The Life of a Virus

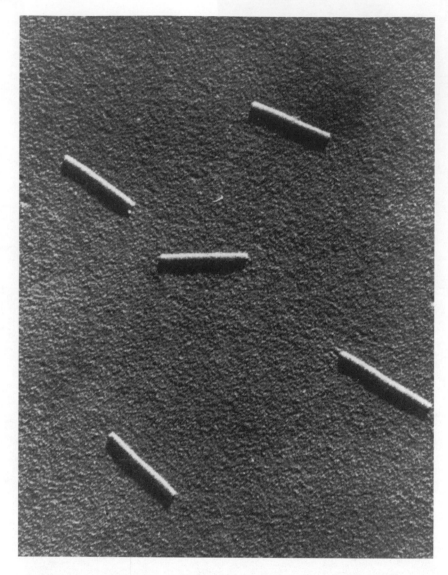

Electron micrograph of a uranium-shadowed preparation of tobacco mosaic virus (TMV), ×42,000. Five rod-shaped virus particles are shown, each measuring approximately 300 × 18 nanometers. Micrograph taken in the Virus Laboratory in the 1950s. (Reprinted courtesy of the Bancroft Library, University of California, Berkeley.)

The Life of a
Virus

*Tobacco Mosaic Virus as an
Experimental Model,
1930–1965*

Angela N. H. Creager

THE UNIVERSITY OF CHICAGO PRESS
Chicago & London

ANGELA N. H. CREAGER is associate professor of history at Princeton University. She is the recipient of a National Science Foundation CAREER Award and a National Science Foundation Science and Technology Studies Scholars Award in support of her work in the history of biomedical research.

The University of Chicago Press, Chicago 60637
The University of Chicago Press, Ltd., London

11 10 09 08 07 06 05 04 03 02   1 2 3 4 5

ISBN: 0-226-12025-2 (cloth)
ISBN: 0-226-12026-0 (paper)

Library of Congress Cataloging-in-Publication Data

Creager, Angela N. H.
    The life of a virus : tobacco mosaic virus as an experimental model, 1930–1965 /
Angela N. H. Creager.
        p.   cm.
    Includes bibliographical references and index.
    ISBN 0-226-12025-2 (cloth : alk. paper)—ISBN 0-226-12026-0 (paper : alk. paper)
    1. Tobacco mosaic virus—Research—History.   2. Stanley, Wendell M. (Wendell
Meredith), 1904–   3. Virology—History.   I. Title.
    QR402 .C74   2002
    579.2′8—dc21   2001035855

For Bill

Contents

Preface

A decade ago, I was a graduate student doing laboratory work in Stanley Hall, at the University of California, Berkeley. Like most of my peers, I did not know much about who Stanley was, nor what he had done to merit having a building named after him. However, whenever I looked out of the fourth floor windows at sunset, I imagined—wistfully—what it must have been like to work there before the massive computer building blocked its spectacular view of the San Francisco Bay. I also wondered what strange institutional politics had led to the creation, on opposite ends of the same campus, of two departments whose scientific agendas seemed nearly indistinguishable: molecular biology and biochemistry. (I happened to work for the only professor with appointments in both departments, which merged during my years there.) A few years later, having given up laboratory work to pursue the history of biology, I was writing an article on the development of a particular line of research at Berkeley. No one, I realized, had done an institutional study of postwar biology there that I could draw on as background to my story. So I ventured into the archives to try to pull together a bit of institutional history. There I encountered Wendell Stanley, the Nobel Laureate who came to Berkeley in 1948 to recreate biochemistry on campus, and whose inability to manage the biochemists led to the formation of an additional molecular biology department in 1964. Stanley's style of leadership, I found, derived not from a brilliant scientific vision nor from political savvy, but from his promotion of a particular object of research, the virus. His single-minded advocacy of virus research helped him acquire a small research empire in Berkeley and a prominent role as a scientific statesman. The more I read of Stanley, the more interested I became in viruses—and in Stanley's favorite virus, tobacco mosaic virus, or TMV. Given this plant virus's significance to biology, and to molecular biology in particular, I was surprised that so little had been written on it. I decided to tell the story of TMV, the tobacco pathogen turned laboratory model.

In the course of working on this book, I have benefited from the support of many organizations and individuals. The research and early writing associated with this project were largely funded through an NSF Science and Technology Studies Scholars Award (SBR 94-12291; 1994–98). A faculty research grant enabled me to conduct further archival research, and Princeton's Philip and Beulah Rollins Bicentennial Preceptorship afforded me additional time away from teaching to write. An NSF CAREER Award (SBE 98-75012; 1999–2003) provided summer support and research assistance during the final stages of work on the manuscript. In collecting materials for the book, I relied on the help of numerous librarians and archivists. I thank William Roberts, Laura Lassleben, and David Farrell at the Bancroft Library, Scott de Haven at the American Philosophical Library, Robert Gelles at the National Foundation Archives, and Thomas Rosenbaum and Lee Hiltzig at the Rockefeller Archive Center. David Goodman at Princeton's Biology Library helped me locate some of my more obscure published sources. For sharing unpublished materials and papers, I am grateful to Soraya de Chadarevian, Nathaniel Comfort, Jean-Paul Gaudillière, Elihu Gerson, Lily Kay, Evelyn Fox Keller, Nick Rasmussen, Hans-Jörg Rheinberger, María Jesús Santesmases, and Susan Spath. I also thank my diligent research assistants, Eric Ash, Elizabeth Merchant, Gail Schmitt, and Suman Seth, for their immense help.

Many colleagues and friends read chapters of the book and offered their comments. For their thoughtful suggestions, I thank Vincanne Adams, Thomas Brock, Peter Brown, Seymour Cohen, Laura Engelstein, Shel Garon, Jean-Paul Gaudillière, Gerry Geison, Bryan Harrison, Sally Smith Hughes, Lily Kay, Evelyn Fox Keller, Manfred Laubichler, Mary Morgan, Andrew Mendelsohn, Hans-Jörg Rheinberger, Jan Sapp, Suman Seth, Judy Swan, Jonathan Weiner, Norton Wise, and Evelyn Witkin.

Others offered insightful comments in response to papers I presented, including Olga Amsterdamska, Rachel Ankeny, Thomas Broman, Soraya de Chadarevian, Clifford Geertz, Charles Gillispie, Jim Griesemer, Bill Jordan, Harmke Kamminga, Dan Kevles, Ursala Klein, Robert Kohler, Susan Lindee, Lisa Lloyd, Harry Marks, Mike Mahoney, Phil Nord, Lynn Nyhart, Karen Rader, Joan Scott, Fred Tauber, Daniel Todes, and others who attended seminars I gave at the Institute for Advanced Study, Johns Hopkins, the University of Pennsylvania, the University of Wisconsin, the California Institute of Technology, Boston University, Princeton University, and Indiana University.

Collaborations with both Jean-Paul Gaudillière and Judy Swan provided ongoing and stimulating discussions while I was writing this book; I thank them for their insights. I am also indebted to discussions with colleagues and graduate students in Princeton's Program in History of Science, especially those involved with Gerry Geison's "Materials" workshop and with the more recent workshop, which I organized with

Norton Wise and Elizabeth Lunbeck, on "Model Systems, Cases, and Exemplary Narratives." Tom Vogt and my lunch pals in the interdisciplinary Reality Check—Vincanne Adams, Charlie Gross, Ben Heller, Hope Hollocher, Alison Jolly, Rena Lederman, Emily Martin, Gideon Rosen, Tracy Shors, Lee Silver, Michael Strauss, Norton Wise—encouraged me to envision biologists and social scientists as part of my intended audience. Howard Schachman, who advised my dissertation research in his laboratory, became a lively historical informant about Stanley. Karen-Beth Scholthof, John Shaw, and Milton Zaitlin welcomed me into the community of plant pathologists in addition to providing their astute comments on my chapters.

A few other intrepid souls read the penultimate version of the entire book and offered valuable responses; for their time and thoughtfulness I thank George Ayoub, David Cole, Joshua Lederberg, and two anonymous referees for the press. Susan Abrams, my editor, has been a marvelous guide and sounding board as I completed the book. I am grateful to Nicholas Murray for his meticulous job on the copyediting. The remaining weaknesses and any errors are, of course, my own.

Finally, there are more personal debts to acknowledge. Countless phone conversations with Beth Streeter and Emily Thompson sustained me as I worked on the manuscript. I thank my parents, William and Janice Hooper, my parents-in-law, Nance and Barbara Creager, my grandmother, Viola Mitchell, my brother and siblings-in-law, Will and Gena Hooper and Martha and David Barnes, and my cousin, Susan Mitchell, for their support and keen interest in this book. My sons Elliot and Jameson have endured my work on this project for most of their young lives; they and their little sister, Georgia, have kept me smiling throughout. My final and most unrepayable debt of gratitude is to my husband, Bill Creager, who read chapter drafts, helped with figures, cared for our children, and made life sweet. I dedicate this book to him.

Princeton
September 2000

Abbreviations for Archival Sources

APS   American Philosophical Society, Philadelphia

NFA   National Foundation Archives, White Plains, NY

RAC   Rockefeller Archive Center, Pocantico Hills, NY

RF   Rockefeller Foundation

RU   Rockefeller Institute/University

Barker papers   Horace A. Barker papers, CU-467, Bancroft Library, University of California, Berkeley

Delbrück papers   Max Delbrück papers, California Institute of Technology Archives, Pasadena

Jones papers   L. R. Jones papers, University of Wisconsin–Madison, Steenbock Archives

Records of the Chancellor   Records of the Chancellor, University Archives, CU-149, Bancroft Library, University of California, Berkeley

Records of the President   Records of the President, University Archives, CU-5, Bancroft Library, University of California, Berkeley

Scientific Reports   Scientific Reports of the Laboratories to the Board of Scientific Directors, Rockefeller Institute of Medical Research

Stanley papers   Wendell M. Stanley papers, 78/18c, Bancroft Library, University of California, Berkeley

Note on citation style: In references to holdings in the Rockefeller Archive Center, the number codes refer to the record group and series in which the document is located. For all citations, carton and box numbers are labeled as such.

"Where Tobacco Mosaic Has Led Us"

Much of the remarkable advance made in the last few years in our knowledge of the viruses has been due to the stimulus of W. M. Stanley's work on the virus of tobacco mosaic, which he was the first to show could be obtained as a crystalline protein material.
"Where Tobacco Mosaic Has Led Us," *Lancet,* 10 May 1941

At the beginning of the twentieth century, bacteriologists appeared triumphant in their ability to identify infectious disease agents by filtering them, culturing them in cell-free media, and viewing them in microscopes.[1] Following an 1892 report that the plant pathogen for tobacco mosaic disease could not be ascertained using any of these methods, however, reports appeared in the scientific literature of other anomalous "filterable viruses" (infectious agents that passed through bacteria-retaining filters). The growing list of invisible pathogens included viruses associated with human illnesses such as yellow fever, rabies, and paralytic polio, as well as with numerous diseases afflicting animals, insects, and even bacteria. Medical and agricultural researchers, physicians, and biologists devoted decades of effort to understanding these pathogens and developing therapies and vaccines against them.

By 1950 the nature of the virus was no longer a mystery. Viruses were known to be macromolecules, genetic units, parasites that depend on their hosts for metabolism and reproduction. But a funny thing happened on the road to this knowledge. The viruses that most shaped this emerging portrait were not the most dangerous pathogens, but those examples, however innocuous to humans, that made good laboratory subjects. Researchers constructed general knowledge about viruses based on a few that, by reason of historical precedence or biological robustness, were intensively studied as representatives of the rest.

1. The classic history is Bulloch, *History of Bacteriology*. For an excellent recent account of the emergence of bacteriology, see Mendelsohn, *Cultures of Bacteriology*.

The significance of these laboratory exemplars extended far beyond disease control. Some archetypal viruses became principal experimental tools in the emerging field of molecular biology, transforming the understanding of life itself.[2] Approaches developed to study viruses proved useful in understanding other biological entities, such as genes and cell organelles. The similarities between viruses and elements of cellular genomes (especially those implicated in cancer) called into question the strict divide between heredity and infection. In short, virus research became as important to biology as it was to medicine and agriculture.

This book focuses on the experimental life of one of virology's most influential representatives, tobacco mosaic virus, or TMV. As the first virus identified as a nonbacterial pathogen, TMV was central to virology from the beginning. But the virus did not attain public visibility until 1935, when Wendell Stanley obtained TMV in pure form, as needle-shaped crystals. The isolation of the virus as a simple chemical brought Stanley worldwide recognition and established his research career. In 1946, this plain-spoken chemist from Indiana received the Nobel prize for his work with TMV, and his laboratory continued to be a major site for virus research during the postwar period. Stanley presented TMV as a model object, not only for other viruses but for life itself, as its simplest incarnation—a reproducing molecule. Although every history of virology includes mention of Stanley's crystallization of TMV, this book reassesses his role in the longer history of TMV research in an attempt to trace out the experimental life of his model virus in the decades following its purification.

The story of TMV as a model virus opens a window on several major developments in twentieth-century biology. TMV was a protagonist in debates about the origin of life, a crucial tool in research on biological macromolecules, and an unlikely major player in the development of commercial instrumentation in science. In the midcentury campaigns to fight poliomyelitis and cancer, the state of knowledge about TMV helped to justify large-scale funding of virus research and provided a pragmatic guide for the investigation of human pathogens. And on the emerging scene of molecular biology in the postwar decade, TMV figured among the most preferred objects of investigation.

This book examines these larger changes—in instrumentation, science funding, and postwar biology—through a critical history of TMV research from the early 1930s to the early 1960s. At first glance, this approach may seem ambitious. However, a detailed history of experimentation on one particular virus can offer more than an incidental view on turning points in twentieth-century biology. The uses of model objects in research

2. For an overview, see Levine, *Viruses*.

reveal the otherwise inconspicuous connections between biological experimentation and activities usually relegated to the domains of technology, politics, medicine, and agriculture. Rather than interpreting shifts in experimental practice as mere responses to these contextual factors, I emphasize the ways in which key scientific achievements set in motion trends ranging far outside the laboratory walls.

This approach benefits from the insights of other historians who have emphasized the materials and tools of experimentation. From Robert Kohler's history of the fruit fly as an instrument for genetics to Peter Galison's history of detection instruments in physics, recent scholarship in twentieth-century science has been attuned to the importance of the concrete and tangible elements of research.[3] This focus has proven particularly fruitful for understanding the life sciences, where variability and complexity often confound attempts to derive universal laws, and generalizations usually remain tied to specific experimental demonstrations.[4] The experimentalist in biology strives for the *typical* in his organism or setup, but leaves behind any dreams of a final theory.[5] Thus a search for sweeping theoretical revolutions in the experimental life sciences would fail to turn up many of the dramatic changes of this past century— changes in how biologists visualize entities, how they organize their work, and how they create new forms of life.[6] Some commentators judge the gradual accumulation of more details and examples in the absence of grand conceptual change to amount to the "end of science."[7] This book is based on a very different premise: that key conceptual shifts in biology have often been incremental, and embedded in specific experimental examples.

3. Kohler, *Lords of the Fly;* Galison, *Image and Logic.* The wider literature addressing science's "material culture" is too vast to cite here; a few examples, besides others cited below, are Clarke, "Research Materials"; Geison and Creager, "Research Materials"; Rader, "Mice, Medicine, and Genetics"; Carson, "Minding Matter/Mattering Mind"; Thompson, "Dead Rooms and Live Wires"; and Lenoir, "Models and Instruments." See also the collection edited by Van Helden and Hankins, *Instruments;* Wise, "Mediating Machines"; and Smith and Wise, *Energy and Empire.*

4. See Geison and Laubichler, "Varied Lives of Organisms."

5. Steven Weinberg contends that finding unifying physical laws is the ambition of physics (*Dreams of a Final Theory*); not all physicists agree (see, e.g., Anderson, "More Is Different"; Langer, "Non-Equilibrium Physics").

6. The so-called recombinant revolution is perhaps the best example of a technological transformation that did not necessarily overturn the scientific understanding of its field, molecular genetics. Rather, it accelerated the acquisition of more detailed knowledge and brought sweeping change to methods used in all of the life sciences in the form of cloning techniques. On visualization, see Rasmussen, *Picture Control;* on organization, see Clarke and Fujimura, *Right Tools.*

7. Horgan, *End of Science;* Stent, *Coming of the Golden Age.*

Systems for Research

Attempts to understand the material and experiential character of research have prompted historians and philosophers to give analytic attention to the language scientists themselves use in describing their work. Two
such terms are especially important to the general picture offered here of
virus research and biological experimentation: *experimental system* and
model system.

First, discussions of laboratory research on TMV are often framed in
terms of the *experimental system*. This phrase, drawn from the parlance of
biologists, refers not only to the material being investigated (e.g., infected
tobacco leaves), but to all of the methods and instruments involved in
handling, isolating, and visualizing the object of study.[8] The experimental
system, in other words, encompasses the assemblage of things and techniques that together generate results. As Hans-Jörg Rheinberger emphasizes in his epistemological analysis of molecular biological research, a
productive experimental system is constantly changing, generating novel
results and new representations of the object being studied.[9] His depiction
of the dynamic character of experimentation provides a useful perspective
for analyzing the many small shifts in the production and visualization of
TMV, which often resulted in conceptual changes to the general category
of viruses.

In order to recapture experimental dynamics of this kind, the historian
must focus on research at close range. My account of experimental
systems using TMV draws principally on research at one laboratory,
Stanley's. Stanley's chemical purification of TMV in 1935 set in motion a
prominent program of biochemical and biophysical research on viruses;
this scientific trajectory provides the core narrative of this book. The aim
in focusing on experimentation associated with a particular scientist is
not so much biographical as epistemological. To concentrate on the ma-
teriality and specificity of biological research raises the question of how
experimental systems produce general knowledge. This account examines
how the picture of TMV generated in one laboratory was reconciled with
research results on the same object in laboratories elsewhere, and how it
was experimentally related to work on other viruses.

Over time, TMV retained its important place in virus research, in
molecular biology, and in the wider world of biomedicine because researchers viewed it as a model in two senses. It was intensively studied

---

8. Kohler, "Systems of Production" and *Lords of the Fly;* Rheinberger, "Experiment,
Difference, and Writing" and *Toward a History of Epistemic Things.* See chapter 8 for a
fuller discussion of *experimental system* as a historiographical category.

9. Rheinberger refers to this as the "epistemic object" (*Toward a History of Epistemic
Things*, esp. 28–31).

as a representative virus, with the expectation that knowledge gleaned from TMV could be generalized (provisionally) to other viruses. Second, it showed by example how other viruses and biological objects could be studied. Throughout the book, I use the term *model system* to refer to these two aspects of TMV's utility.

Like *experimental system, model system* is part of the biologist's idiom. The best-known model systems—standardized organisms such as the laboratory mouse and the fruit fly—are investigated by an entire community of biologists. Model systems become prototypes within which key biological questions are defined and resolved, useful precisely because they have already been so well studied. TMV was a model system in these respects, studied and discussed by a large contingent of biochemists, plant pathologists, and other agricultural and medical researchers.[10]

Historical studies tend to focus on this first feature of model systems: their role as standard prototypes. Robert Kohler's account of the construction of the "instrument" *Drosophila melanogaster* presents the fruit fly as the defining technology through which chromosomal genetics was worked out.[11] Kohler's story emphasizes how the fruit fly, once crafted into a scientific tool, shaped the behaviors and practices of its researchers. In his account, alternative biological systems were displaced rather than reconfigured by the dominant fly. The inbred mouse, another prominent model organism, became the central prototype through which biologists extended the study of chromosomal genetics to mammals. The motivations for standardizing the mouse were different from those animating fly research; cancer research provided the major impetus for making the mouse into a genetic tool.[12] But in both cases, the simple fact that these creatures were so intensively studied made them more usable than other candidates for advancing research. In this sense, model systems have a self-reinforcing quality.

TMV also became a widely studied research object whose initial advantages as a stable and abundant laboratory material were enhanced by the sheer magnitude of knowledge that accumulated about this plant virus. My analysis of TMV, however, will stress the other (second) aspect of model systems mentioned above—the way they serve as *exemplars* for studying and understanding other entities and organisms. A model system is productive not only because of the amount of information available

10. Of course, TMV was not the only well-studied virus. At various times during the twentieth century, fowl plague virus, bacteriophages (bacterial viruses), influenza virus, poliovirus, adenovirus, papilloma virus, and several other pathogens were model systems in virology. Nevertheless, TMV research tended to lead the way in biochemical and structural studies of viruses.

11. Kohler, *Lords of the Fly.*

12. Rader, *Making Mice;* Löwy and Gaudillière, "Disciplining Cancer."

about it, but also because scientists can draw useful analogies between it and other less-understood objects or systems. Researchers summoned TMV as an example of how one could gain knowledge about other viruses, often in ways that linked basic laboratory research with research aimed at disease control in animals, humans, and crops.

These analogies had concrete consequences, prompting the extension of experimental techniques and representations from TMV research to new situations. For example, the characterization of TMV as a macromolecular particle inspired efforts to obtain and understand other viruses, including the agents of papilloma and influenza, using the same physical-chemical instruments. Exemplars in this sense are transient, not fixed: as the experimental systems using TMV changed, the kind of model TMV provided also shifted. Thus the purification in 1956 of infectious ribonucleic acid (RNA) from TMV prompted the isolation of infectious nucleic acid from Coxsackie virus and poliomyelitis virus, associated with a new understanding of virus infectivity as mediated by nucleic acids. In these examples, experimental techniques and conceptual changes moved together, based on the referent of TMV as a model virus.

Developments in TMV research in turn drew on other experimental systems as models. John Northrop's crystallization of three digestive enzymes and his demonstration that all three were proteins motivated Stanley to try to crystallize TMV as a protein. After Stanley succeeded in isolating TMV, the driving analogy between enzymes and viruses faded from view. Two decades later, Frederick Sanger's heroic protein sequencing of insulin provided a different model to Stanley's laboratory researchers, who began to pursue the complete chemical characterization of TMV protein in terms of amino acid sequence.

The history of research on TMV reflects a series of very specific analogies of this sort between it and other viruses and objects. Some of the analogies were to conceptual entities, for which TMV could provide a material referent. For example, similarities between viruses and genes (postulated units of heredity) guided researchers to use TMV to understand the genetic properties of self-duplication and mutation. In the 1940s, newly apparent resemblances between viruses and cytoplasmic particles further shaped the investigation of genes through viruses, in ways that are barely acknowledged in standard histories of genetics. Through emphasizing the role of such analogies and exemplars in research, I want to recover these concrete connections, in which common references to particular model systems linked (however temporarily) research programs involving different biological, medical, and agricultural objects.

In these ways, my analysis of TMV distinguishes the experimental system, as more local, from the model system, as more communal. The resulting account offers not only a history of experimentation on TMV (in terms of the experimental system) but also points to many of the ways in which results and representations achieved with TMV were adapted—using TMV as a model—to other research situations. This process illustrates one way in which biological knowledge generated in the highly local environment of a research laboratory "travels" and becomes generalized. Behind biological principles and experimental approaches often stand specific model systems which inspire the investigation of other organisms and materials along similar lines.

Model systems, I argue, are exemplars. For a historian of science to use the term *exemplar* at all is to call up Thomas Kuhn's depiction of scientific change through paradigm shifts. To the degree that Kuhn stressed the importance of analogies between exemplars and new problems, my approach indeed draws from his insights.[13] Yet my use of *exemplar* in this book departs from Kuhn's in key respects, not least that his emphasis was on the role of exemplars in scientific pedagogy. In the 1969 postscript to *The Structure of Scientific Revolutions*, Kuhn asserted, "The paradigm as shared example is the central element of what I now take to be the most novel and least understood aspect of this book. Exemplars will therefore require more attention than the other sorts of components of the disciplinary matrix."[14] Kuhn proceeded to offer three examples of "what I mean by learning from problems to see situations as like each other, as subjects for the application of the same scientific law or law-sketch."[15] To the degree that Kuhn's vision relies on the centrality of theories, and even *laws*, his schema is problematic when applied to biomedical research, where the kinds of symbolic generalizations that characterize physics are rare.[16]

Even more important, whereas Kuhn emphasized the role of theory and problem solving in terms of conceptual work, I emphasize the centrality of experimentation and materials at the leading edge of scientific change. What scientists take from exemplars such as TMV is not necessarily a way to *see* the world differently as a way to *handle* or materially configure the world differently. And even then, such experimental exemplars are not static. TMV was a central model system in various ways for thirty years—it was actually a cluster of possible models and templates based on the experimental systems of Stanley and others. Rather than

---

13. For an astute analysis of the role of analogies in Kuhn's approach and its indebtedness to Wittgenstein—and through him, to Lichtenberg—see Cedarbaum, "Paradigms." Of course, Kuhn was not the only scholar to stress the role of analogies in science; for one classic view, see Hesse, *Models and Analogies*.

14. Kuhn, *Structure of Scientific Revolutions*, 187.

15. Ibid., 190.

16. See Schaffner, *Discovery and Explanation*.

seeing exemplars (or models, for that matter) as fixed or rigid, my account analyzes them as changeable points of reference amid the day-to-day decisions that are constitutive of research.[17] It is there, I argue, that we will discern how model systems shaped the direction of research across fields of investigation.

A fuller historiographical discussion of experimental systems and model systems is offered at the conclusion of the book (chapter 8). The remainder of this chapter sketches out the contours of the book's narrative, indicating how the story of TMV revises our understanding of the history of biomedical research in this century.

TMV: From a Crystal to a Code

When Stanley began his work on TMV, this pathogen was already established as a good representative virus for laboratory research (chapter 2). But scientists disagreed about whether viruses constituted a unique category of infectious agents (separate from bacteria) and whether viruses were living or nonliving entities. The ramifications of these issues extended beyond the problems posed by viral diseases in medicine and agriculture. The prospect of self-reproducing units smaller than cells resonated with speculations that colloids or enzymes might be the fundamental matter of living substances. Disputes about the "nature of the virus" appeared experimentally resolvable—if a virus could be completely isolated from its host and other contaminants. TMV seemed an ideal material for such a task. Bacteriologists, viewing the virus as an ultramicroscopic microbe, sought to cultivate TMV in lifeless media. Chemists attempted to purify it as a molecule.[18]

Stanley's 1935 publication in *Science*, announcing that he had isolated TMV as needle-shaped crystals of pure protein, attracted worldwide attention because it demonstrated that viruses could be recovered as chemical entities (chapter 3).[19] One writer for the *New York Times* compared Stanley to another chemist whose contributions to medicine were legendary: Pasteur.[20] Stanley's evidence that TMV was "simply" a chemical—but one that could reproduce itself—fueled speculation about the role of viruses in the origin of life.[21] Purified TMV also served as a

17. For more on the temporality of scientific activity, see Pickering, *Mangle of Practice*.

18. On these two approaches, see Helvoort, "What Is a Virus?"

19. Stanley, "Isolation of a Crystalline Protein." Lily Kay has provided historical analysis of this contribution in "Stanley's Crystallization."

20. In 1937 Stanley was recognized by a prize from the American Association for the Advancement of Science; the comparison of Stanley to Pasteur was part of the news story of this award (*New York Times*, 2 Jan 1937, 1:7).

21. Podolsky, "Role of the Virus."

tangible representative of viruses, which could now be defined in terms of proteins rather than disease symptoms.

Stanley's celebrated feat followed years of working with TMV to establish a productive experimental system. The resulting image of a crystalline virus appeared stable and decisive, but a close look at Stanley's research reveals a marked contrast between the "public" icon of the immutable crystal and the more unstable "private" life of TMV in the laboratory.[22] Other researchers challenged Stanley's initial claims, and Stanley himself kept tinkering with his preparation method, producing purified virus with altered characteristics.[23] A modified portrayal of TMV accommodated Stanley's critics and his changing results: the virus was not a pure protein but contained nucleic acid as well; its needle-shaped precipitates were not true three-dimensional crystals; and TMV did not reproduce autocatalytically. Yet these corrections served to consolidate the chemical approach to virus research at just the moment, ironically enough, when Stanley's chemical practices changed even more dramatically. In 1936, the ultracentrifuge displaced the precipitation procedures and once again revised the conception of TMV.

TMV was the first virus obtained in a form pure enough to be analyzed in an ultracentrifuge (chapter 4). Physical chemists developed analytical ultracentrifuges in the 1920s and 1930s as tools to gain information about the size and shape of materials by viewing them as they sediment in solution, spinning at high speeds. TMV's properties in the ultracentrifuge proved more complex than Stanley expected: the virus particles behaved as macromolecules much larger than most proteins. In 1936 biophysicist Ralph Wyckoff found that the ultracentrifuge also provided an alternative means to *purify* TMV; one could efficiently separate the virus from host contaminants in plant sap by sedimentation. This observation completely altered the way TMV was prepared in the laboratory. Almost immediately, researchers followed the example of TMV and began isolating other viruses in the ultracentrifuge, a use of the machine its original inventor did not intend (or even endorse). During the war, Stanley further adapted his centrifuge-based method of isolating viruses to develop a new kind of influenza vaccine that was mass-produced for the army and subsequently for civilian use. Thus TMV served as a model for applied research on influenza virus.

The use of the ultracentrifuge in virus research was part of a broader pattern of technological innovations in interwar biology. The Rockefeller Foundation funded the instrument's development for biological use

22. My attention to this contrast is indebted to Geison, *Private Science*.
23. Stanley's most significant critics were F. C. Bawden and N. W. Pirie; see Bawden et al., "Liquid Crystalline Substances."

through Warren Weaver's grants program, which encouraged physicists, chemists, and mathematicians to collaborate with life scientists.[24] Historians of science have paid significant attention to the role of Weaver (who coined the term *molecular biology*) in encouraging this "technology transfer" into biology and, with it, a preoccupation with life at the submicroscopic level.[25] However, the effects of this program on scientific practice have only begun to be appreciated.[26] Along these lines, the development of the ultracentrifuge illustrates the consequences of specific materials and model systems in shaping these new instruments. The utilization of ultracentrifuges by virus researchers spurred the development of a version of the machine that could prepare as well as characterize macromolecules.

The adaptation of the ultracentrifuge for virus research also traces out the long-range outcome of the Rockefeller program: In the end, the successful commercialization of the ultracentrifuge after World War II attenuated the collaborations Weaver set in motion. The contributions of physicists and engineers were "black-boxed" into standardized, industrially produced machines that biologists could purchase and operate themselves. This shift from collaboration to commercialization marked the development of several other new instruments as well, as company-produced electron microscopes, electrophoresis equipment, scintillation counters, and spectrophotometers came on the market.[27]

During the same time period that virus research benefited from this expanded technological infrastructure, new sources of financial support for virus research came into play (chapter 5). The National Foundation for Infantile Paralysis (NFIP), best known for its yearly fundraiser, the March of Dimes, granted millions of dollars to virus researchers (including Stanley) in the 1940s and 1950s. The confidence that laboratory research would vanquish polio arose from widespread public perception that scientists had helped win World War II. By 1950, the federal government, too, began funding basic research on a large scale in the name of improving the health of citizens (having failed to enact legislation providing universal health care).[28]

24. See Kohler, "Instruments of Science," ch. 13 of *Partners in Science*.

25. Ibid.; Kohler, "Management of Science"; Abir-Am, "Discourse of Physical Power"; Kay, *Molecular Vision of Life*.

26. For exceptions, see Kay, "Laboratory Technology and Biological Knowledge"; Rasmussen, *Picture Control*. Accounts of the Rockefeller Foundation's role in encouraging innovation in instrumentation include Kohler, *Partners in Science*, and Zallen, "Rockefeller Foundation and Spectroscopy Research."

27. See, for example, the works by Rasmussen and Kay cited in note 26 above, and Rheinberger, "Putting Isotopes to Work."

28. Starr, *Social Transformation*.

Two generations of historians have analyzed the legacy of World War II for postwar science. The core literature focuses on two important issues: debates over the formation of the National Science Foundation, and the postwar interests of the government (and the military) in the growth of the physical sciences.[29] The case of virus research reveals that a different set of actors and concerns shaped the emergence of biomedical research policy. Before there was consensus in Congress about the merits of government-sponsored science after the war, voluntary health agencies championed the notion that researchers should continue to be mobilized to solve pressing medical problems. Their publicity presented the agents of dread diseases as the new enemies against which teams of scientists should be deployed. The NFIP and American Cancer Society (ACS) did not restrict their support of science to applied research or therapeutic development: proceeds from their disease-based fund-raising campaigns included generous grants to university laboratories to support basic biomedical research. Symbolically, public expectations that scientists could win the war against disease provided a powerful legitimation of the value of laboratory research. More tangibly, model systems such as TMV provided experimental exemplars for the study of human pathogens. For example, Stanley undertook the purification and biochemical characterization of poliovirus, drawing on the same approaches and techniques developed in TMV research.

The resources that the polio foundation (and the ACS) directed at virus research fell on fertile biological ground; scientists already viewed viruses as good experimental tools for understanding life at the most elemental level. Since the early twentieth century, similarities between viruses and genes—both self-reproduce and both accumulate and perpetuate mutations—had intrigued life scientists. For biologists interested in understanding the physical basis of heredity rather than concentrating on genes as operational entities (as located through chromosomal mapping), viruses seemed an even more promising experimental tool than *Drosophila*.

Stanley's isolation of TMV further stimulated these hopes; in 1936 H. J. Muller referred to Stanley's crystalline virus as "a certain kind of gene."[30] Stanley's biochemical and biophysical characterization of TMV, including studies of mutant strains of the virus, aimed at discerning the chemical basis for mutations. His crystallization of TMV also inspired Max

---

29. See, for example, Dupree, "Great Instauration"; Kevles, "National Science Foundation" and *Physicists;* Forman, "Behind Quantum Electronics"; Schweber, "Mutual Embrace"; and, more recently, Galison, *Image and Logic,* ch. 4. I present a fuller historiographical account of the debates over federal funding of science in chapter 5.

30. Muller, "Physics in the Attack," 213.

Delbrück to work with the "T" coli bacteriophages (viruses that infect bacteria) to develop a quantitative experimental system to study virus multiplication. Thus the NFIP's investment in basic research intersected with intensifying biological interest in viruses, particularly as vehicles for studying heredity.

In the 1940s, Delbrück's experimental system for studying virus reproduction assimilated the standard terms and techniques of genetics— mapping and crossing of mutants. Delbrück also recruited many other researchers to adopt the phage system as a tool for biological research. The members of the "phage group" he inspired contributed to the emerging identity of molecular biology, both through their scientific contributions and through their sense of singular group affiliation.[31] Beginning with the accounts of Delbrück's students and followers, histories of molecular biology have tended to stress the centrality of bacteriophages and the phage group at the expense of other contributors and experimental materials.[32]

The story of TMV research, juxtaposed to the work on phages, makes visible two features of the history of molecular biology that are obscured by a central focus on Delbrück as disciplinary father and the bacteriophage as central model system. First, reaching back to the above discussion, the tremendous resources available in the 1940s through the 1960s to virus researchers, including early molecular biologists, derived from the relevance of their work to medical concerns, especially polio. This context for the development of molecular biology is rarely addressed in the historiography of the field.[33]

Second, the linear narrative from Delbrück's experimentation with bacteriophage in the 1930s to the identification of DNA as the hereditary material in the 1950s overlooks the way the virus-gene analogy veered in the 1940s (chapter 6). Many biologists focused on apparent similarities between viruses and other cellular entities (e.g., microsomes, chloroplasts,

31. For a sociological analysis, see Mullins, "Development of a Scientific Specialty."

32. Central to the literature on the role of phage research in the emergence of molecular biology is the Festschrift to Max Delbrück: Cairns, Stent, and Watson, *Phage and the Origins*. See chapter 6 for more complete references. For the responses to this book by other contributors to molecular biology eager to assert ownership over "ancestor spots" (in David Berol's nice phrasing; see *Living Materials and the Structural Ideal*, 2), see Kendrew, "How Molecular Biology Started," and the Festschrift for Jacques Monod: Lwoff and Ullman, *Origins of Molecular Biology*.

33. This point is indebted to Jean-Paul Gaudillière and Ilana Löwy's more general emphasis on the construction of postwar bio*medicine*. For exceptions to my generality about neglect: Kevles, "Renato Dulbecco" and de Chadarevian and Kamminga, *Molecularizing Biology and Medicine*. The more standard sources are Olby, *Path to the Double Helix;* Judson, *Eighth Day of Creation*. Some recent accounts challenge Olby and Judson, but do not take up the embeddedness of molecular biology in the larger world of biomedical research (e.g., see Kay, *Molecular Vision of Life*).

and "plasmagenes") that appeared to share the genetic properties and constitution of viruses without being infective. The growing list of analogues to nuclear genes suggested the existence of a general mechanism for the self-reproduction of cytoplasmic particles, including viruses.

These analogies motivated a new conception of viruses, which were no longer viewed as necessarily exogenous from the host, either materially or genetically. But what was fruitful for rethinking the virus was not so helpful for discerning the gene. The fact that viruses, most cytoplasmic particles, and chromosomes were all composed of a combination of nucleic acid and protein (nucleoprotein) led researchers to undervalue early evidence that genes could inhere in simple DNA (deoxyribonucleic acid). The focus on nucleoproteins as such was largely replaced in the 1950s by efforts to differentiate the roles played by DNA, RNA, and protein in gene expression and duplication—as elements of the cell's "library of specificities."[34] The formerly perceived similarities between viruses, genes, and cytoplasmic particles were rapidly forgotten as molecular biologists took credit for recognizing the hereditary "secret of life": DNA.[35] Nonetheless, examining the analogies between the many different entities and experimental systems considered to have "genetic" properties in the 1940s helps account for both the changing understanding of viruses and the slow appreciation for the special role of nucleic acids.

In 1960, Stanley's chemical approach to virus research culminated in the publication from his laboratory of the complete amino acid sequence of the TMV coat protein (chapter 7).[36] However, the meanings associated with TMV as a macromolecule had shifted decisively since the 1940s. Stanley's earlier representation of the rod-shaped virus particle as a unitary chemical object was slowly undermined by the increasing attention to the functional pieces of the virus, both protein and nucleic acid.

Crucial to this transition was the "reconstitution experiment" of Heinz Fraenkel-Conrat and Robley Williams at Stanley's Virus Laboratory in Berkeley. They dismantled the virus into its two constituents, nucleic acid and protein, and then reassembled these into infectious virus particles, indistinguishable from the original TMV. This experiment, published in 1955, was instrumental in leading TMV researchers to acknowledge (rather belatedly) the crucial role of nucleic acid in infectivity and

34. Nanney, "Epigenetic Control Systems," 712.
35. The exception is viral genes, which can be carried by RNA (as in the case of TMV). The more general formulation is to say that the carriers of hereditary information are nucleic acids.
36. See Tsugita et al., "Complete Amino Acid Sequence." See also Anderer et al., "Primary Structure." Stanley's group, unlike the latter, attempted to distinguish the amino acids glutamic acid from glutamine, and aspartic acid from asparagine, in the protein coat sequence. Corrections were subsequently made to both sequences, however.

heredity. Because one could correlate chemical changes in the protein and the nucleic acid portions of the virus, TMV rapidly became a central tool in efforts at cracking the genetic code (the pattern showing how nucleic acid sequence specifies protein sequence). Representations of the Berkeley TMV, like many other experimental systems in molecular biology, took an "informational" turn.[37] But by the mid-1960s, the development of more manipulable experimental systems (cell-free translation systems) overtook the experimental use of TMV to resolve the genetic code.[38] TMV's laboratory career again shifted gears, as researchers began to explore its use as a model for macromolecular self-assembly.[39]

During the 1950s and 1960s, TMV and bacteriophage embodied different approaches to "molecular" explanation and provided competing exemplars in the field. As science journalist Greer Williams noted in 1959, "These two viruses, TMV and phage, have been the greatest rivals in fundamental research into the nature of midget microbes."[40] Research on bacteriophage guided scientists toward visualizing other viruses— including polioviruses and tumor viruses—using molecular genetic techniques.[41] TMV inspired the conceptualization of viruses and other subcellular entities in biochemical terms, drawing on chemical techniques as well as technologies for visualizing macromolecules. The greater emphasis on bacteriophage in the conventional "origins" stories about molecular biology reflects, in part, the turf wars between biochemists and self-declared molecular biologists of the 1950s and 1960s. In the view of many biochemists, molecular biology was not scientifically distinctive and certainly not revolutionary—it was, at most, a successor name for biochemistry.[42] Erwin Chargaff encapsulated this viewpoint in his famous complaint, "molecular biology is essentially the practice of biochemistry without a license."[43]

These disciplinary disputes have provided an important historiographical nexus in historical work on the postwar reconfiguration of the life sciences.[44] The consolidation of molecular biology encroached on other

37. Kay, *Who Wrote the Book of Life?;* Rheinberger, *Toward a History of Epistemic Things.*

38. Kay, ibid.; Rheinberger, ibid.

39. For an overview, see Klug, "Tobacco Mosaic Virus Particle."

40. Williams, *Virus Hunters,* 476.

41. On the modeling of animal viruses on bacteriophage at Caltech, see Kevles, "Renato Dulbecco"; and Creager and Gaudillière, "Experimental Arrangements."

42. See Fruton, *Skeptical Biochemist.*

43. Chargaff, "Amphisbaena," 176.

44. See, for example, Gaudillière, *Biologie moléculaire;* Abir-Am, "Politics of Macromolecules," and "Themes, Genres, and Orders of Legitimation"; and the essays and comments gathered by de Chadarevian and Gaudillière under the theme, "The Tools of the Discipline: Biochemists and Molecular Biologists" for a special volume of the *Journal*

fields as well, such as microbiology, immunology, cell biology, and genetics, and it assimilated much of what went by the name of biophysics in the 1940s.[45] Focusing attention on the prominent model systems in these overlapping fields is a valuable way to explore how postwar disciplinary boundaries were transgressed, contested, and redrawn.

Commitment to particular model systems sometimes became implicated in disciplinary ruptures. In Berkeley, where Stanley worked hard to integrate biochemistry and virus research, the subsequent divorce between the biochemists and the virologists in the late 1950s created an opening for a new department of molecular biology centered around Stanley's Virus Laboratory.[46] The departmental biochemists were never as committed to the importance of viruses as Stanley's group. On the other hand, model systems could be shared across disciplinary and departmental lines, providing points of mediation between researchers working in different traditions. In Cambridge, England, during the same period, hemoglobin provided an important nexus for unifying the efforts of crystallographers, biochemists, and geneticists at the Medical Research Council's Molecular Biology Unit.[47] The emphasis on TMV in this book is meant to encourage further studies of the successive model systems and modes of experimentation that shaped the growth of molecular biology.

Stanley, surveying recent developments in biology in 1959, offered this assessment of his own contribution: "The discovery that a self-reproducing microbe could be treated like a chemical substance, and be even crystallized, engendered a storm of philosophical controversy, but after the dust raised by the arguments whether viruses are 'organisms' or 'molecules,' 'living' or 'dead,' had settled, the new discipline, called 'molecular biology' nowadays, had come into being."[48] The following chapters take issue with his synopsis; the story of TMV as an experimental model was much more circuitous and complex than is conveyed by the settling of dust on a new scientific order. The point of *The Life of a Virus* is surely not to replace the master narrative of molecular biology with a different teleology or another founding virus. Rather, the history of a particular research object shows the interrelatedness of change in disparate sites— we see how an unlikely plant virus made cameo appearances in many

of the History of Biology: de Chadarevian and Gaudillière, "Tools"; Creager, "Stanley's Dream"; de Chadarevian, "Sequences, Conformation, Information"; Rheinberger, "Comparing Experimental Systems"; Gaudillière, "Molecular Biologists, Biochemists, and Messenger RNA"; Kay, "Biochemists and Molecular Biologists"; Burian, "Tools."

45. Rasmussen, "Midcentury Biophysics Bubble," and *Picture Control*.
46. See Creager, "Stanley's Dream."
47. de Chadarevian, "Following Molecules."
48. Summary, 9 Mar 1959, Stanley papers, carton 22, folder Biographical data.

different venues, from speculations about the origin of life to rousing pamphlets for the March of Dimes and discussions of the genetic code. In the 1990s TMV has become a commercial vector, enabling a biotechnology company to grow therapeutic human enzymes in Kentucky tobacco fields.[49] Thus the unexpected turns in the laboratory career of TMV have not ceased; indeed, its most profitable years may have just begun.

49. Turpen, "Tobacco Mosaic Virus and the Virescence of Biotechnology."

2

Viruses Enter the Laboratory

Little is known about most of these filterable viruses. They appear to be
of various natures, and the only property common to them is minuteness.
"Filter-Passing Viruses," *Encyclopedia Britannica*, 1929

Through the first third of the twentieth century, discovery claims for filterable viruses proliferated—viruses seemed to inhabit all living things, from
bacteria through insects, animals, humans, and plants. In his 1927 review of
"Filterable Viruses," Thomas Rivers covered seventy-one disease groups
attributed to viruses, citing 410 publications.[1] Researchers from a variety of orientations—epidemiological, clinical, veterinary, pathological, as
well as chemical—were actively pursuing knowledge about viral diseases
and their causes. Yet in the laboratory these seemingly ubiquitous infectious agents evaded captivity. As late as 1932 the category of viruses was
still principally defined in terms of negative attributes: inaccessible by
filtering, invisible in the microscope, and unculturable in media.[2]

For the many medical and agricultural scientists interested in understanding viruses, the absence of experimental evidence positively defining
or visualizing these agents posed a challenge. Although the presence of
viruses could be detected through the infectivity of filtered sap or serum,
in the context of an infection it seemed impossible to answer questions
about the virus itself, such as whether viruses were enclosed in mem-
branes or possessed metabolic activity. Thus, as the literature on viral
diseases expanded rapidly, researchers complained that the nature of the
virus remained mysterious. As Frederick Twort observed in 1923, "Since
... they have never for certain been either seen or grown, it is impossible
to define their nature, and in carrying out experiments on these viruses it
is unwise to *assume* that they represent very minute varieties of ordinary
bacteria."[3]

1. Rivers, "Filterable Viruses: A Critical Review."
2. Rivers, "Nature of Viruses," 423.
3. Twort, "Ultramicroscopic Viruses," 351; emphasis in original.

This predicament led researchers in many fields to look for ways to isolate viruses as objects for further study. If a virus could be separated from its host, then it could be subjected to bacteriological and biochemical analysis to determine its chemical composition and structure, its nutritional requirements, its means of reproduction, and its mechanisms for infecting a cell. Investigations along these lines could shed light on whether viruses should be considered animate or inanimate entities. Such research might also advance therapies for viral diseases. This effort to understand viral diseases in ontological terms, rather than only as pathological processes, drew on the bacteriological understanding of diseases in terms of causative organisms.[4] At the preeminent center for bacteriological research in the United States, the Rockefeller Institute for Medical Research, founding director Simon Flexner encouraged attempts to investigate viruses as autonomous entities.

Two distinct ways of isolating a virus seemed most promising, and both approaches were attempted at the Rockefeller Institute.[5] One could use the techniques of bacteriology to attempt the cultivation of viruses in artificial media (containing nutrients but no cells). Guided by the framework of germ theory, many pathologists regarded viruses as authentic bacteria that were simply smaller than other known microbes. These expectations were kept alive by a steady stream of publications by researchers claiming to have cultivated filterable viruses.[6] For example, Peter Olitsky, a prominent bacteriologist at the Rockefeller Institute, asserted in *Science* that he had artificially cultured the virus of tobacco mosaic (TMV).[7] This claim was especially noteworthy because TMV, as a highly stable and transmissible infectious agent, already served as an important representative of filterable viruses. Olitsky's work was strongly contested by plant pathologists; his paper was followed by three others whose authors were unable to reproduce his results.[8] As Rivers observed, "Unequivocal

4. Bulloch, *History of Bacteriology;* Mendelsohn, *Cultures of Bacteriology,* ch. 2.

5. On the two approaches to studying viruses, see Helvoort, "What Is a Virus?"

6. Simon Flexner was among the most influential supporters of the idea that filterable viruses could be cultivated in lifeless media; in 1915 he and Hideyo Noguchi claimed to have artificially cultivated poliomyelitis virus. As late as 1931, Eagles and McClean (of the Lister Institute) published a claim to have cultivated vaccine virus in cell-free medium. Several other viruses, such as fowl plague virus, could be made to grow in media with blood or cells present. See Rivers, "Filterable Viruses: A Critical Review," 228–29; Benison, *Tom Rivers,* 110–11, 116, 145; and Waterson and Wilkinson, *Introduction to the History of Virology,* 33.

7. Olitsky, "Experiments on the Cultivation of the Active Agent of Mosaic Disease" (1924). The longer version is Olitsky, "Experiments on the Cultivation of the Active Agent of Mosaic Disease in Tobacco and Tomato Plants" (1925).

8. Mulvania, "Cultivation of the Virus of Tobacco Mosaic by the Method of Olitsky"; Purdy, "Attempt to Cultivate an Organism from Tomato Mosaic"; Goldsworthy, "Attempts

confirmation of Olitsky's work...would settle one of the most important problems in the whole field. [However], in general it can be said
that no worker has proved that any of the etiological agents of the [viral]
diseases in [the] table...is susceptible to cultivation in the absence of
living cells."[9]

An alternative way to achieve the goal of isolating a virus was through
the use of chemical methods. The decline in the adequacy of protoplasmic
theory at the end of the nineteenth century motivated new approaches
to understanding life, such as the biochemical axiom that each chemical
change in the cell is mediated by an individual enzyme.[10] As an increasing
number of enzymes were being purified in the early 1900s through precipitation and centrifugation, researchers began to extend these methods
to isolate viruses.[11] At the Rockefeller Institute, the analogy between
enzymes and viruses prompted just such a chemical approach to virus research. John Northrop used the methods that had enabled him to crystallize digestive enzymes in an attempt to purify bacteriophage; his colleague
Wendell Stanley had even more success using Northrop's precipitation
methods to isolate TMV. The prospect that viruses might be retrieved as
chemical entities buoyed scientific expectations that the physical sciences
would provide the tools for resolving fundamental problems in biology
and medicine.

Bacteriologists and chemists offered quite different conceptions of
viruses as entities—microbial organisms versus molecules—that served
as two poles in a broader discussion from the 1890s to the 1930s about the
nature of viruses. In the course of recounting these scientific debates, this
chapter includes many well-known stories about the earliest discoverers
of viruses; however, I want to emphasize two central issues in this retelling.
First, as mentioned above, the techniques for detecting viruses and the desire to conceptualize viruses as *objects* were part of the legacy of medical
bacteriology, especially at research sites like the Rockefeller Institute for
Medical Research. But since the first filterable virus—TMV—was identified in a plant, not in an animal or human, agricultural researchers had
equal claim from the outset to the domain of viruses. Indeed, the comparative nature of virus research propelled a biological approach to these
disease agents. Since viruses apparently infected all living things, the issue

---

to Cultivate the Tobacco Mosaic Virus." Note that tomato is another host for TMV. In his
oral history, Olitsky blamed his technician for the observation of apparent multiplication
of tobacco mosaic virus in artificial culture; see Olitsky papers, RAC RU 450OL4, box 1,
folder 3, "Oral History," 28–29.

9. Rivers, "Filterable Viruses: A Critical Review," 228.

10. Kohler, "Enzyme Theory," esp. 181, 185.

11. On early efforts at purifying enzymes, see Fruton, *Molecules and Life,* "From Diastase
to Zymase."

of understanding their *nature* became an important scientific question early on.

Second, the "virus" was not a scientifically precise entity whose characteristics became more and more obvious as research on various viruses proceeded. Rather, the term *virus* designated a domain of etiologic agents whose occupants proliferated and shifted as new observations were made in the field and the laboratory. Veterinarians, physicians, epidemiologists, bacteriologists, and plant pathologists found and conceptualized viruses in somewhat different ways. Dealing with viruses in terms of their etiological or pathological effects was often sufficient for practitioners to find ways to treat patients, livestock, or crops suffering from viral diseases. At the same time, researchers from many quarters endeavored to discern the typical features of viruses. To this end, particular examples became canonical referents and systems for defining viruses in general. This chapter relates how TMV became one such laboratory prototype by the 1920s, not only among agricultural researchers but in the medical domain as well.

Mosaic Disease and Its Filterable Agent

In late-nineteenth-century Europe, the etiological puzzle of viral diseases came into view as part of the response to a devastating blight afflicting tobacco. In 1886, Adolf Mayer, a German bacteriologist working in Holland, sought to standardize the labeling of this tobacco disease. He described the leaf-mottling pathology, gave it the name tobacco mosaic disease, and showed that it was infectious. In 1892, Dmitri Iosifovich Ivanovskii, following up Mayer's investigations, was surprised to observe that the infectivity of tobacco sap from mosaic-disease plants remained even after he passed it through a bacteria-retaining filter.[12] Mayer's colleague, Martinus Willem Beijerinck, a Dutch microbiologist, also found that the agent of tobacco mosaic disease passed through a sterilizing filter. He, unlike Ivanovskii (whose work he did not know), argued in 1898 that the agent was not simply an ultramicroscopic bacterium, but something fundamentally different.[13] Thus TMV became the earliest "filterable virus" discovered, although only Beijerinck used the term *virus* in his paper.

Every history of virology treats the work of these three nineteenth-century researchers as the point of embarkation for modern virus research. In fact, scientists still disagree about which one of the three

12. Ivanovskii, "O dvukh bolezniakh tabaka." A partial translation by J. N. Irvine is available as an appendix to Hughes, *Origins and Development*.
13. Beijerinck, "Über ein Contagium vivum fluidum."

should be considered the father of virology.[14] My aim in more fully recounting their contributions is not to attempt to settle the genealogy of
virology, but to emphasize the still-tentative status of viruses as a distinct
class of organisms at the turn of the century. One way to organize the
myriad observations about viruses into a systematic framework was to
privilege particular viruses as model systems. The earliest experiments on
tobacco mosaic were frequently cited, and often reinterpreted, as TMV
became a standard referent for viruses generally.

At the time of the earliest studies of tobacco mosaic disease, the criteria
for what constituted a virus were far from settled. In part, this ambiguity
reflected the complex history of the term. From the first century A.D., the
word *virus* was associated in Latin with disease, and connoted poison,
venom, or slime.[15] In 1658, Athanasius Kircher used *virus* in connection
with plague, and the connotation of infection was increasingly carried in
eighteenth- and nineteenth-century usages of the term, adapted from the
Latin into vernacular languages.[16] Edward Jenner referred to "cow-pox
virus" in his 1798 publication of the method for smallpox vaccination,
establishing the precedent for the subsequent association of *virus* with
vaccination (and with long-term immunity after infection).[17]

A dozen years after Charles-Emmanuel Sédillot's coining of the word
microbe in 1878, Louis Pasteur made his famous declaration: "Tout virus
est un microbe."[18] *Virus* in the age of the germ theory of disease served
as a catch-all for infectious agents, particularly those not yet identified bacteriologically.[19] The reporting of "filterable viruses" associated
with animal and plant diseases, following Beijerinck's use of *virus* for
the agent of tobacco mosaic, gave the term *virus* a more specific meaning. Viruses were now operationally (albeit negatively) defined by filters
and microscopes, standard bacteriological implements. But Ivanovskii's
and Beijerinck's observations also challenged the scientific hegemony of
medical bacteriology. Because the first representative of the category of

14. See, for example, Lustig and Levine, "One Hundred Years of Virology"; and Bos,
"Embryonic Beginning of Virology."

15. Hughes, *Virus,* appendix A, 109–14. As Hughes points out, the ancient and modern
meanings of *virus* (as poison and agent of infection) meet in a quotation from the Roman
encyclopedist, Cornelius Aulus Celsus, c. 50 A.D.: "Especially if the dog is rabid, the virus
should be drawn out with a cupping glass" (109).

16. Ibid., 111.

17. Jenner, *Inquiry into the Causes and Effects of the Variolae Vaccinae;* Waterson and
Wilkinson, *Introduction to the History of Virology,* 5–6.

18. Waterson and Wilkinson, *Introduction to the History of Virology,* 10; Sédillot, "De
l'influence des découvertes de M. Pasteur"; Pasteur, "La rage."

19. Hughes, *Virus,* 112. Hughes also traces the way in which the old association of virus
with poison or infectious exudate continued to be carried into the early twentieth century.

FIGURE 2.1 Picture of mosaic-infected tobacco leaves. The captions following are from the source. "*Fig. 1:* A young tobacco leaf, in the first stage of the disease, with a moderate amount of virus. The dark-green spots [here, dark gray] are visible next to the vein. *Fig. 2:* A mildly diseased tobacco leaf in the second stage of the disease with a few... spots that were produced through the premature necrosis of the tissues. *Fig. 3:* A vari-colored tobacco leaf of a plant that had become vari-colored through the mixed infection of the virus with *Bacillus anglomerans. Figs. 4 and 5:* Small, malformed tobacco leaves, produced by the introduction of large amounts of virus into the stem." (Photograph and captions from Beijerinck, "Über ein Contagium vivum fluidum" [1898], plate 2. Beijerinck's original plate was in color. Reprinted from *Phytopathological Classics* 7 [1942], the American Phytopathological Society, St. Paul, MN.)

"filterable viruses" was observed in a plant, virus research was the purview of agricultural as well as medical researchers.

A closer look at the agricultural origins of TMV research points to explanatory tensions within plant pathology, which sought to incorporate bacteriological theory and practices while retaining its long-standing emphasis on the role of environmental factors in disease. In 1879, Mayer, who headed the Agricultural Experiment Station in Wageningen, the Netherlands, examined samples of diseased tobacco leaves sent to him by a local chapter of the Dutch Agricultural Society. He coined the term *die Mosaikkrankheit,* the *mosaic disease,* to refer to the blight (see fig. 2.1). Mayer and his coworkers discerned that elevated temperatures and nutritional deficiencies, such as lack of essential elements nitrogen and potassium, were not responsible for the disease. Rather, the mosaic disease could be transmitted from plant to plant by sap extracts; Mayer therefore argued that it was due to an infectious agent. However, he looked in vain for the agent among nematodes, animal parasites, and plant fungi. Microbes could be isolated from diseased tissue, but none transmitted the mosaic disease. Moreover, sap extract from diseased plants could be filtered through a single layer of paper and retain the ability to transmit infection. In 1886, Mayer speculated:

> One should realize that a definite capacity to infect, as has been proved in our case, may be determined either by an unorganized [enzymatic] or an organized ferment [i.e., a cell]. It is true that the former would be rather unusual as a cause for a disease, and also that an enzyme should reproduce itself is unheard of. Yet this situation has been taken under consideration . . . [F]iltered extract has about the same effect (the percentage of diseased plants is somewhat smaller) as the original.[20]

His openness to ferments as infectious, even self-multiplying, agents would have raised the eyebrows of medical bacteriologists. Nonetheless, Mayer held that his results could also be compatible with a bacterial etiology, even though he had been unable to isolate the causative microbe.[21] In fact, the attribution of plant disease to bacteria was itself rather novel. Plant pathology was so dominated by mycologists, who had identified fungi responsible for many of the common agricultural diseases, that many pathologists did not believe that plants were susceptible to bacterial infection.[22]

Other European universities were also responding to the concerns of tobacco farmers. In 1887, the dean of natural science at the University of St. Petersburg sent a Russian instructor of plant physiology and anatomy, Ivanovskii, to the Ukraine to study a tobacco disease causing substantial damage. Ivanovskii and his companion V. V. Polovtsev published a set of papers associating this pox disease of tobacco with overused lands, and they advocated crop rotation as a possible solution. Ivanovskii's subsequent Ph.D. dissertation, "On Two Diseases of Tobacco," differentiated between Mayer's mosaic disease and the pox disease he observed with

20. Mayer, "Über die Mosaikkrankheit des Tabaks," translated and reprinted as "Concerning the Mosaic Disease of Tobacco," 22.

21. Mayer reported that although the agent could pass through one layer of filter paper, two layers screened out the agent, a result that current plant pathologists find perplexing.

22. On mycology, see Ainsworth, *Introduction to the History of Plant Pathology*. As Hughes observes, Anton de Bary, the founding father of plant mycology, declared in 1885, "According to present experience, parasitic bacteria are after all only of slight importance as contagia of plant diseases. Most contagia of the numerous infectious diseases of plants belong to other animal and plant groups, principally. . . to the true fungi" (*Virus*, 16).

Polovtsev.[23] After conducting more fieldwork on tobacco diseases in the Crimea region, Ivanovskii moved in 1891 to the laboratory of botany at the Academy of Sciences in St. Petersburg, where he became engrossed in studies of alcoholic fermentation. However, Ivanovskii continued publishing on tobacco diseases, and is best known for his short 1892 paper to the Academy, "Concerning the Mosaic Disease of the Tobacco Plant," in which he confirmed and extended Mayer's observations. Much of the paper concerns his ongoing argument that Mayer's *Mosaikkrankheit* is actually two entirely separate diseases, pox and mosaic, rather than two stages of the same disease. But he also offered a new observation and interpretation concerning the transmission of mosaic disease: "I have found *that the sap of leaves attacked by the mosaic disease retains its infectious qualities even after filtration through Chamberland filter-candles* [see fig. 2.2]. According to the opinions prevalent today, it seems to me that the latter is to be explained most simply by the assumption of a toxin secreted by the bacteria present, which is dissolved in the filtered sap."[24] Inspired by Emile Roux's and Alexandre Yersin's isolation in 1888 of diphtheria toxin, Ivanovskii contended that an infectious toxin could account for tobacco mosaic disease.[25] However, he also cautioned that his investigations were incomplete because he had been unable to culture the bacterium.

Independently, Beijerinck, who had observed some of Mayer's initial experiments on tobacco disease while a docent at the agricultural school at Wageningen, took up the problem of mosaic disease in 1887 from his chair of microbiology at the Delft Polytechnic School. Beijerinck's studies focused on abnormal growth and its physiological effects, particularly as enacted through ferments (enzymes responsible for fermentation). Given his interest in metabolism, he was inclined to view tobacco mosaic disease as a physiological aberration, bearing an analogy to the overgrowth of crown gall.[26] Beijerinck, unaware of Ivanovskii's papers, presented evidence that the infectious agent of tobacco mosaic disease was filterable, and furthermore that it could diffuse through agar. He also established that the agent multiplied in dividing tissues, so that infection in series could

23. Ivanovskii, "O dvukh bolezniakh tabaka"; for biographical information, I have relied on Lechevalier, "Dmitri Iosifovich Ivanovski."

24. Ivanovskii, "Über die Mosaikkrankheit der Tabakspflanze" (1892), translated and reprinted as "Concerning the mosaic disease of the tobacco plant," 30; emphasis in source. Chamberland filters were developed by Chamberland in 1885 to obtain "physiologically pure" water. See Bos, "100 Years of Virology."

25. Roux and Yersin, "Contribution à l'étude de la diphthérie." See also Lustig and Levine, "One Hundred Years of Virology."

26. "The behavior of the virus in connection with the growing tissues reminds one of similar relationships in gall formation, for the cecidiogenen substance also can affect only growing parts" (Beijerinck, "Über ein Contagium vivum fluidum," 39 in English translation). On Beijerinck's research career, see Theunissen, "Beginnings of the 'Delft Tradition' Revisited," 205.

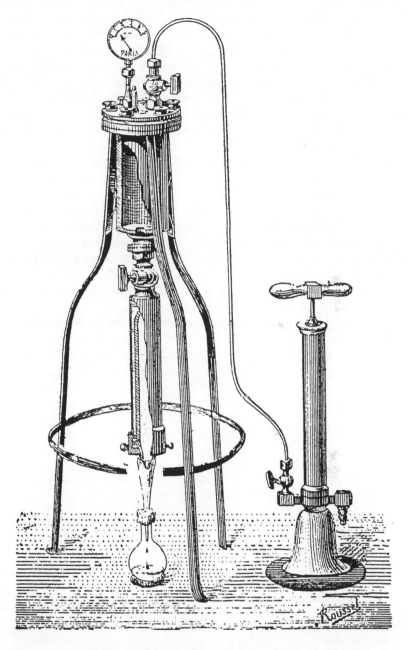

FIGURE 2.2 Drawing of Chamberland filter. The filter is connected to a hand-pump, and is like those used at the Institut Pasteur toward the end of the nineteenth century. (From Duclaux, *Traité de microbiologie* [1898], 1: 103.)

be obtained. In Beijerinck's view, the diffusibility of the substance attested to its liquid or dissolved nature, and contradicted the assumption that an infectious agent must be corpuscular, or cellular, to reproduce itself.[27]

Beijerinck inaugurated a new category for such diffusible agents of infection: "contagium vivum fluidum," an infectious living fluid. As Beijerinck's students have noted, he used this term only in the title, and employed the term *virus* throughout the paper.[28] However, *contagium vivum fluidum* caught the attention of many readers, including Ivanovskii, who criticized it as a contradiction in terms.[29] Beijerinck's contagium conception borrowed substantially from protoplasmic theory; he would as late as 1916 oppose the chemical view of zymase, arguing instead that the enzyme was "an essential, microscopically visible part of the yeast protoplasm," which acts as a "suspensoid."[30] His view of ferments and viruses thus placed them between mere molecules and cells, having autonomy and some of the characteristics of life.

Apparently irritated that Beijerinck had failed to cite him, Ivanovskii remarked in his 1902 (second) dissertation that the Dutch researcher "filtered the sap of diseased plants through a porcelain filter and stated that the sap, sterilized in this fashion, retained its infectivity. The author does not know that I had already established this fact a long time ago."[31] Ivanovskii repeated Beijerinck's experiments to argue strenuously against the soluble character of the infective agent, showing that particles of ink could pass through agar as readily as tobacco mosaic virus.[32] Accordingly, in his view, Beijerinck had not established that the virus was not a small, particulate bacterium.

Like other cases of co-discovery in science, a comparative perspective reveals differences in both the commitments and experimental intentions

27. Hughes, *Virus*, 49. According to the students who wrote his biography, Beijerinck used *dissolved* and *liquid* interchangeably in describing the virus in solution; see Iterson, Dooren de Jong, and Kluyver, *Martinus Willem Beijerinck*, 119.

28. Iterson, Dooren de Jong, and Kluyver, *Beijerinck*, 119. To be more precise, Beijerinck also used *contagium vivum fluidum* in a subtitle, and used *contagium* several times as a noun in the paper, although not nearly so frequently as the term *virus*.

29. Contributing to the attention attracted by Beijerinck's notion of a *contagium vivum fluidum* was the fact that in 1898, *Nature* published a summary of his argument after he presented a paper to the Royal Academy of Sciences Amsterdam ("Prof. Beijerinck, on a contagium vivum fluidum"). On Ivanovskii's objections, see note 31.

30. Iterson, Dooren de Jong, and Kluyver, *Beijerinck,* 117–18; Beijerinck, "Nachweis der Violaceusbakterien." On the background to protoplasmic theory, see Geison, "Protoplasmic Theory."

31. Ivanovskii, *Mozaichnaia bolezn' tabaka* (On the mosaic disease of tobacco), doctoral dissertation at University of Warsaw, 1902, quoted in Lechevalier, "Dmitri Iosifovich Ivanovski," 140.

32. Ivanovskii, "Über die Mosaikkrankheit der Tabakspflanze" (1899).

of Beijerinck and Ivanovskii. Beijerinck's interest in using diffusion studies in agar to assess the liquid nature of the virus has been attributed to his friendship with physical chemist Jacobus van't Hoff.[33] Ivanovskii's regard for the environmental factors in tobacco disease reflected his practical work in soil science and fertilizer development. Whereas Beijerinck abandoned experimental work on tobacco mosaic after 1898, his controversial interpretation seemed to have motivated Ivanovskii to return to studies of tobacco mosaic, a subject that might have otherwise remained dormant.[34] In fact, for both scientists, work on tobacco mosaic disease was tangential to their main programs of research in microbiology. It was only as other observations of filterable pathogens began to accumulate in the first decade of the twentieth century that the priority dispute between Beijerinck and Ivanovskii became consequential.[35]

In 1897, Friedrich Loeffler and Paul Frosch, both employed in Koch's Institut für Infektionskrankheiten in Berlin, were appointed to head a commission on foot-and-mouth disease in Germany. Working on the basis of Behring's discovery in 1890 of antitoxic immunity, Loeffler and Frosch inoculated healthy calves with filtered vesicle lymph, which they surmised might contain antitoxin. Instead, they found that the infective agent itself was transmitted.[36] The fact that the disease could be propagated through serial inoculations with increasing dilutions of the lymph demonstrated that the agent could not be a conventional toxin, but must be multiplying within the host. Shortly thereafter, similar observations of a filterable agent for bovine pleuropneumonia were made by pathologists at the Pasteur Institute, although this agent was visible (barely) in the light microscope and could be cultured under very special conditions.[37] These

33. Helvoort, "What Is a Virus?" 560.

34. On Ivanovskii's return to the problem of tobacco mosaic diseases, see Lechevalier, "Dmitri Iosifovich Ivanovski." On the legacy of Beijerinck's work in microbiology, see Amsterdamska, "Beneficent Microbes"; Spath, "C. B. van Niel's Conception"; and Theunissen, "Martinus Willem Beijerinck."

35. Indeed, as Ton van Helvoort has pointed out ("What Is a Virus?" 561, n. 12), prominent virus researchers a few decades later at the Rockefeller Institute were still divided as to whom to credit with the founding of virology: Louis Kunkel credited Ivanovskii's earlier paper, whereas Thomas Rivers honored Beijerinck as the true father of virus research because he treated the virus as a nonbacterial agent; see Benison, *Tom Rivers*, 115–16.

36. Loeffler and Frosch, "Berichte der Kommission"; Waterson and Wilkinson, *Introduction to the History of Virology*, 30. For a valuable summary of the political and institutional context of Loeffler and Frosch's work and the German historiography on Loeffler, see Schmiedebach, "Prussian State and Microbiological Research."

37. The agent of bovine pleuropneumonia "could be grown in special collodion sacs implanted in the peritoneum of live guinea pigs" (Waterson and Wilkinson, *Introduction to the History of Virology*, 33).

observations of filterable viruses at the bacteriological centers in Paris and Berlin greatly increased the perceived significance of the earliest experiments in Holland and Russia. Filterable infectious agents began to attract the attention of physicians and medical researchers. As noted by veterinarian John M'Fadyean in 1908,

> It is true that the experiments of Iwanowski and Beijerinck had previously shown that the virus of the spotted disease of the tobacco plant was filtrable, but that discovery does not appear to have attracted wide notice, and its importance was certainly not immediately apprehended. The discovery of Loeffler and Frosch, on the other hand, at once attracted general attention, and gave a great impetus to the investigation of the nature of the virus in those diseases which had hitherto baffled investigations conducted on ordinary bacteriological lines.[38]

Giuseppe Sanarelli's 1898 isolation of an invisible virus from the lymph of rabbits infected with myxomatosis, even though it could not be filtered, strengthened the apparent generality of these infectious agents.[39]

The reports of these four infectious agents in 1898—for tobacco mosaic, foot-and-mouth disease, bovine pleuropneumonia, and rabbit myxomatosis—provided conflicting criteria for differentiating viruses from other germs. The discoverers of TMV disagreed over whether the agent was a small microbe or a dissolved substance. Bovine pleuropneumonia could be seen in a microscope, whereas the rabbit myxomatosis agent was initially unfilterable. Which agents fell into the "filterable" category depended on several technical specificities: whether the Chamberland or Berkeland filters were used, which size of either filter was employed (e.g., Berkeland W or V), from what kind of solution the agent was filtered (watery or albuminous), and what bacterium was used as a control. Furthermore, accumulating evidence that some yeast and bacteria could pass through these filters, whereas some chemical compounds such as strychnine could not, challenged assumptions that filtering was a simple, size-based discrimination.[40] Nonetheless, new reports of viruses being published, such as M'Fadyean's identification of a viral agent for African horse-sickness in 1900, reinforced the links between culturability, filterability, and visibility in the microscope.[41]

38. M'Fadyean, "Ultravisible Viruses," 66.

39. Sanarelli, "Das myxomatogene Virus."

40. M'Fadyean ("Ultravisible Viruses") discusses the problem of different solutions, and Remlinger ("Les microbes filtrants") offers a good discussion of the different bacterial controls used by the various researchers.

41. M'Fadyean, "African Horse-Sickness," 15.

Reports of viral etiologies for fowl plague, yellow fever, and rabies added further support to the generality of viral diseases.[42] Moreover, a pattern of citations emerged among the growing number of publications that established the precedence of earlier observations. In his report on the viral nature of fowl plague in 1902, Centanni cited Beijerinck's 1898 article,[43] and M'Fadyean's report on African horse sickness in 1900 cites Loeffler and Frosch's work on foot-and-mouth disease.[44] A series of important review articles on viruses in French, English, and German further solidified the cast of representative viruses from plants, animals, and humans. In 1903, Emile Roux included ten known "invisible viruses" in his review of the field; three years later his compatriot Paul Remlinger enumerated nineteen "microbes filtrants."[45] M'Fadyean's review, "The Ultravisible Viruses," published in 1908, specified sixteen viruses; his set differed somewhat from Remlinger's. Two German reviewers in 1911, Loeffler and Doerr, counted twenty viruses;[46] the following year Harvard's bacteriologist Wolbach enumerated thirty viral diseases.[47] Despite reviewers' different interpretations of the nature of viruses and their varying list of the pertinent discoveries, their citations reveal an emerging canon.

Table 2.1 lists the eight disease agents that reviewers tended to mention as the first observed viruses; TMV became the standard earliest discovered example. In his review, M'Fadyean remarked, "One can see that the diseases caused by ultravisible viruses form a remarkably heterogeneous group. When compared with one another, they exhibit differences almost as great as one finds among the diseases caused by visible bacteria."[48] At the same time, this author first articulated a criterion for

42. The first researchers to publish on fowl plague virus were Lode and Gruber, "Bakteriologische Studien"; Centanni, "Die Vogelpest"; and Maggioria and Valenti, "Über eine Seuche von exsuditivem Typhus bei Hühnern." (Centanni and Maggioria and Valenti first reported their results in Italian in 1901; see Wilkinson, "Development of the Virus Concept....Part 1. Beginnings.") On yellow fever as a viral disease, see Reed and Carroll, "Etiology of Yellow Fever." On rabies, see Remlinger, "Le passage du Virus rabique."

43. Centanni, "Die Vogelpest," 198; Wilkinson, "Development of the Virus Concept. ...Part 1. Beginnings," 216.

44. M'Fadyean, "Ultravisible Viruses," 16.

45. Roux, "Sur les microbes dit 'invisibles'"; Remlinger, "Les microbes filtrants"; Waterson and Wilkinson, *Introduction to the History of Virology,* 49.

46. Loeffler and Doerr, "Über filtrierbares Virus," 4–6.

47. Wolbach, "Filterable Viruses: A Summary."

48. M'Fadyean, "Ultravisible Viruses," 239. In 1936, Rivers observed that "the term virus used in connection with an infectious agent has lost its old indefinite meaning and has acquired a new significance similar in exactness to that borne by the words bacterium and spirochete" ("Viruses and Koch's Postulates," 2).

Table 2.1 Earliest Reported Observations of Filterable Viruses

Researchers	Year	Disease	Passes through Bacteria-Retaining Filters?	Culturable?	Visible?	Regarded as a Bacterium?
D. Ivanovskii;	1892	Tobacco mosaic	Yes	No	No	Yes
M. W. Beijerinck	1898					No
F. Loeffler and P. Frosch	1898	Foot-and-mouth	Yes	No	No	Yes
E. Nocard, E. Roux, et al.	1898	Pleuropneumonia	Yes	Yes	Yes	Yes
G. Sanarelli	1898	Rabbit myxomatosis	No, then yes	No	No	No
J. M'Fadyean	1900	African horse-sickness	Yes			Yes
E. Centanni; A. Maggiora and G. L. Valenti; A. Lode and J. Gruber	1901	Fowl plague	Yes	No	No	"True virus" (Maggiora and Valenti); "Something like an enzyme" (Lode and Gruber)
W. Reed and J. Carroll	1902	Yellow fever	Yes	No		
P. Remlinger	1903	Rabies	Yes (but only under certain conditions)	No (Pasteur)	No (although Negri bodies were visible)	

viruses other than technical failures to culture, visualize, or filter out the infective agents: "Another character common to all the ultravisible organisms is that they appear to be *obligatory parasites*."[49] This characteristic distinguished viruses fundamentally from bacteria.

Given the growing interest in the "nature of the virus," particular viruses began to be intensively investigated as models of their class of infectious agents. Suitability for laboratory study was of foremost concern. Eugenio Centanni noted about fowl plague in 1902: "Fowl plague virus is not inferior to these other viruses [the agents of foot-and-mouth disease, bovine pleuropneumonia and African horse-sickness] with regard to its wide distribution and its lethal effect; it is far superior with respect to convenience of study, requiring only small, easily available animals, whereas for the other agents investigated so far the choice has been restricted almost exclusively to cattle and horses. This does not permit the exhaustive investigations required on these still mysterious viruses."[50] Fowl plague was one of the first viruses to be subjected to chemical isolation in the laboratory. In 1912, Mrowka, a staff veterinarian at the German naval base at Tsingtao, published his account of precipitating the virus of fowl plague from infective serum and centrifuging the "globulin" away from cellular constituents.[51]

Other researchers promoted their own preferred subjects as candidate exemplars for viruses. Félix d'Herelle's publications on bacteriophage in 1917 and 1921 drew attention to their amenability to laboratory study.[52] H. J. Muller rapidly promoted the analogy between "d'Herelle bodies" and genes, the former being particularly susceptible to investigation: "They are filterable, to some extent isolable, can be handled in test tubes, and their properties, as shown by their effects on the bacteria, can then be studied after treatment."[53] In 1923, B. M. Duggar and Joanne Karrer Armstrong observed that the cause of mosaic diseases was the problem that most engaged and confounded plant pathologists, and TMV was

49. M'Fadyean, "Ultravisible Viruses," 241; emphasis added.

50. Centanni, "Die Vogelpest," 201, as translated and quoted in Wilkinson and Waterson, "Development of the Virus Concept. . . . Part 2. The Agent of Fowl Plague," 53.

51. Mrowka, "Das Virus der Hühnerpest ein Globulin"; Waterson and Wilkinson, *Introduction to the History of Virology,* 43. Two years later P. Andriewsky furthered these efforts by using ultrafiltration methods to determine the size of fowl plague virus: see "L'ultrafiltration et les microbes invisibles." On the continuing importance of fowl plague virus as a model system in animal virology, see Schäfer, "Structure of Some Animal Viruses."

52. d'Herelle, "Sur un microbe invisible," and *Le bactériophage.* There was considerable controversy over whether bacteriophages were genuinely viruses; see Helvoort, "Controversy."

53. Muller, "Variation Due to Change," 48. I take up the analogy between genes and viruses in chapter 6 (where I quote more extensively from this article).

the candidate among plant viruses most likely to provide the answer.[54] The "nature of the virus" remained an open question through the early 1930s, even as fowl plague virus, bacteriophage, and TMV provided some of the leading (if competing) models for its resolution.

Infectious Agents between Organisms and Molecules

Bacteriology provided the tools and concepts for identifying viruses as possibly distinct infectious agents, but viruses also challenged the reigning microbial understanding of disease, defying identification by Koch's postulates.[55] And viruses were not the only conundrum bacteriologists faced. The identification of "healthy carriers" in epidemics of tuberculosis and cholera led to a reevaluation of the germ theory of disease in the early twentieth century.[56] Differential susceptibility became a new focus of research for epidemiologists and medical pathologists. Plant pathologists, on the other hand, had a tradition of emphasizing the role of environment in disease. As mentioned above, much of the success of plant pathology in the late nineteenth century had been due to research in mycology.[57] Environmental factors play an important role in opportunistic fungal diseases of plants. Consequently, plant pathologists tended to also consider the role of environment for other diseases, even as they assimilated bacteriological approaches and explanations.

In particular, plant pathologists, unlike their medical counterparts, did not necessarily extend the organismal view of pathogens to a more microscopic realm to account for mosaic diseases. Viral diseases could also be conceptualized as environmentally induced disruptions of endogenous factors rather than as due to strictly exogenous, infectious agents. In 1899 the American plant pathologist A. F. Woods challenged the infectious etiology of tobacco mosaic, contending that Beijerinck's contagious fluid consisted of accumulated oxidase and peroxidase, which cause mottling of the leaves by enzymatically destroying chlorophyll.[58] In 1902 Woods

54. Duggar and Armstrong, "Indications," 191.
55. Rivers, "Viruses and Koch's Postulates."
56. I am indebted on this point to the excellent article by Mendelsohn, "Medicine and the Making of Bodily Inequality."
57. See Ainsworth, *Introduction to the History of Plant Pathology*. He argues that there was a national element to the status of mycology in plant pathology: "The early American work on bacterial diseases of plants was not universally accepted, particularly in Germany where scientific work was then in the ascendant and American work held in low esteem" (70). Erwin Frink Smith did a great deal to establish bacteriology in plant pathology; see Rodgers, *Erwin Frink Smith*.
58. Woods, "Destruction of Chlorophyll by Oxidizing Enzymes," especially 751–52. Woods was following upon the successes of his colleague L. R. Jones in accounting for plant diseases through enzyme disturbances (Campbell, Peterson, and Griffith, *Formative Years of Plant Pathology*, esp. 192). As Robert Olby has described Woods's theory, "In the

further elaborated his enzymatic explanation.[59] Although F. W. T. Hunger offered a critique of Woods's theory in 1905, Hunger himself accounted for mosaic disease through a toxic ferment produced through poor growth conditions and nutrition.[60]

There were exceptions to this trend. Defending Beijerinck's earlier conception of the virus as an infectious agent in 1915, H. A. Allard mobilized his studies of the effect of dilution upon the infectivity of filtered virus of tobacco mosaic disease to argue that the disease was not of a spontaneous origin or due to unfavorable conditions, but resulted from a contagious agent which multiplied in its host.[61] Allard contrasted this theory of "parasitic origin" with Woods's alternative theory of "enzymic origin."[62]

Historians have conventionally followed Allard in drawing similar distinctions between exogenous (usually bacteriological) versus endogenous (such as enzymatic) conceptions of viruses.[63] While this differentiation is very useful at a heuristic level, it does not adequately account for the range of theories about viral diseases, which included notions of viruses described in terms of ferments or toxins, as well as environmental and microbial etiologic agents. Researchers frequently invoked colloids to account for the activity of such subcellular particles; early on, Mrowka, who investigated fowl plague, argued in 1912 for "the colloid nature of the virus" by analogy with bacterial antigens and antitoxins, "which are generally regarded as colloidal in nature."[64] This trend registered a prevailing interest in circumventing the dichotomy between organisms and molecules. Colloidal chemistry provided a framework for conceptualizing entities which are active if not alive, and particulate if not corpuscular—agents between enzymes and bacteria.[65]

---

necrotic spots chlorophyll was changed to xanthophyll, owing to a local accumulation of oxidizing enzymes" (*Path to the Double Helix*, 149).

59. Woods, "Observations on the Mosaic Disease of Tobacco."

60. Hunger, "Untersuchungen und Betrachtungen über die Mosaik-Krankheit der Tabakspflanze," and "Neue Theorie zur Ätiologie der Mosaikkrankheit des Tabaks."

61. Allard, "Effect of Dilution." In addition, Allard provided a more direct refutation of Woods's claims by demonstrating the agent's ability "to destroy completely the peroxidase present without destroying the infective principle" ("Some Properties of the Virus of Mosaic Disease," 666). The same point is made by Wilkinson, "Development of the Virus Concept. . . . Part 3. Lessons of the Plant Viruses," 119.

62. Allard, "Effect of Dilution," 299.

63. Helvoort, "What Is a Virus?"; Waterson and Wilkinson, *Introduction to the History of Virology*; Wilkinson, "Development of the Virus Concept. . . . Part 3. Lessons of the Plant Viruses." Hughes (*Virus*) divides theories into microbial and nonmicrobial groups. This distinction was similarly important to bacteriophage research, for which the observations of lysogeny brought an explicitly hereditary component to the virus as an exogenous agent.

64. Mrowka, "Das Virus der Hühnerpest ein Globulin," quoted in Waterson and Wilkinson, *Introduction to the History of Virology*, 43.

65. Morgan, "Strategy of Biological Research Programmes."

Enzymes, colloids, and ultravisible or filterable viruses all inhabited this no-man's-land between molecules and cells, and hinted tantalizingly at a level of vitality more organized than protoplasm, the amorphous substance that had guided much thinking about the "physical basis of life" in the late nineteenth century.[66] The ascent of the discipline of biochemistry, with its emphasis on the dynamic nature of metabolism and the centrality of enzymes, provided both research tools for such questions and a large audience for their answers.[67] The nature of enzymes was hotly contested until the 1920s, with the adsorption models of colloidal chemists providing the reigning explanation for enzyme action.[68] Just as Eduard Buchner's 1897 isolation of the enzyme zymase called into question the distinction between unorganized (chemical) and organized (cellular) ferments, so the persistently nonbacterial nature of filterable viruses provided a new reason to question the adequacy of cell theory—despite the demise of a rigorous protoplasmic alternative.[69] The philosophical ramifications of such questions were extensive: In 1908 Archibald Macallum argued that the apparent vitality of both colloidal suspensions and viruses provided a new window onto the possibility of spontaneous generation, and in 1917 Leonard Troland offered his speculative theory of enzyme self-replication through autocatalysis, with its explanatory relevance to the chemical nature of heredity.[70]

In the 1920s, continuing interest in the notion that colloids might be responsible for the fundamental features of life reflected a high regard for the physical sciences as a source of biological explanation. While biochemists, colloidal chemists, physiologists, and cytogeneticists viewed the problems of biology in different, sometimes even incommensurable, terms, they shared a faith that key questions about the nature of life could be answered by applying the (right) methods and tools of physics and chemistry. Moreover, many life scientists suspected that answers to their questions resided just beyond the realm accessible to the light microscope. The nature of subcellular biological entities—enzymes, colloids, proteins, antibodies, and genes—invited endless speculation and disagreement. The prominent biologist E. B. Wilson offered an update on T. H. Huxley's fifty-year-old protoplasmic theory in 1923 and emphasized the biological centrality of the submicroscopic realm to life: "We have now

66. Geison, "Protoplasmic Theory of Life."
67. Kohler, "Enzyme Theory."
68. See Edsall, "Proteins as Macromolecules"; and Fruton, *Molecules and Life,* esp. 156. Many chemists believed that enzymes were low-molecular-weight catalysts adsorbed onto a carrier, such as a protein (viewed here as a nonspecific colloid or colloidal aggregate).
69. Kohler, "Reception of Eduard Buchner's Discovery."
70. See Farley, *Spontaneous Generation Controversy;* Macallum, "On the Origin of Life"; Troland, "Biological Enigmas."

arrived at a borderland, where the cytologist and the colloidal chemist are almost within hailing distance of each other—a region, it must be added, where both are treading on dangerous ground."[71] Among the various residents of this borderland, viruses were particularly attractive subjects for investigation by physicochemical tools on account of their ability to self-reproduce. The *Biology and Medicine* volume of Jerome Alexander's encyclopedic *Colloid Chemistry* in 1928 included several articles on the colloidal nature of protoplasm; among them were "Filtrable Viruses" by Charles Simon and "Bacteriophage, a Living Colloidal Micell," by Félix d'Herelle. At issue in Simon's and d'Herelle's articles was whether viruses or bacteriophage could be considered animate, and if they were also particulate, how colloidal chemistry might be used to elucidate their nature.[72] Defining viruses, then, would have implications for many areas and disciplines, but researchers first needed to overcome the seeming intractability of isolating them.

TMV as a Model Virus

In the midst of these wide-ranging speculations about viruses as agents, plant researchers in the 1910s and 1920s sought to advance their understanding of mosaic diseases by using a variety of laboratory methods. TMV was a particularly suitable choice for research because of its relatively high transmissibility in comparison with other plant viruses, and because it was stable over time in filtered sap. Tobacco mosaic virus also has, despite its specific name, an extensive host range, including tomato, pepper, bean, petunia, and eggplant. One could thus straightforwardly compare rates and symptoms of virus infection in different species and in hybrids of closely related species.[73] Beyond these practical considerations, tobacco is, of course, an important economic crop in the United States and was therefore a priority for plant pathological research. (In Britain, by comparison, plant virus research tended to be directed toward potato pathogens.)[74] A handful of plant pathologists in the United States, predominantly affiliated with the U.S. Department of Agriculture (USDA) or state agriculture experiment stations, took up research on TMV. Their principal scientific contributions can be summarized along four lines of

71. Wilson, *Physical Basis of Life,* 26. This first William Thompson Sedgwick memorial lecture was named after T. H. Huxley's 1868 lecture in Edinburgh, "On the Physical Basis of Life."

72. The widely perceived promise of colloid chemistry to rationalize "the nature of protoplasm" is reflected in Nordenskiöld's 1928 survey, *History of Biology,* 595.

73. See, for example, Allard, "Specific Mosaic Disease"; Johnson and Grant, "Properties of Plant Viruses."

74. Wilkinson, "Development of the Virus Concept. . . . Part 3. Lessons of the Plant Viruses," 128.

research: virus infectivity, virus genetics, host resistance to infection, and immunochemical studies of the virus.

Investigations of virus infectivity resembled contemporary biochemical studies of enzyme activity under various conditions. Allard, mentioned above for his insistence on the contagious nature of TMV, assessed the effects of heat, germicidal agents, and chemical reagents on viral infectivity, publishing his results in a series of articles in the USDA's *Journal of Agricultural Research*.[75] A few years later, Maurice Mulvania conducted similar studies assessing the effects of light rays, heat, dialysis, and exposure to bacteria on the infectivity of TMV, publishing his work in the other major journal for research on plant diseases, *Phytopathology*.[76] The susceptibility of TMV to some of these inactivating agents led him to speculate that the virus might be a simple protein or enzyme.

A second fruitful line of research focused on genetic approaches. In 1926, H. H. McKinney first isolated a yellow-mottled variant of TMV and demonstrated the heritability of its distinct features.[77] McKinney also gathered naturally occurring strains of TMV from the Canary Islands, West Africa, and Gibraltar, and on this basis suggested in 1929 that the virus could, like an organism, mutate. "A satisfactory interpretation of the yellow and green mosaic associations cannot be given at this time because so few data are available. . . . It seems entirely possible that viruses may become altered locally in the plant, thus producing mutations, to use this term in its broadest meaning."[78] Several years later James Jensen corroborated McKinney's field observations through his own experimental studies of TMV strains, and documented the reversion of a yellow variant to the original mosaic strain.[79] In the 1920s and 1930s, genetic studies of plant mosaic viruses were potentially more useful for understanding general features of viruses than bacteriophage genetics, where lysogeny—the appearance of infected cultures without any contamination from an exogenous agent—greatly complicated the perceived interplay between infection and heredity.[80]

75. Allard, "Effects of Various Salts," "Some Properties of the Virus," and "Mosaic Disease of Tomatoes and Petunias."

76. Mulvania, "Studies on the Nature of the Virus."

77. McKinney, "Virus Mixtures." On McKinney's contributions, see Lederman and Tolin, "OVATOOMB," 242. In 1925 Eubanks Carsner had first reported on the attenuation of virus strains in plants, although he does not explicitly discuss it in terms of mutations: see "Attenuation of the Virus." In addition to his genetic studies, McKinney tried to use centrifugation and filtration to purify TMV; see "Quantitative and Purification Methods."

78. McKinney, "Mosaic Diseases," 576.

79. Jensen, "Isolation of Yellow-Mosaic Viruses."

80. Lederman and Tolin, "OVATOOMB." I take up virus genetics in chapter 6.

Plant pathologists also sought to understand susceptibility or resistance to mosaic disease in terms of the genetics of the host. Resistant strains of various crop plants were sought, motivated in large part by the need to control disease. Along these lines, McKinney observed that infection of tobacco with a mild strain of the mosaic virus (one which had been attenuated by passing it through other hosts or growing the host at elevated temperatures) protected the plant from susceptibility to other strains.[81] This phenomenon, termed "induction of immunity," seemed to be general to plant viruses and suggested the possibility of developing a live-vaccine-like method of protecting crops from severe viruses.[82] The mechanism of this immunity, however, was puzzling to plant pathologists, who could never detect antibodies in plant tissues. Nonetheless, the existence of such immune effects provided an important link with medical studies of viruses, for one of the defining features of viral diseases in humans was the lasting immunity upon recovery.[83]

Immunological methods could also be used as a tool to probe virus structure. Following upon Mayme Dvorak's success in raising rabbit antibodies to potato leaf extracts from healthy and mosaic-diseased plants, Helen Purdy of the Boyce Thompson Institute made a "study of the immunologic reactions with a tobacco mosaic virus to an antiserum from rabbits . . . with the hope of throwing more light upon the nature of the virus."[84] Her study revealed that TMV isolated from different hosts (from petunia to pepper) exhibited similar serological reactions, and that anti-TMV serum reacted differently to filtered sap from plants infected with tobacco mosaic, cucumber mosaic, and ringspot viruses.

The accumulating record of publications on TMV left unresolved questions about its mechanisms of infection and reproduction. Even so, plant virus research was widely viewed as important, for both disciplinary and scientific reasons. Plant pathology was increasingly recognized as an academic field of research. In 1933, the American Council of Education assembled a list of forty-eight disciplines characterized by graduate study, and plant pathology was one of twenty sciences listed.[85] In the

81. McKinney, "Inhibiting Influence."
82. See Lederman and Tolin, "OVATOOMB," 242.
83. See Rivers, "Some General Aspects," 7. Louis Kunkel's further observations of the immunity conferred on plants through virus infection were frequently cited: see Kunkel, "Studies on Acquired Immunity."
84. Purdy, "Immunologic Reactions with Tobacco Mosaic Virus," 702; and "Specificity of the Precipitin Reaction"; Dvorak, "Effect of Mosaic on the Globulin of Potato."
85. Letter from R. M. Hughes to Louis O. Kunkel, 22 May 1933, Kunkel papers, RAC RG450K963, box 4, folder 3. The others I have counted as scientific disciplines (excluding mathematics and engineering fields) are anatomy, animal nutrition, astronomy, bacteriology, biochemistry, biophysics, botany, chemistry, entomology, genetics, geology, human nutrition, pathology, physics, physiology, plant physiology, psychology, soil science, and zoology.

previous decades, leadership in the field had begun to pass from German centers to American institutions, marked by the founding of the American Phytopathological Society in 1908, which commenced publication of *Phytopathology* in 1911.[86] The federal government's support of plant research conducted at geographically scattered agricultural experiment stations as well as at the facilities of the USDA Bureau of Plant Industry built up a sizable network of laboratories for plant science.[87]

Within the field of plant pathology, viral diseases attracted keenest interest. One 1923 review offered this assessment of the field: "Among plant pathologists there is today no topic of more engaging interest and no problem more difficult than that of the nature of the causal agency in mosaic and allied plant diseases."[88] They singled out tobacco mosaic disease as the key representative for studies of viruses. A dozen years later, in a retrospective on research on plant virus diseases, Louis Kunkel cited the publication of three thousand articles on plant viruses between 1910 and 1935.[89] Kunkel's review of research on plant viruses in Thomas Rivers's volume, *Filterable Viruses,* also highlighted the central place of mosaic diseases in plant virology: 162 different species of plants had been observed to be susceptible to mosaic diseases; by comparison only 24 viral diseases outside of the mosaic disease group had been reported. Plant viruses, and especially those causing mosaic diseases, were viewed as relevant for understanding the fundamental biology of viruses. As John Caldwell noted in 1933, "During the past few years, as the symptomatology of plant virus diseases has been increasingly clarified, more attention has been directed to the study of the nature of the virus."[90]

Virus Research at the Rockefeller Institute for Medical Research

Interest in TMV as a laboratory material was not confined to agricultural researchers. In the 1920s and 1930s, bacteriologists and chemists at the Rockefeller Institute for Medical Research began to investigate TMV in their search to discern the general features of viruses. The reasons for this medical interest in TMV derived from the strong emphasis on virus research at the Rockefeller Institute as well as a growing interest there in comparative approaches to disease, encompassing even those afflicting plants.

86. On the founding of the society, see Ainsworth, *Introduction to the History of Plant Pathology,* 217–18. On the rise of the discipline in the United States, see Campbell, Peterson, and Griffith, *Formative Years of Plant Pathology.*

87. On the USDA, see Paul and Kimmelman, "Mendel in America," esp. 285–93.

88. Duggar and Armstrong, "Indications," 191.

89. Kunkel, "Virus Diseases of Plants."

90. Caldwell, "Physiology of Virus Diseases in Plants," 100.

The Rockefeller Institute was established in 1901 through the philanthropy of John D. Rockefeller Sr. as a center for experimental medical research, modeled on the Pasteur Institute. But whereas the European laboratories built in honor of Pasteur, Koch, and Ehrlich were centered on bacteriology (and hence, from the American perspective, "instruments for extending the personalities of these men of genius"),[91] the Rockefeller Institute was intentionally developed to support a wider range of research areas in medical science.[92] Under the directorship of Simon Flexner, laboratories were organized around divisions of pathology and bacteriology, experimental surgery, physiology, and chemistry, with units of cancer research, biophysics, chemistry, and chemical pharmacology being added over the years. From the outset, the physical sciences were held in high esteem as a source of tools and of standards for experimental rigor in medical research. Flexner contended that the Institute researchers' reliance on chemistry and physics further distinguished the Rockefeller Institute from its European counterparts.[93]

In keeping with the broad and interdisciplinary ideals for research, Flexner espoused a culture of scientific cooperation. As Lily Kay has argued, this vision of research had a strong resonance with the Rockefellers' business ideology.[94] Flexner himself remarked, "The Institute is favorably

91. Simon Flexner, "Developments of the First Twenty Years of the Rockefeller Institute and an Outlook for Further Growth," Scientific Reports, RAC RU 439, box 3, vol. 14, 1925–26, 4.

92. In Flexner's words, "From the beginning, the organization of the Rockefeller Institute departed essentially from these models: It was not built about an outstanding personality" (ibid.). These contrasts between the Rockefeller Institute and its European counterparts are Flexner's own justifications for the American model, and they greatly oversimplify Pasteur's and Koch's enterprises. For a historical assessment, see Weindling, "Scientific Elites and Laboratory Organisations." Weindling emphasizes the orientation of both Pasteur's and Koch's institutes to the production of innovative therapies and public health, with links to the state or the military.

93. "The Rockefeller Institute departed from the earlier models also because between 1886, when the Pasteur Institute was founded, and 1904, when the first laboratories of the Rockefeller Institute were opened in New York, the scientific outlook on medical research was undergoing rapid changes. Bacteriology as such, the cornerstone of the three European institutions, was, without losing its important position, becoming more closely linked with pathology, was beginning to draw vitally on chemistry, and presently was to draw significantly on physics also. These changes reflected the shift in viewpoint of the biological processes in general—whether physiological or pathological—which were becoming more and more closely related to the new knowledge being derived from fundamental discoveries in organic and physical chemistry and in physics.... This wider conception of medical research has been a guiding factor in the development of the Institute" (Simon Flexner, "Sketch of the First Twenty-five Years of the Rockefeller Institute for Medical Research," Scientific Reports, RAC RU 439, vol. 18, 1929–30, 15–16). Again, Flexner's views on the Rockefeller Institute's distinctive features are admittedly self-serving.

94. Kay, *Cooperative Individualism and the Growth of Molecular Biology;* and *Molecular Vision of Life,* 26–27. See also Abir-Am, "'New' Trends," in which she refers to the

circumstanced for bringing a kind of purposeful combined or cooperative effort to bear on certain problems, instead of leaving the separate efforts more or less to chance undertakings. This desirable relationship is oftener secured in industrial research than in university research."[95] As compared with universities, where researchers were physically separated into departmental buildings, the Institute was designed "to bring the workers into the most intimate possible personal and scientific relationships."[96] In such an open arrangement, boundaries were still required to assure fair credit. "An unwritten rule at the Institute has been that of independence of subject or field of research. No two men are encouraged to work on precisely the same problems, or in exactly the same fields."[97]

Flexner's particular interest in virus research derived in part from his own work on poliomyelitis, which he and Paul Lewis found to be carried by a filterable virus in 1909.[98] Flexner was eager to see breakthroughs in virus research occur at the Institute, even as he tended to view "filter-passing microbes" as not fundamentally different in nature from disease-causing bacteria.[99] Under his directorship, viral diseases were intensively studied by the Institute's pathologists, physiologists, and chemists. In 1911, F. Peyton Rous found that chicken sarcoma virus could be transmitted by a filterable virus, an observation that attracted worldwide attention.[100] At the Department of Animal Pathology (located an hour away in the more rural Princeton, New Jersey), Carl TenBroeck and Richard Shope focused research efforts on viral agents of hog cholera, equine encephalitis,

---

Rockefeller Foundation as the "extra-mural" program of Rockefeller philanthropy in research support, as compared to the "intra-mural" Rockefeller Institute. It is difficult to find archival evidence of such a coordinated science policy.

95. Flexner, "Developments of the First Twenty Years," 44 (see citation in note 91 above). On industrial research in the United States, see Reich, *Making of American Industrial Research;* and Carlson, "Innovation and the Modern Corporation."

96. Flexner, "Developments of the First Twenty Years," 45.

97. Ibid.

98. Flexner and Lewis, "Nature of the Virus of Epidemic Poliomyelitis." Some accounts credit Karl Landsteiner and his assistants with the discovery that the etiological agent of poliomyelitis was a virus (see Paul, *History of Poliomyelitis,* 98–100). The difficulty in assigning priority is due to the several ways of identifying a disease as viral—Landsteiner and Popper showed that a bacteria could not be cultivated from the infectious extract, whereas Simon Flexner and his coworker showed that the agent was filterable. See Landsteiner and Raubitschek, "Demonstriert mikroskopische Präparate," and Landsteiner and Popper, "I. Übertragung der Poliomyelitis acuta auf Affen."

99. Flexner valued Rivers's eminence as a virologist enough to tolerate Rivers's public criticisms of his belief that viruses were bacterial in nature. On Flexner's and Rivers's disagreement over the cultivation of viruses, see Benison's oral history, *Tom Rivers,* 110–11.

100. Rous, "Transmission of a Malignant New Growth." On the reception of Rous's work on chicken sarcoma virus, see Rous, "Challenge to Man"; Corner, *History of the Rockefeller Institute,* 110–11; and Helvoort, "Century of Research."

rabbit papilloma, and swine influenza. Flexner recruited Thomas Rivers from Johns Hopkins in 1923 with the mandate of developing a vibrant virus research program at the institute-affiliated hospital.[101]

In 1925, the USDA recognized the expertise of virus researchers at the Rockefeller Institute when they appointed Peter Olitsky to lead a commission on foot-and-mouth disease in Europe. Questions about the "nature of the virus" were intertwined with the pressing need for disease control. Olitsky wrote at the outset of the commission's efforts that "questions were often posed as to whether filter-passing viruses of the type of foot and mouth diseases were inanimate chemicals, or multiplying, living bodies; or whether they were fluid (as postulated by [Beijerinck's] theory of *contagium vivum fluidum*) or particulate, that is, corpuscular."[102] He concluded that the virus was corpuscular, although it behaved as a chemical, and that it might well be a living body.

The utility of TMV as a good model for viruses in general was not lost on the Rockefeller Institute bacteriologists. Olitsky's research on the bacteriological cultivation of TMV had begun in 1924, and after returning from Europe he ordered greenhouses for the Rockefeller Institute in order to study the disease in tobacco as well as tomato plants.[103] Flexner justified the extension to plant disease in practical terms: "The class of filter-passers has a very large significance for pathology and for human and animal diseases. They also produce large losses among cultivated plants—tobacco, tomato, potato, etc. The Institute's studies in the field have been significant. Dr. Olitsky is perhaps a leading investigator in it."[104]

Olitsky subsequently collaborated with enzyme chemist Northrop to determine the physical properties of partially purified TMV. He also worked with Frederick L. Gates, in the division of biophysics, subjecting vesicular stomatitis virus to inactivation by ultraviolet light. Virus research was not only at the core of the Rockefeller Institute's scientific interests, but it provided opportunities for the kind of interdisciplinary

101. Benison, "History of Polio Research," 318. Rivers's influence went beyond the researchers in the Department of the Hospital; as he put it, "It is true that I proselytized for virology among young investigators, and that I did try to get people in the Institute laboratories, as opposed to the hospital laboratories, to work on things I was interested in" (Benison, *Tom Rivers*, 131). For summaries of the ongoing research at the Department of Animal Pathology, see Corner, *History of the Rockefeller Institute*.

102. Olitsky, Scientific Reports, RAC RU 439, box 3, vol. 14, 1925–26, 78.

103. On Olitsky's early positive results on the cultivation of TMV, see his letters to Flexner, 6 Apr, 4 Jun, 4 Aug, and 24 Aug, all 1924, Flexner papers, APS F365, folder Olitsky, Peter K. #4; on the greenhouses, telegram Flexner to Olitsky, 15 Nov 1927, Olitsky papers, RAC RU 450OL4, box 4, folder 1, "Correspondence 1912–34."

104. Flexner, "Developments of the First Twenty Years," 28 (see citation in note 91 above).

collaborations that Flexner esteemed. Unfortunately, Olitsky's strictly bacteriological investigations of TMV—resulting in his claim that he had cultured the virus in cell-free media—were rapidly criticized by established plant pathologists.[105] And the combination of medical and physicochemical approaches was not necessarily harmonious. Gates remarked that his demonstrations of the chemical reactivity of vaccine virus and bacteriophage had failed to move his collaborators to admit the chemical nature of viruses.[106]

A Branch of Plant Pathology at Princeton

In 1926 the Board of Scientific Directors began considering a plan to formally extend a broadly comparative approach to studying disease by adding a Division of Plant Pathology to the Department of Animal Pathology in Princeton, New Jersey.[107] Louis Kunkel, who specialized in the study of viral diseases in plants, quickly emerged as an attractive choice for the position of director of the new unit.[108] Kunkel was already familiar to the Institute researchers, having been one of the few outside authors chosen to write the chapter on plant viruses for Thomas Rivers's edited collection *Filterable Viruses,* first published in 1928. Kunkel's laboratory had also provided one of the refutations of Peter Olitsky's claim to have artificially cultured TMV.

Flexner first contacted Kunkel to express interest in his work in January 1930, and negotiations for the move, new staff, and laboratory facilities were resolved the following December. Flexner stressed to Kunkel the unique opportunity at the Rockefeller Institute for comparative virus research: "It is so exceptional that problems in plant pathology can be studied alongside related problems in animal and human pathology."[109] Kunkel's acceptance letter to Flexner voiced similar hopes: "With your help and advice and with the cooperation of our colleagues in animal

105. Olitsky, "Experiments on the Cultivation of the Active Agent of Mosaic Disease" (1924), and "Experiments on the Cultivation of the Active Agent of Mosaic Disease in Tobacco and Tomato Plants" (1925); Mulvania, "Cultivation of the Virus of Tobacco Mosaic by the Method of Olitsky"; Purdy, "Attempt to Cultivate an Organism from Tomato Mosaic"; Goldsworthy, "Attempts to Cultivate the Tobacco Mosaic Virus."

106. Gates, Scientific Reports, RAC RU 439, box 3, vol. 15, 1926–27, 94–95. Gates had other personal as well as scientific ties to the Rockefeller Institute; his father, Frederick Taylor Gates, was President of the Board of Trustees and had advised John D. Rockefeller in the establishment of his philanthropies since 1891; see Corner, *History of the Rockefeller Institute,* 19 ff.

107. Corner, *History of the Rockefeller Institute,* 313–15.

108. For some personal reminiscences on achievements in virus research at the Princeton branch of the Rockefeller Institute, see Schlesinger, "Virus Research at the Princeton Rockefeller Institute."

109. Flexner to Kunkel, 3 Dec 1930, Kunkel papers, RAC RU 450K963, box 1, folder 1.

pathology, I am sure we shall succeed in prying into many of the secrets of plant diseases."[110] In November 1931 ground was broken for the new plant pathology buildings in Princeton; the Rockefeller Institute was already funding Kunkel's research at his current institution. In preparation for his new position, Kunkel embarked on scientific exploration of European laboratories that year, fully paid for by the Institute.[111]

Kunkel's background reflected the diverse institutional settings for research on plant diseases in the early part of the century. Having taken his Ph.D. at Columbia in 1914, Kunkel also trained in Germany for a year before taking a post as pathologist with the USDA in their Bureau of Plant Industry. After six years of work on potato wart as well as on corn and sugar mosaic diseases, Kunkel joined the Experiment Station of the Hawaiian Sugar Planters' Association as associate pathologist. This agribusiness setting proved to be very conducive to research. Within four months, Kunkel had identified an aphid as the hitherto mysterious vector of sugar cane mosaic, and he went on to publish a number of histological and cytological studies of diseased sugar cane. These findings brought Kunkel acclaim and new opportunities. In 1923, he accepted a position as pathologist at the new Boyce Thompson Institute for Plant Research in Yonkers, New York. Kunkel assembled a diverse research team at his new laboratory, including Francis Holmes and Helen Purdy (later Beale), both of whom made important contributions to the TMV field. This institute had been founded by Colonel William Boyce Thompson, a wealthy New York banker whose hobby was growing prize aster flowers. Kunkel did not neglect the patron's practical concerns. He devoted significant time to investigating a disease of asters known as "yellows," demonstrating that it was caused by a virus and carried by leaf-hoppers.

Kunkel moved into his new Princeton laboratory in 1932. Two aspects of his group's research agenda there are striking. First, although Kunkel had been asked to set up a division of plant pathology that included studies of bacterial and fungal diseases, he focused the effort entirely on his specialty, mosaic viruses.[112] Holmes, William C. Price, and Philip R. White, who had all been part of Kunkel's team in Yonkers, came with him to Princeton. They continued studying the genetics of resistance to tobacco mosaic disease, isolating strains of the virus, and seeking to develop assay and tissue culture techniques.[113] New postdoctoral fellows Herbert T. Osborn and Ernest L. Spencer began studies of the insect transmission

110. Kunkel to Flexner, 24 Jan 1931, Kunkel papers, RAC RU 450K963, box 1, folder 1.
111. Kunkel papers, "News Item for Publication, April 6, 1931," RAC RU 425, box 2, folder 5 (1930–47); and RAC RU 450K963, box 1, folder 1.
112. Corner, *History of the Rockefeller Institute,* 315.
113. Scientific Reports, RAC RU 439, box 5, vol. 21, 1932–33, 369–77.

of pea mosaic as well as nutritional studies of tobacco mosaic. James H. Jensen and H. H. Thornberry, graduate students in plant pathology, took up projects on mutations of TMV and quantitative studies of virus filtration. Together they composed perhaps the largest single group in the country of researchers focused on the study of mosaic viruses, especially TMV.

Second, Flexner specifically cultivated Kunkel's decision to include chemical approaches to plant virus research. Kunkel actually had two members of his group at the Boyce Thompson Institute, Carl G. Vinson and A. W. Petre, who were attempting the chemical purification of TMV. Nonetheless, the director there complained to a Wisconsin plant pathologist in 1932 that he had never been able to interest Kunkel in chemical approaches; he expected that problems concerned with the "chemical nature of viruses and with their isolation and purification" would not be touched by Dr. Kunkel at Princeton.[114] His prediction proved altogether wrong.

Flexner requested a reprint of Vinson and Petre's paper as he corresponded with Kunkel about the Princeton situation, and he planted a seed for new collaborative efforts in this direction after Kunkel's first luncheon seminar at the Institute:[115]

> The next time you are at the Institute I should like you to talk with Dr. [P. A. T.] Levene. He has some ideas on the chemistry of the viruses which I think may appeal to you. If you do, you can, I believe, help in the collection of proper material to be studied. You know of course Dr. Levene's wide knowledge and experience in dealing with minute and delicate chemical substances.[116]

The collaboration never actually got off the ground, for it turned out that the quantity of TMV Levene needed for his proposed study would have required Kunkel to grow between two and ten thousand infected tobacco plants.[117] However, Flexner specifically encouraged Kunkel to write Harvard's James Conant to ask him to recommend a young chemist to take up the project of investigating TMV in the new Division of Plant

114. Letter from William Crocker to L. R. Jones, 5 May 1932, Jones papers, folder "Correspondence, 1932." According to George Corner, after moving to Princeton, Kunkel continued supplying Vinson in Yonkers with tobacco mosaic virus for chemical studies (Corner, *History of the Rockefeller Institute,* 317).

115. Flexner to Kunkel, 26 Jan 1931, Flexner papers, APS F365, folder Kunkel, L. O.

116. Flexner to Kunkel, 5 Nov 1931, Flexner papers, APS F365, folder Kunkel, L. O.

117. Although Kunkel claimed that growing and inoculating this number of tobacco plants would pose no great difficulty or cost, Flexner, upon reviewing Kunkel's letter, warned Levene not to ask for another budget increase; see Kunkel to P. A. T. Levene, 17 Nov 1931, Kunkel papers, RAC RU 450K963, box 1, folder 1; Flexner to Levene, 27 Nov 1931, marked "personal"; Simon Flexner papers, Microfilm Copy at RAC, reel #9, correspondence to P. A. Levene, 1934–45.

Pathology.[118] Conant recommended his postdoctoral student, Samuel Kamerling, but Wendell Stanley, who was already at the Rockefeller Institute as a postdoctoral fellow with physiologist W. J. V. Osterhout, also expressed interest.[119] In the end, Kunkel accepted Stanley to launch the group's chemical work on TMV.[120]

Thus Stanley did not initiate, but rather inherited, an emphasis at the Rockefeller Institute on the chemical study of viruses, including TMV. This was part of a general program geared toward the use of chemical and biophysical techniques to understand disease agents. But more specifically, the chemical isolation of a virus such as TMV offered a way to circumvent the apparent futility of using bacteriological techniques. Thomas Rivers summarized the predicament facing virus researchers in 1928 in the introduction to *Filterable Viruses:* "How, then, is progress to be made in the study of virus diseases? It will be difficult for the best-trained workers and doubly difficult for those poorly trained to progress rapidly in this field. As long as viruses resist cultivation on simple media, just so long will it be necessary to study them in the host, in the tissues of the host, or in the emulsions and filtrates of the tissues."[121] Chemistry provided an alternative set of techniques for recovering and characterizing viruses as objects.

Conclusions

The origin of virology is most often narrated as the history of a concept, that of the virus as a nonbacterial pathogen.[122] This chapter has instead emphasized the materials and laboratory examples that shaped the emerging meaning of the category "filterable virus." Due both to the historical precedents set by the earliest observations of viruses and the fact that some pathogens proved more obliging in the laboratory than others, a handful of "filterable viruses" contributed disproportionately to the characterization of this class of pathogens. The titles given several of the

118. Kunkel to Conant, 8 Apr 1932, Kunkel papers, RAC RU 450K963, box 6, folder 2.

119. Kunkel to Conant, 18 May 1932, and response, 19 May 1932, Kunkel papers, RAC RU 450K963, box 2, folder 3. Osterhout took on Conant's student Samuel Kamerling as Stanley's replacement.

120. In a letter to Flexner years later, Stanley recalled the transfer in somewhat different terms: "I shall never forget the occasion when I was called into your office and asked if I'd like to go to Princeton to work on viruses—and my stumbling answer due to the fact that, although I didn't know what a virus was, I did want to get out of New York and go to Princeton." Because Stanley did a minor field in bacteriology while in graduate school and had studied mosaic diseases (see the next chapter), this statement of his ignorance (which has often been quoted) may be a bit disingenuous; see Stanley to Flexner, 6 Nov 1940, Flexner papers, APS F365, folder Stanley, W. M.

121. Rivers, "Some General Aspects," 22.

122. Hughes, *Virus*.

contributions to Thomas Rivers's 1928 *Filterable Viruses* suggest the important role of such examples: "Virus Diseases of Man As Exemplified by Poliomyelitis," "Virus Diseases of Mammals As Exemplified by Foot-and-Mouth Disease and Vesicular Stomatitis," and "Virus Diseases of Fowls As Exemplified by Contagious Epithelioma (Fowl-Pox) of Chickens and Pigeons." Moreover, scientific and medical reviewers from the earliest years of the century grouped these exemplars together in an effort to ferret out the general features of viruses as a class of pathogens.

TMV was viewed as a particularly promising candidate for laboratory experimentation and one that might illuminate the common features of viruses. Its stability and resilience made TMV especially suitable for investigation. As two agricultural researchers noted in 1933, "Tobacco mosaic retains its infectivity for years in dried material or in expressed sap; it is highly infectious and can be diluted to one in a million and still cause infection; it has a thermal death-point of 88°C for 10 min. heating; and it has considerable resistance to chemicals."[123]

If the advantages of TMV were pragmatic, the rewards of its domestication in the laboratory went beyond the control of tobacco disease. As H. H. Dale observed, "The problems presented by the nature and behaviour of the viruses cannot fail to raise questions of the greatest interest to anyone concerned with general physiological conceptions. What is the minimum degree of organisation which we can reasonably attribute to a living organism? What is the smallest space within which we can properly suppose such a minimum of organisation to be contained? Are organisation, differentiation, separation from the surrounding medium by a boundary membrane of special properties, necessary for the endowment of matter with any form of life?"[124] Viruses, in sum, appeared to have some properties of life and yet did not seem to be living organisms as defined by bacteriology; consequently their study opened general questions about whether their vital properties—and indeed all vital properties—might be explained in terms of physics and chemistry.

123. Samuel and Bald, "On the Use of the Primary Lesions," 71.
124. Dale, "Biological Nature of the Viruses," 599.

3

Crystals at the "Threshold of Life"

It has astonished the scientific world that a single molecule can be the
causative organism of a disease. How can a crystal be made up of living
molecules?

Barclay Moon Newman, 1937

When Wendell Stanley transferred from the Rockefeller Institute's Department of General Physiology to the rural Department of Animal and
Plant Pathology in 1931, he was joining a major center of comparative
virology. It was from his hands at Princeton that the virus received its
most stunning identification in 1935, namely, as a crystalline molecule.
In the midst of ongoing debates about the nature of the virus, Stanley's
announcement in *Science* on 29 June of that year, "The Isolation of a
Crystalline Protein Possessing the Properties of Tobacco-Mosaic Virus,"
was widely hailed in both the scientific and popular media. Almost three
decades later, George Corner, the official historian of the Rockefeller
Institute, asserted that "no discovery made at the Rockefeller Institute,
before or since, created such astonishment throughout the scientific world
as did this."[1] By claiming to have isolated the virus as a crystal, Stanley vindicated a physicalist approach to biology and medicine. For researchers
at the Rockefeller Institute, Stanley's feat of obtaining a virus in "pure"
form marked the triumph of chemical over bacteriological methods in the
quest to apprehend viruses.

While Stanley's professed crystallization of TMV was the symbolic
culmination of his work, his establishment of a productive *experimental
system* using tobacco mosaic began two years earlier, when he entered
Louis Kunkel's laboratory and set out to repeat the published experiments
of Kunkel's previous associate Carl Vinson. This chapter focuses on
TMV in terms of Stanley's experimental system—on the sequence and

---

1. Corner, *History of the Rockefeller Institute,* 320. Commentary on Stanley's 1935 paper
is common in the history of biology; most prominently, Lily Kay has analyzed the reasons
for and impact of Stanley's isolation of TMV as a protein in "Stanley's Crystallization."

arrangement of materials, techniques, instruments, and methods for visualizing the virus as an object. The diseased tobacco plant did not determine these experimental manipulations, but neither was the shaping of the system unconstrained by the plant material. Only in the process of taking the system in particular directions, however, do its "inherent" possibilities and limitations become evident.

An experimental system manifests multiple decisions by the researcher about how to isolate and visualize the object or phenomenon of study; in this sense the system registers and reflects the laboratory habits, culture, and training of the researcher.[2] The manner in which systems epitomize individual decisions and local resources is discernible in the relatively unusual case of replication of experiments. In the case at hand, Vinson and Stanley worked under the same person (Vinson had been at the Boyce Thompson with Kunkel) and ostensibly on the "same" object, TMV, yet their systems for experimentation were noticeably distinct. Vinson was a plant pathologist conducting chemical experiments, whereas Stanley was a chemist working on plant materials. Without reifying disciplinary identity, one can note the chemical way in which Stanley ordered his materials, attending more carefully than had Vinson to solution conditions such as pH and composing his results in terms of the graphical representation conventional to biochemistry.

Stanley's choices were not the only ones that mattered: his project on TMV had been specifically cultivated by Simon Flexner, and he benefited from the prevailing high regard for chemical approaches at the Institute.[3] More concretely, colleagues at the Institute contributed expertise and materials. Stanley's accomplishments relied on a quantitative assay for virus activity developed in the Kunkel laboratory, as well as on the guidance and pure enzymes of John Northrop, whose laboratory was only one hundred yards away from Stanley's bench.[4] In fact, Northrop's crystalline enzymes proved to be key models as well as reagents in Stanley's crafting of TMV as a chemical substance.

Stanley's public presentation of the virus as a crystalline protein dramatized rather than resolved the question of whether viruses are alive. The

2. On experimental systems, see Kohler, "Systems of Production"; Rheinberger, "Experiment, Difference, and Writing," and *Toward a History of Epistemic Things*. While my analysis is indebted to Rheinberger's brilliant insights into experimental practice, I (along with Jean-Paul Gaudillière) seek to ascribe more agency to the researcher than Rheinberger has; see Creager and Gaudillière, "Meanings in Search of Experiments," esp. 3–5; and Gaudillière, "Wie man Labormodelle."

3. Flexner's continuing interest in Stanley's progress is documented in his correspondence with John Northrop during May 1935 (Flexner papers, APS F365, folder Northrop, John H. #3).

4. Stanley, "Some Chemical, Medical, and Philosophical Aspects of Viruses," 146.

image of the virus as a living crystal evoked the familiar analogy of crystallization to biological growth and reproduction, further destabilizing the boundary between life and nonlife.[5] At the least, the needle-like paracrystals of TMV conveyed the immobility and ontological certainty of viruses as chemicals. But Stanley's continuing experimentation after the initial purification shows how uncertain and shifting was the virus as a laboratory entity. Even before the longer account of the crystallization of tobacco mosaic virus was published, Stanley tinkered with his method of isolation, re-forming the virus as a chemical object. Researchers who tried to repeat Stanley's crystallization obtained material with different characteristics than he had reported, and an exchange of letters records the intense cross-Atlantic negotiations over the chemical nature of the virus.

In the end, Stanley's system and his authority as a chemist were resilient enough to absorb corrections, and his sheer productivity in generating further results was taken as support of his early claims. A close analysis of the experimental life of TMV following Stanley's 1935 announcement reveals a contrast between the crystalline icon in the press and the changing referent in the laboratory. Moreover, the representation of TMV as a crystal had a life of its own in the late 1930s, playing into heated debates about the origin of life and reinforcing expectations that physics and chemistry might further illuminate the nature of life.

Establishing an Experimental System in Kunkel's Laboratory

Wendell Stanley had been trained as an organic chemist by Roger Adams at University of Illinois during the 1920s, when its division of organic chemistry "flourished on the surging postwar demand for industrial chemists."[6] The division's close ties with industry did not sully the academic reputation of Illinois's researchers; Adams, as chair from 1926 to 1954, has been credited with creating the most outstanding department of chemistry of its time.[7] As a graduate student, Stanley received a minor in bacteriology, and the notes he took in a course on that topic included a survey of research on filterable viruses, including "tobacco plant mosaic disease."[8] His first twelve research articles concerned diphenyl

5. Lorch, "Charisma of Crystals in Biology." Among the many advocates of the biological significance of crystallization, Lorch (456) includes this quotation from Herbert Spencer's *Principles of Biology:* "Perhaps the widest and most familiar induction of Biology is that organisms grow.... Crystals grow.... Growth is indeed a concomitant of Evolution.... The essential community of nature between organic growth and inorganic growth is ... that they both result in the same way." See also Haraway, *Crystals, Fabrics, and Fields.*

6. Kohler, *From Medical Chemistry to Biochemistry,* 266. On the rapid growth after 1910 of industrial jobs for organic chemists, see Thackray et al., *Chemistry in America.*

7. Tarbell, Tarbell, and Joyce, "Students of Ira Remsen and Roger Adams."

8. Stanley papers, carton 20, folder "Class Notes, University of Illinois," 6.

compounds and other synthetic chemicals assayed for their bactericidal action against *Mycobacterium leprae,* the disease agent held responsible for leprosy.[9] Thus, although Stanley did not formally train in biochemistry, about which he later expressed some regret to one of his professors,[10] his interests as a chemist were from the outset medical. After completing his Ph.D., Stanley remained an extra year in Adams's laboratory to prepare sufficient amounts of the antileprosy compounds to enable their clinical testing.[11] Stanley then accepted a National Research Council postdoctoral fellowship to spend two years working in Munich with Heinrich Wieland (who had just received the Nobel Prize for Chemistry in 1927), a choice which reflected the high regard for German structural chemistry in his home department. He returned to the United States in 1931 and began postdoctoral work in W. J. V. Osterhout's laboratory in the Department of General Physiology at the Rockefeller Institute. His work on membrane permeability to ions in the marine organism *Valonia* resulted in four more publications and introduced Stanley to plant physiology.

Upon moving to Princeton to join Kunkel's Department of Plant Pathology, Stanley met John Northrop, a distinguished chemist who was also stationed there. Northrop began his career at the Institute as an assistant to Jacques Loeb, and just after Loeb's death in 1924, Northrop became a member of the Department of General Physiology. However, he soon petitioned to move his group to the rural quarters of the Department of Animal Pathology, which had been in Princeton since 1915, because he "detested city life."[12] Northrop's group had been working for several years to purify and crystallize digestive enzymes, and in 1930 bacteriologist Alfred Krueger joined his laboratory to extend the same methods to the study of bacteriophage. Thus, as Stanley began his work as the chemist in Kunkel's group, he had recourse to local expertise about purifying and assaying enzymes.

Stanley began in the fall of 1932 by repeating what had already been published on the chemistry of TMV. In doing so, Stanley set himself in direct competition with Carl G. Vinson, whose 1927 note in *Science* first reported his partial success in isolating tobacco mosaic virus with lead acetate.[13] After Kunkel moved to Princeton, Vinson and coworker Petre continued to work on TMV at the Boyce Thompson Institute. Their subsequent articles and abstracts detailed their various strategies in attempting

9. List of Publications of Wendell Meredith Stanley, 308x S789c, Bancroft Library, v. 1.

10. "Recently I have regretted very much the fact that, while at Illinois, I thought I was so busy with organic chemistry that I had no time for biochemistry" (Letter from Stanley to W. C. Rose, 30 Jul 1935, Stanley papers, carton 11, folder "Rose, William C.").

11. Shriner, "William H. Nichols Medalist," 751.

12. Corner, *History of the Rockefeller Institute,* 173.

13. Vinson, "Precipitation of the Virus."

to purify TMV.[14] Stanley utilized their published methods for precipitation of the virus with lead acetate and safranin, and he also repeated their
reported investigations of the effect of digestive enzymes—catalysts that
break down proteins—on the integrity of the virus.

By using different controls and attending carefully to pH, Stanley
offered a very different interpretation of the reported inactivation of
TMV. Vinson and Petre had published observations that pepsin, emulsin,
and yeast extract did not affect TMV, but that trypsin, papain, and pancre-
atin inactivated the virus. Stanley focused on pepsin and trypsin, both of
which had been purified and investigated in John Northrop's laboratory.
Repeating Vinson's result, Stanley showed that trypsin (obtained in crystalline form from Moses Kunitz in Northrop's lab) completely inactivated
the virus at neutral pH. However, further experiments revealed that this
effect was *not* due to proteolytic action of the enzyme on the virus, but to
the interaction of trypsin with the plant leaves.[15] Thus the trypsin studies,
while demonstrating elegantly the importance of controls in biochemical experimentation, revealed nothing about the chemical nature of the
virus.

The pepsin studies, however, did yield a positive result. In contrast
to Vinson's report that pepsin had no effect on TMV, Stanley found that
pepsin (both the commercially available form from Parke-Davis and crystalline pepsin, courtesy of Northrop's researcher Roger Herriott) effected
a nearly complete inactivation of TMV at low pH. At neutral pH, as in
Vinson's experiments, pepsin had no effect on the infectivity of TMV, but
this was because pepsin is only active in acidic conditions. Stanley inferred
from the inactivation of TMV at low pH that pepsin was specifically cata-
lyzing the breakdown of the virus into constituent polypeptides. Stanley
concluded that the virus was either itself a protein or closely associated
with a protein.[16]

In both the trypsin and pepsin studies, Stanley reworked Vinson and
Petre's published observations in such a way as to corroborate and
account for their results while undermining their interpretation. Stanley
thus established himself as the more credible chemical researcher even
before he attempted to follow their steps further and purify the virus.

14. Vinson's other publications at this time included the following: Vinson and Petre,
"Mosaic Disease of Tobacco," "Mosaic Disease of Tobacco. Part 2," and "Progress in Freeing the Virus"; Lojkin and Vinson, "Effect of Enzymes"; and Waugh and Vinson, "Particle
Size of the Virus." Kunkel sent copies of Vinson's papers to Stanley upon appointing him to
the laboratory (Letter from Kunkel to Stanley, 8 Jun 1932, and Stanley's reply to Kunkel,
13 Jun 1932, Kunkel papers, RAC RU 450K963, box 14, folder 6).

15. Scientific Reports, RAC RU 439, box 5, volume 21, 1932–33, 371–72; Stanley, "Chemical Studies on the Virus of Tobacco Mosaic. Part 1."

16. Stanley, "Chemical Studies on the Virus of Tobacco Mosaic. Part 2," 1289.

In this regard, Stanley benefited not only from Vinson's deficiencies as a chemist, but also from his weaknesses as an author. In comparison with Vinson's papers, which offer seemingly undigested accounts of actual experiments and qualified interpretations, Stanley's experimental findings are tightly constructed as rational investigative narratives that provide definitive answers to carefully framed questions.[17] In the introduction to his first paper on TMV, the study of trypsin's effects, he opens with a clear problem: Vinson has reported that trypsin inactivates TMV, whereas Caldwell has found the inactivation to be reversible upon heating, a seemingly paradoxical result:

> If the virus were actually split or digested by trypsin, as first suggested by Vinson, reactivation by heat would be improbable. Furthermore, if the inactivation were due to proteolysis, one would expect pepsin, rather than trypsin, to be an active proteolytic agent. Trypsin usually fails to digest native proteins, whereas pepsin is usually able to digest both native and denatured proteins. It was decided, therefore, to reinvestigate the action of trypsin and of pepsin on the virus of tobacco mosaic. The present paper records the results of a study of some effects of trypsin upon this virus.[18]

These investigations are not original—they are repeating published work—but Stanley defended them on the basis of their superior experimental logic.

The differences between Vinson's and Stanley's science are not restricted to how they presented their results; although working on the "same" virus, Stanley established a distinct and more productive experimental system. Stanley employed the same elements Vinson had in his investigations, such as diseased tobacco plants, chemical reagents, and benchtop centrifuge, yet he altered their usage in important ways. In particular, Stanley privileged the pH meter as a key implement for rationalizing the results of his studies of virus inactivation. This choice was not the result of his keener scientific insight so much as recent experience and local resource. In Stanley's studies with Osterhout of the accumulation of electrolytes in the marine plant *Valonia,* acid-base considerations proved critical to the development of their model for physiological regulation.[19] In fact, Duncan MacInnes and Malcolm Dole had developed a pH meter at the Rockefeller Institute specifically for Osterhout's investigations of

17. Creager and Swan, "Fashioning the Virus." I thank Judy Swan for allowing me to draw on our joint work in writing this section.
18. Stanley, "Chemical Studies on the Virus of Tobacco Mosaic. Part 1," 1055.
19. Osterhout and Stanley, "Accumulation of Electrolytes. Part 5." Stanley probably brought a sense of the importance of attending to pH with him to Osterhout's laboratory, reflecting his chemist's training.

the electrical properties of plant cells.[20] Publications from the Northrop laboratory similarly showed careful attention to pH in their work on digestive enzymes. As Stanley turned to the chemical reactivity of TMV, he constructed elaborate experiments that included multiple time-interval and pH points to discern the effects of enzymes and chemicals on infectivity (see fig. 3.1). By making pH a variable worthy of study, Stanley recast his experiments concerning acid-base effects as the key controls showing that Vinson and Petre's results were flawed not because their enzymes were bad but because they had not attended to the conditions of their use.[21]

Stanley also utilized the findings of other members of the Kunkel laboratory to make his system more sensitive than Vinson's in two respects. First, he employed colleague Francis Holmes's local lesion method in bean and tobacco plants to assay active virus concentration. This method allowed for far more quantitative results than Vinson and Petre assembled.[22] Ordinarily, plant pathologists transmitted the disease experimentally by the "needle-prick" method, followed (when successful) by the appearance of systemic infection in a tobacco plant.[23] In 1928 Holmes published a new technique developed for quantifying plant virus, analogous to the plaque titration of bacteriophage advanced by Félix d'Herelle.[24] In mosaic-diseased *Nicotiana glutinosa,* a less common tobacco species, infection resulted in the appearance of visible lesions rather

20. Cranefield, "Glass Electrode." The pH meter was not available as a commercial instrument until 1936; the device established Arnold Beckman's career as an instrument maker.

21. See Stanley, "Chemical Studies on the Virus of Tobacco Mosaic. Part 2." According to Bryan Harrison, subsequent researchers have not confirmed Stanley's elegant result: "Others have claimed that TMV particles are resistant to pepsin, or are only digested when partially denatured" ("Chemical Nature of Tobacco Mosaic Virus Particles," 71).

22. Whereas Vinson and Petre had quantified the concentration of virus in their extract by assaying the number of plants per ten inoculated tobacco plants that showed systemic infection, Holmes's lesion-counting methods enabled Stanley to increase the precision of the titer measurements by at least a factor of ten. There are two aspects to this difference: one is that the needle-puncture method used by Vinson and Petre was quite ineffective in transmitting TMV; the other is that Holmes's new method involved counting lesions around the inoculation site rather than counting diseased plants, which provided a much more sensitive measure of infection.

23. McKinney, "Quantitative and Purification Methods." This article details a method for obtaining close to 100 percent infection of needle-inoculated plants, which had been difficult to achieve with tobacco mosaic.

24. d'Herelle, *Le bactériophage* (1921); the book was issued the next year in English (*The Bacteriophage*). Holmes did not specifically cite d'Herelle's work, but rather mentioned bacteriological plating methods, asserting that his method "is as helpful in the study of tobacco mosaic virus as Koch's plate method is in the study of bacterial cultures" ("Local Lesions," 54).

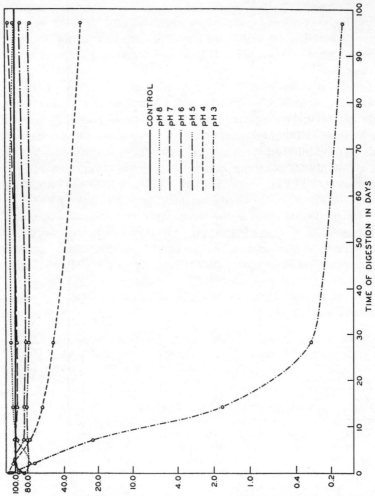

AVERAGE OF NUMBER OF LESIONS OBTAINED WITH DIGESTION MIXTURES
(EXPRESSED AS PERCENTAGE OF NUMBER OF LESIONS OBTAINED WITH CONTROLS)

TIME OF DIGESTION IN DAYS

CONTROL
pH 8
pH 7
pH 6
pH 5
pH 4
pH 3

than mere mottling of the leaves (see fig. 3.2).[25] The number of these lesions, which became visible at the site of infection, could be quantitatively correlated with the concentration of the infectivity of the sap. Holmes's assay thus enabled "much more accurate quantitative work" on TMV than was previously possible, but even so it was not widely adopted. Other researchers claimed that Holmes's publication of the technique "did not present many of the details which would be involved in any practical application of it," so that Stanley's direct access to Holmes in Kunkel's laboratory was probably critical to his success with the lesion-counting technique.[26]

Second, Stanley conducted experiments on various mutant strains of the virus being isolated and characterized in Kunkel's Princeton laboratory, which allowed him to see whether effects of pH or proteolytic enzymes were strain-specific. To this end, his institutional location allowed him to incorporate one novel element into the system: the exquisitely pure proteolytic enzymes being prepared in Northrop's laboratory. But it was not (principally) the higher-grade material that generated the improved system. In his first papers on TMV, Stanley not only reproduced but overturned published results. His use of the experimental system both raised and resolved controversies early on, suggesting its capacity for mediating future conflicts.

Given his skills and the resources at hand, Stanley might have used his productive experimental system to investigate any number of questions about TMV. One early exploratory study concerned the effects of sound waves on the infectivity of the virus.[27] But Stanley's other efforts built largely on his chemical expertise. In 1934, Stanley tested the effects of pH

25. Holmes, "Accuracy in Quantitative Work," and "Local Lesions."
26. Samuel and Bald, "On the Use of the Primary Lesions," 70. It is unclear why Vinson and Petre did not exploit Holmes's innovation, since they all worked in Kunkel's group at the Boyce Thompson Institute at the time Holmes developed the assay. Karl Maramorosch, who also worked in the Kunkel laboratory, says that Kunkel told him Vinson and Petre simply "paid no attention to the work of the young parasitologist Holmes and were well familiar with the needle inoculation technique" (personal communication, 13 Sep 1999).
27. Stanley, "Action of High Frequency Sound Waves."

FIGURE 3.1 Pepsin inactivation of TMV. "The graph shows the effect of different hydrogen-ion concentrations on the inactivation of TMV with pepsin at 37°C. The lines showing the average results obtained at pH 5, 6, 7, and 8 fall very close to the line for the controls, thus indicating little or no inactivation of the virus. The line showing the results obtained at pH 3 drops sharply, thus indicating a rapid inactivation of the virus." (Graph and caption reprinted, with permission, from Stanley, "Chemical Studies on the Virus of Tobacco Mosaic. Part 2," *Phytopathology* 24 [1934]: 1269.)

FIGURE 3.2 Local lesions on a tobacco leaf. These spots result from application
of a dilute solution of TMV, enabling a quantitative measure of virus infectivity
for the solution. (From Knight and Fraser, "Mutation of Viruses," 75. Reprinted
courtesy of the Virus Laboratory and the Bancroft Library, University of
California, Berkeley.)

and 110 chemical reagents (from aniline to zinc chloride) on infectivity,
resulting in two more long papers for *Phytopathology*.[28] To the extent
that Stanley was using these various agents to probe the virus chemically
as a protein, the papers were modeled on Northrop's work characterizing
proteolytic enzymes. But his attempts to further purify the virus were
overtaking the myriad chemical studies on infectivity, and here Vinson
and Petre's publications guided the way. After his first note in *Science*
announcing his successful precipitation of TMV, Carl Vinson published
two subsequent papers on improved methods for isolating the virus, none
of which, however, produced TMV in pure form.[29]

Stanley was not the only virus worker to pick up Vinson's lead,
and his competitors' reports reveal how contestable was Stanley's early
conjecture that the virus was a protein. In 1933, E. Barton-Wright and
Alan McBain treated TMV with potassium phosphate and acetone; they
isolated a crystalline powder containing no protein or nitrogen, which
nevertheless proved infectious.[30] Several weeks later John Caldwell

28. Stanley, "Chemical Studies on the Virus of Tobacco Mosaic. Part 3," and "Chemical
Studies on the Virus of Tobacco Mosaic. Part 4." On the timing of these studies, see "Report
of Dr. Kunkel and Associates," Scientific Reports, RAC RU 439, box 6, volume 22, 1933–
34, 216 and volume 23, 1934–35, 281–82. Stanley's purification of the virus as a crystalline
material was not reported until volume 24, 1935–36.
    29. Vinson, "Further Purification of the Virus," and "Purification of the Virus."
    30. Barton-Wright and McBain, "Possible Chemical Nature of Tobacco Mosaic Virus."
Vinson responded in *Science*: "Possible Chemical Nature of Tobacco Mosaic Virus."

responded in *Nature* that upon repeating Barton-Wright and McBain's procedures, he found their "virus" crystals to be composed almost exclusively of the potassium phosphate used to precipitate the virus activity. "That the crystals have no specific relation to the virus is easily demonstrable," Caldwell asserted.[31] Barton-Wright and McBain stuck by their asserted observations in response, but failed to produce a supporting publication.[32] Vinson also responded to Barton-Wright and McBain's claim that they had isolated "infectious crystals that lacked nitrogen, presumably to reassert his priority. However, his statement revealed an unwillingness to commit himself to an interpretation about the nature of TMV:

> Under our conditions and using plants of *Nicotiana Tabacum*, var. Turkish, I still find nitrogen present in infectious preparations obtained as described above. The nitrogen content, however, is not high, especially when the diseased plants have been grown during the short, gloomy days of midwinter....
>
> Under our conditions, purified virus preparations which seemed to contain the major portion of the original virus have not yet been obtained free of nitrogen. This is not stated, however, as argument against the possibility of nitrogen-free preparations having been obtained by others under their conditions.[33]

In contrast to Vinson's ambivalence, Stanley claimed that his studies with trypsin and pepsin had "indicated quite strongly that tobacco mosaic virus was protein in nature.... It was obvious that the methods of protein chemistry so successfully used by Northrop and associates in their work on enzymes might prove useful in work with this virus."[34]

Stanley's "Crystals" of TMV

Stanley's isolation of TMV in chemically pure form, recounted below, is his most widely cited—and least explained—contribution to virology. Scientists and historians have conventionally attributed Stanley's success to his easy access to Northrop's expertise and methods with enzyme crystallization, painting a picture of Stanley's project as fundamentally derived from Northrop's.[35] In effect, this argument simply shifts credit from

31. Caldwell, "Possible Chemical Nature of Tobacco Mosaic Virus."
32. Barton-Wright and McBain, "Possible Chemical Nature of Tobacco Mosaic Virus (response to Caldwell)."
33. Vinson, "Possible Chemical Nature of Tobacco Mosaic Virus."
34. Stanley, "Biochemistry and Biophysics of Viruses," 472.
35. For example, in James, *Nobel Laureates in Chemistry,* the author of Stanley's biographical sketch, Albert Costa, asserts, "He decided to investigate TMV with John Northrop's newly devised methods for preparing crystalline enzymes" (302). In her account, Lily Kay is more emphatic: "A skilled technician, he emphasized physicochemical manipulations but did little innovative thinking. His research approach, its weaknesses and

a junior colleague (Stanley was thirty when he announced his crystallization) to a senior colleague, leaving unexamined the ways in which Stanley
adapted other available chemical methods in his work on TMV, and
neglecting entirely the ways in which Northrop then attempted to benefit
in his own work from Stanley's success. Rather than a unidirectional transfer of concepts and methods, there was an interplay between three lines of
work: Northrop's crystallization of digestive enzymes, Stanley's purification of TMV, and Northrop and Krueger's work on the autocatalytic properties of bacteriophage. I emphasize the way that each of these experimental systems was used as a model or guide for work on the other objects.
Northrop's experimental system for purifying crystalline enzymes as proteins served as a model for Stanley's experimental work on his own system;
in turn TMV, configured as a purified protein, became the exemplar for
chemical work on viruses. Northrop was just as committed to the analogy
between enzyme and virus as Stanley, and attempted, unsuccessfully, to
extend Stanley's success by isolating bacteriophage as a pure protein.

To obtain crystalline digestive enzymes from swine and bovine pan-
creas and stomach, Northrop subjected an extract of the starting material
to repeated fractionation with salts. The use of salts (such as sodium
chloride and ammonium sulfate) to precipitate proteins from solution
dated back to the nineteenth century; Northrop's refinement of such
precipitation methods in protein purification represented the state of
the art in the early 1930s.[36] Because different cellular proteins exhibit
varying degrees of solubility in salt solutions, one could enrich the concentration of a particular protein by adding a specific amount of one salt
(at a particular pH and temperature), and then separating the insoluble—
thus precipitated—proteins from those still in solution. Northrop and
his coworkers used this method of fractional precipitation to purify
pepsin, trypsin, and chymotrypsin as crystalline—and thus chemically
pure—proteins.[37] These achievements not only constituted an important
technical advance in protein chemistry, but also demonstrated that these
enzymes were composed entirely of protein, not other substances,

---

strengths, rested firmly on the scientific reputation of others, on his successful extension
of their work, and on his managerial skills in the scientific and public domains.... By
applying many of Northrop's concepts and techniques, Stanley seemed to have proved
that the same molecular mechanisms that governed the action of enzymes also determined
the reproduction of viruses" ("Stanley's Crystallization," 451).

36. See Fruton, *Molecules and Life,* 114 ff., on Willy Kühne's pioneering studies of polypeptide solubility in various salt solutions and Prosper-Sylvain Denis's use of salting-out
methods on serum constituents.

37. Northrop advocated the use of repeated (fractional) crystallization as a means
of further purification; see Northrop, "Crystalline Pepsin. Part 1"; Northrop and Kunitz,
"Crystalline Trypsin. Part 1"; Kunitz and Northrop, "Crystalline Chymo-Trypsin and
Chymo-Trypsinogen. Part 1."

Table 3.1 Comparison of Northrop's and Stanley's Precipitation Protocols

Northrop's Method for Obtaining Crystalline Chymo-Trypsin	Stanley's Method for Obtaining Crystalline TMV
Take pancreas from freshly slaughtered cattle; immerse in N/4 sulfuric acid. Grind into suspension; leave at 5°C overnight.	Take frozen, ground, infected tobacco leaves (4000 kg); extract twice with disodium phosphate (5000 l).
Bring to 0.4 saturation with ammonium sulfate; discard precipitate.	Celite extraction, leaving "sparkling clear brown liquid."
Bring to 0.7 saturation with ammonium sulfate; leave in cold room for 48 hours.	Add sulfuric acid to pH 5. Add ammonium sulfate to 0.4 saturation.
Collect precipitate and redissolve in 3 volumes water and 2 volumes ammonium sulfate; discard precipitate and repeat previous step.	Refilter precipitate through celite; repeat above step.
	Precipitate with lead subacetate at pH 8.
Crystallization of filter cake with ammonium sulfate at pH 5.0.	Crystallization with ammonium sulfate at pH 4.5.

SOURCE: Adapted from information in Kunitz and Northrop, "Crystalline Chymo-Trypsin and Chymo-Trypsinogen. Part 1" (1935) and Stanley, "Chemical Studies on the Virus of Tobacco Mosaic. Part 6" (1936).

confirming John Sumner's 1926 crystallization of the enzyme urease as a protein.[38]

Stanley's experiments treating TMV with pepsin and trypsin (which specifically digest protein substrates) had led him to believe the virus was also a protein. He began trying to purify TMV by reworking and optimizing Vinson and Petre's published methods. Specifically, Stanley's protocol included as a first step a potassium phosphate treatment and as the last step a lead acetate precipitation, both modified from Vinson and Petre.[39] Stanley then incorporated Northrop's procedure for salt precipitation, modeling TMV on Northrop's enzymes by treating his virus preparation similarly with ammonium sulfate (see table 3.1).[40] Thus Northrop's experimental system served as a key referent for Stanley in his attempts to render the virus chemically.

38. Sumner, "Isolation and Crystallization." On the resistance to accepting Sumner's evidence by scientists adhering to a colloidal explanation of enzyme activity, see Fruton, *Molecules and Life,* 158.

39. Compare Stanley, "Isolation of a Crystalline Protein" (especially the sixth paragraph on 644), with Northrop, "Isolation and Properties of Pepsin and Trypsin," 241. On Stanley's adaptation of Vinson and Petre's lead acetate precipitation, see Stanley, "Chemical Studies on the Virus of Tobacco Mosaic. Part 5."

40. Northrop, "Crystalline Pepsin. Part 1." Vinson and Petre had also published attempts to use ammonium sulfate to precipitate TMV, but it was Northrop's protocol that Stanley drew upon.

Stanley's tinkering with a combination of these available purification methods paid off handsomely. Late in the spring of 1935, Stanley obtained needle-shaped crystals from his tobacco mosaic virus preparations, and by summer he had sent an account to *Science* magazine.[41]

> A crystalline material, which has the properties of tobacco-mosaic virus, has been isolated from the juice of Turkish tobacco plants infected with this virus.... Although it is difficult, if not impossible, to obtain conclusive positive proof of the purity of a protein, there is strong evidence that the crystalline protein herein described is either pure or is a solid solution of proteins.[42]

Ten successive reprecipitations of the needle-shaped crystals yielded a material with a constant isoelectric point and a regular X-ray diffraction pattern, suggesting that it was a protein relatively free of impurities (see fig. 3.3). Stanley's chemical analyses of the material showed it to contain 20 percent nitrogen and 1 percent ash. Immunological reactivity confirmed the virus titer obtained through Holmes's local lesion assays on Golden Cluster bean plants, which were susceptible to infection by a billion-fold dilution of the highly purified material. In the *Science* account, Stanley reported the accumulation of only ten grams of active crystalline protein, but the longer version published in *Phytopathology* made clear the large scale of operation required to obtain this material: 4,000 kg of infected fresh Turkish tobacco plants were processed into 5,000 liters of extract, the various precipitations being achieved in five-gallon tinned pails.[43] From the pails of green liquid, and through the labors of many extractions and precipitations, Stanley procured "pure chemical compounds," with the "general properties of a protein."[44]

The immediate impact of Stanley's 1935 announcement is clearly discernible from popular and scientific media, especially as compared with the attention given his early competitors. On the day Stanley's paper appeared in *Science,* the *New York Times* covered his report with a front-page story entitled, "Crystals Isolated at Princeton Believed Unseen Disease Virus" (see fig. 3.4). Another story followed in the next day's paper; this one mentioned in passing that Stanley's achievement was "foreshadowed by the work of Dr. Carl G. Vinson."[45] Six months later, the *New York*

41. Stanley, "Isolation of a Crystalline Protein." For the sake of historical (but not scientific) accuracy, I will follow Stanley and other scientists in using the word *crystalline* to refer to his purified TMV, although I discuss the still-valid objections to this usage below.
42. Ibid., 644.
43. Stanley, "Chemical Studies on the Virus of Tobacco Mosaic. Part 6," 306–10.
44. Ibid., 313, 317.
45. "Crystals Isolated at Princeton Believed Unseen Disease Virus," *New York Times,* 28 Jun 1935, 1: 4; "Life in the Making," *New York Times,* 29 Jun 1935, 14: 2.

FIGURE 3.3 Stanley's crystalline TMV ×675. This photomicrograph was reproduced in several magazine and journal accounts (e.g., *Scientific American*) of Stanley's purification of TMV. (Reprinted courtesy of the Bancroft Library, University of California, Berkeley.)

Times reported on Stanley's announcement at a bacteriology conference that he had recently obtained aucuba mosaic, another strain of TMV, in crystalline form.[46] The *Literary Digest* ran a story on Stanley's achievement less than a month after the paper in *Science,* and *Time* magazine reported on Stanley's work with TMV four times during the following year.[47] (Figure 3.5 is a portrait taken of Stanley during this time.)

46. "Scientist Isolates Virus in Quantity," *New York Times,* 28 Dec 1935, 16: 1.
47. "Virus Apparently Made Visible at Last," *Literary Digest* 120 (13 Jul 1935): 18; "Virus Diseases," *Time,* 23 Mar 1936, 45; "Virus a Protein Molecule," *Time,* 11 Jan 1937, 40; "Viruses Analyzed," *Time,* 5 Jul 1937, 39; "Macro-Molecules," *Time,* 15 Nov 1937.

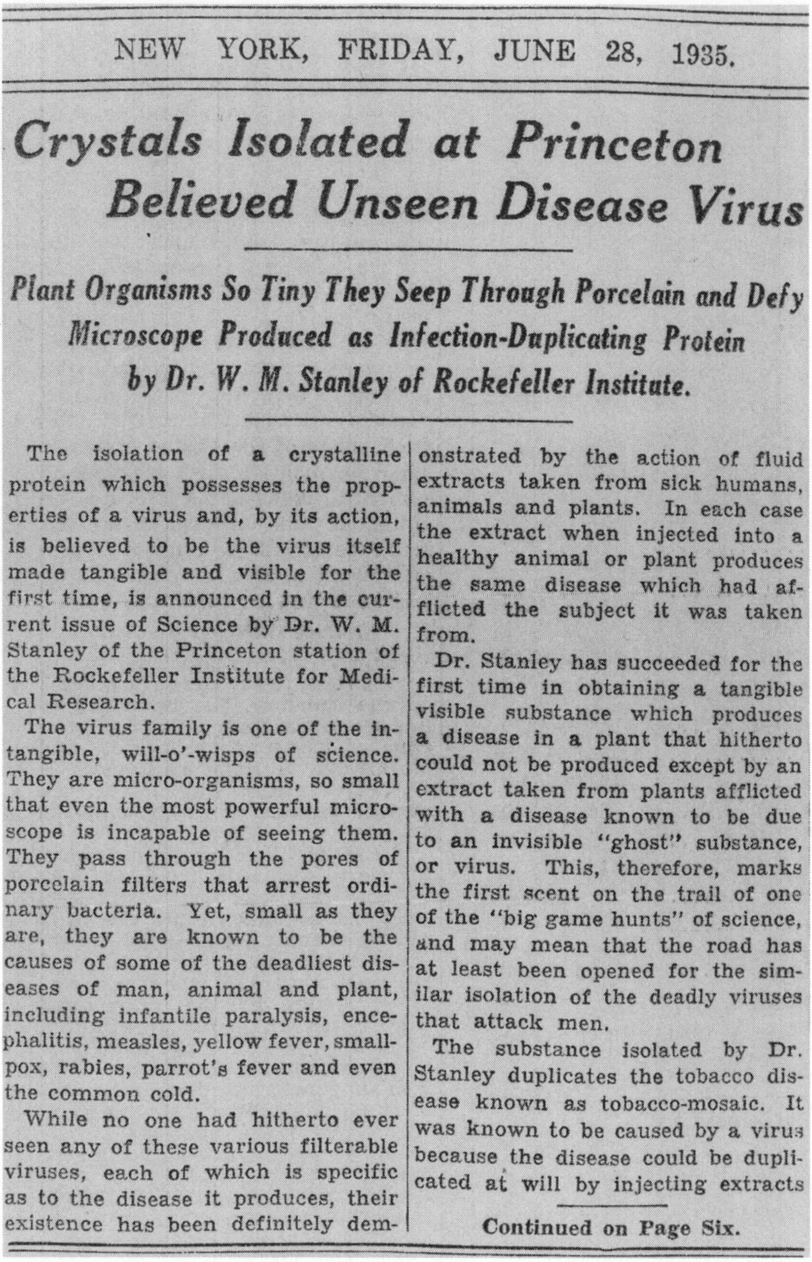

Crystals Isolated at Princeton Believed Unseen Disease Virus

Plant Organisms So Tiny They Seep Through Porcelain and Defy Microscope Produced as Infection-Duplicating Protein by Dr. W. M. Stanley of Rockefeller Institute.

The isolation of a crystalline protein which possesses the properties of a virus and, by its action, is believed to be the virus itself made tangible and visible for the first time, is announced in the current issue of Science by Dr. W. M. Stanley of the Princeton station of the Rockefeller Institute for Medical Research.

The virus family is one of the intangible, will-o'-wisps of science. They are micro-organisms, so small that even the most powerful microscope is incapable of seeing them. They pass through the pores of porcelain filters that arrest ordinary bacteria. Yet, small as they are, they are known to be the causes of some of the deadliest diseases of man, animal and plant, including infantile paralysis, encephalitis, measles, yellow fever, smallpox, rabies, parrot's fever and even the common cold.

While no one had hitherto ever seen any of these various filterable viruses, each of which is specific as to the disease it produces, their existence has been definitely demonstrated by the action of fluid extracts taken from sick humans, animals and plants. In each case the extract when injected into a healthy animal or plant produces the same disease which had afflicted the subject it was taken from.

Dr. Stanley has succeeded for the first time in obtaining a tangible visible substance which produces a disease in a plant that hitherto could not be produced except by an extract taken from plants afflicted with a disease known to be due to an invisible "ghost" substance, or virus. This, therefore, marks the first scent on the trail of one of the "big game hunts" of science, and may mean that the road has at least been opened for the similar isolation of the deadly viruses that attack men.

The substance isolated by Dr. Stanley duplicates the tobacco disease known as tobacco-mosaic. It was known to be caused by a virus because the disease could be duplicated at will by injecting extracts

Continued on Page Six.

FIGURE 3.4 Front-page news story on Stanley's isolation of crystalline TMV. (From the *New York Times,* 28 Jun 1935. Copyright ©1935 by the New York Times Co. Reprinted by permission.)

FIGURE 3.5 Portrait of Wendell Stanley circa 1936. (Reprinted courtesy of the Rockefeller University Archives.)

Whereas broad interest in the "nature of the virus" fueled the popularity of Stanley's achievement, his contributions did little to answer long-standing questions about viruses. Stanley's interpretative stance in the *Science* article was largely restricted to the closing statement, which evoked Leonard Troland's earlier speculations: "Tobacco-mosaic virus is regarded as an autocatalytic protein which, for the present, may be assumed to require the presence of living cells for multiplication."[48] Lily Kay has argued that this assertion played into prevailing assumptions that both viruses and genes were self-reproducing enzymes, a conception both she and Robert Olby have identified as central to "the protein paradigm."[49] While it is indisputable that Stanley drew on current conceptions of enzymes and proteins, his interest in autocatalytic proteins can be traced to a more immediate source. The juxtaposition in Northrop's laboratory of related studies of two different kinds of systems informed a local interpretation that viruses, like digestive enzymes, are generated autocatalytically from precursors. Northrop modeled his studies of bacteriophage on trypsin and pepsin, viewing the multiplication of bacteriophage in bacteria as analogous to the activation by autocatalysis of the enzyme precursors trypsinogen and pepsinogen. Further, the lytic nature of bacteriophage—their ability to destroy infected cells—could be seen as analogous to the protein-destroying activity of the digestive enzymes. These views on autocatalysis as the mechanism for virus reproduction were adopted by Stanley along with Northrop's methods for purifying enzymes as proteins, with one point of disagreement. Northrop did not want to see bacteriophage as a true virus, but rather as an autocatalytic bacterial enzyme which emerged only in diseased cells. However, he was unable to procure pure bacteriophage (much less its precursor) using the same precipitation methods that had yielded crystalline enzymes. As noted in the Institute's 1935–36 Scientific Reports, although "[a] protein which has extremely powerful lytic properties" had been isolated, it was "extremely unstable and denature[d] at the rate of about 20 per cent a day under the most favorable conditions so far found."[50] By contrast, during the same time, Stanley was able to obtain purer and more abundant preparations of TMV.

Bawden and Pirie: Contestation and Confirmation of a Chemical TMV

Stanley's claims, widely cited, did not go uncontested. In December of 1936, F. C. Bawden, N. W. Pirie, J. D. Bernal, and I. Fankuchen reported

48. Stanley, "Isolation of a Crystalline Protein," 645; Troland, "Biological Enigmas."

49. Kay, "Stanley's Crystallization"; Olby, *Path to the Double Helix*. Kay uses the phrase "protein paradigm" in "Laboratory Technology and Biological Knowledge." I address this issue further in chapter 6.

50. Scientific Reports, RAC RU 439, box 6, vol. 24, 1935–36, 128.

that their isolation of the virus following Stanley's methods had yielded a substantially different chemical substance. In the hands of the British researchers, the preparation contained less nitrogen (16.7 percent, as compared with Stanley's 20 percent) and showed traces of elements and compounds that Stanley did not detect, such as sulfur (0.2–0.7 percent), phosphorus (0.5 percent), and carbohydrate (2.5 percent). As they noted, "The last two constituents can be isolated as nucleic acid of the ribose type from protein denatured by heating."[51] In addition, they contended that the crystals had a two-dimensional rather than three-dimensional regularity, so that they should be regarded as liquid crystals or "paracrystals," rather than as true crystals. (According to the standards of X-ray crystallography, Bawden, Pirie, Bernal, and Fankuchen were correct to chastise Stanley and other researchers for continuing to refer to the purified TMV as crystalline. I will, however, continue to use the term *crystalline* where scientists persisted in doing so, even if they were, by present-day standards, incorrect.)

While the important contribution of the British group never garnered quite the same public acclaim as had Stanley's original paper, it has attained a certain canonical status in accounts of virus research. The standard story has been encapsulated as follows for juvenile readers: "Not long after Stanley's great experiment, F. C. Bawden and N. W. Pirie, two researchers at the Rothamsted Experimental Station in England, discovered that Stanley was not quite correct. They found that tobacco mosaic virus is not 100 percent protein, after all. Instead, it contains a small but very important amount of a special kind of acid. This acid is called nucleic acid, because it is also found inside the nuclei controlling living cells."[52]

Stanley's relationship to his British competitors, both in public and in private, was more complicated than is suggested by the narrative of progressive corrections toward true scientific knowledge. Bawden, a plant pathologist by training, and Pirie, a biochemist from F. G. Hopkins's laboratory, had become acquainted in 1927 or 1928 as students at the same Cambridge college. In 1934, they began some collaborative experiments on potato virus X.[53] This potato virus proved difficult to isolate chemically, and in the wake of Stanley's *Science* paper, they switched over to TMV, which was clearly the system of choice.[54] As Pirie later reflected, "Like most plant virus workers we ... then concentrated on tobacco mosaic virus."[55] Trying to recreate Stanley's experimental setup, Bawden

51. Bawden et al., "Liquid Crystalline Substances," 1051.
52. Sullivan, *Pioneer Germ Fighters,* 110.
53. Pirie, "Frederick Charles Bawden, 1908–1972," 23; Pierpont, "Norman Wingate Pirie," 402–3. Dates were obtained from Bryan D. Harrison (personal communication, 1 Mar 1999), and Pirie, "Frederick Charles Bawden, Thirty-eight Years of Collaboration."
54. Bawden, "Musings of an Erstwhile Plant Pathologist," 3.
55. Pirie, "Recurrent Luck in Research," 501.

wrote Kunkel in February of 1936 to ask for seeds for the Golden Cluster bean strain used in the local lesion assay. Stanley himself then contacted Bawden and Pirie in the spring of 1936, responding to their recent paper on potato virus X and sending his own reprints.[56] Stanley mentioned to Bawden that he was going to be in England that summer, and suggested a visit.

The correspondence between Stanley and Bawden was highly congenial from the outset. Stanley viewed Bawden as a friendly adversary, remarking in a letter thanking Bawden for hosting his 1935 visit to Rothamsted, "We should have lots of fun during the next few years."[57] Bawden visited Stanley's laboratory for a few months of research in the late 1930s, and during World War II, Stanley offered to care for Bawden's children to get them away from the threat of bombing in England.[58] As Stanley remarked to Max Delbrück about an upcoming meeting where Bawden would also be in attendance: "You will find that we are compatible as well as combatable."[59]

By contrast, Stanley's relations with Pirie were more strained, and may well have reflected the competition between a trained biochemist and a chemist encroaching on biological problems. Stanley's first letter to Pirie concluded patronizingly:

> I think that you will find, as you continue work on viruses, that the chemists' participation will become increasingly important. It appeals to me as a wonderful opportunity for chemical work. We should be able to lay the foundation for much of the virus work during the next few years. Unfortunately, just at present there is tremendous opposition to the acceptance of a protein as a virus. I hope that eventually we shall be able to bring sufficient evidence to overwhelm the opposition.[60]

Pirie, who had trained in biochemistry with F. G. Hopkins at Cambridge, likely resented Stanley's apparent sense of advantage, perhaps connected to his training as an organic chemist. Pirie took every opportunity to point to scientific failings that reflected Stanley's ignorance of basic biochemistry, from missing the nucleic acid component of the virus to believing that a protein could have a specific volume of 0.646, or contain 20 percent nitrogen and no sulfur.[61] Interestingly, Pirie actually came to

56. Bawden and Pirie, "Experiments on the Chemical Behavior of Potato Virus 'X'."

57. Stanley to F. C. Bawden, 14 Sep 1936, Stanley papers, carton 1, folder Sep 1936.

58. F. C. Bawden to Stanley, 27 Sep 1937, Stanley papers, carton 6, folder Bawden, Sir Frederick Charles; Harrison, "Frederick Charles Bawden," 41.

59. Stanley to Delbrück, 24 Jan 1950, Stanley papers, carton 3, folder Jan 1950.

60. Stanley to N. W. Pirie, 12 May 1936, Stanley papers, carton 1, folder May 1936.

61. As Pirie points out, a protein with 20 percent nitrogen would have to be almost exclusively arginine and glycine, which is highly unlikely; see Pirie, "Recurrent Luck in Research," 502.

the Rockefeller Institute in New York as a visiting scientist in 1937 to do two months of organic chemistry with Karl Landsteiner; perhaps this experience better equipped Pirie to challenge Stanley in the chemical pecking order, although there is no evidence of their interaction while Pirie was in New York.

Stanley, who never commented on Pirie's evident antipathy toward him, felt that scientific recognition of his isolation of TMV would not be secure until it was confirmed by another group.[62] Accordingly, he saw Bawden and Pirie's efforts to replicate his results as eminently desirable, and he did everything he could to enable it. When Pirie reported to Stanley that he and Bawden had succeeded in isolating a substance which seemed to be the same as Stanley's, but that they could not get the material to crystallize, he sent along some of this material to Princeton. Stanley gleaned beautiful crystals from their sample.[63] He also obtained remarkably similar lesion counts upon comparing the British samples to his own. This was especially striking given that, as Pirie noted, he and Bawden may well have isolated a different strain of TMV than Stanley's from their White Burley tobacco plants (see fig. 3.6).[64] In addition, where Bawden and Pirie's results differed from those Stanley had reported, subsequent changes in Stanley's methods helped account for the disparities. By mid-1936 Stanley had altered his protocol for TMV isolation, eliminating the use of lead acetate (which seemed to be damaging to the virus protein) in favor of gentler chemical treatments. The benefits were immediate, with a dramatic increase in yield. At the end of the year Stanley could report on his bountiful result: "Several pounds of crystalline protein from mosaic-diseased tobacco plants have been isolated."[65]

Chemical analysis of material from Stanley's improved purification procedure led him to concede that Bawden and Pirie's percentages of

62. In a letter written to his graduate adviser in May 1936, which explained why he chose to turn down an assistant professorship at Harvard, Stanley commented that "my value to an institution should be materially enhanced, for by next summer my work will have been confirmed. It seems to me that it would be to my advantage and also to their advantage to wait until the work is confirmed and more generally accepted" (Stanley to Roger Adams, 12 May 1936, Stanley papers, carton 1, folder May 1936).

63. N. W. Pirie to Stanley, 30 May 1936, Stanley papers, carton 11, folder Pf–Pla misc.; Stanley to Pirie, 16 Jun 1936, Stanley papers, carton 1, folder Jun 1936.

64. Pirie to Stanley, 30 May 1936, Stanley papers, carton 11, folder Pf–Pla misc. During the past several years, scientific research on isolates of tobamoviruses from sites throughout the world has revealed a surprising lack of diversity for RNA viruses. (In current plant virus nomenclature, tobamovirus is the type species for which tobacco mosaic virus and at least twelve other viruses are definitive species.) Indeed, one may view TMV in terms of one worldwide population. In this respect, TMV was a fortuitous material for enabling scientists at different laboratories throughout the world to replicate one another's results successfully; see Fraile et al., "Century of Tobamovirus Evolution."

65. Scientific Reports, RAC RU 439, box 6, vol. 24, 1935–36, 278.

Dilution of protein in mgs.		10^{-3}	10^{-4}	10^{-5}	10^{-6}
Your preparation	Bean	179.2	124.7	35.5	13.2
Our "		186.3	152.0	46.5	13.1
Your preparation	N.glu-	61.5	42.5	27.8	5.7
Our "	tinosa	63.0	47.9	25.5	7.3

FIGURE 3.6 Stanley's comparison of his TMV preparation with Bawden and Pirie's. The numbers represent average lesion counts of dilutions of the virus preparations assayed by the Holmes local lesion assay in both tobacco (*Nicotiana glutinosa*) and bean (*Phaseolus vulgaris* variety Early Cluster). In all cases, the infectivity of the two TMV samples was strikingly similar. (From Stanley's letter to Pirie, 16 Jun 1936, Stanley papers, 78/18c, carton 1. Reprinted courtesy of the Bancroft Library, University of California, Berkeley.)

nitrogen, ribose, and nucleic acid were more accurate than his first published figures. Increasingly suspicious of the analytical numbers in his *Science* paper, Stanley sent new samples (along with a sample Bawden and Pirie had mailed) in June 1936 to a quantitative analyst recommended by Vincent du Vigneaud, a friend from graduate school.[66] The results that came back to Stanley in August supported the British group's detection of nucleic acid and their determinations of the chemical elements in the isolated virus. Stanley wrote Pirie a conciliatory letter in November 1936:

> I think that I have the P content of the virus protein straightened out. As ordinarily prepared, it contains about 2 to 3% nucleic acid. The nucleic acid may be removed on alkaline hydrolysis, even at about pH 8.5, to give a P-free protein. We evidently hydrolyzed off the nucleic acid from our earlier preparations, for they were subjected to hydrogen-ion concentrations around pH 8.5.[67]

Stanley also informed Pirie of a footnote included in a paper sent to press: "The presence of nucleic acid in certain preparations of tobacco mosaic virus protein was called to the writer's attention by Dr. N. W. Pirie."[68]

66. Stanley to du Vigneaud, 15 Jun 1936 and 22 Aug 1936, Stanley papers, carton 1, folder Jun 1936.

67. Stanley to Pirie, 9 Nov 1936, Stanley papers, carton 1, folder Nov 1936. Stanley's estimate of the percentage nucleic acid was still lower than the British group's and would be further revised to 5–6 percent by 1942; see Cohen and Stanley, "Molecular Size and Shape."

68. Stanley, "Chemical Studies on the Virus of Tobacco Mosaic. Part 8," 329. No doubt Pirie would have appreciated more than this mention in a footnote.

Pirie responded to Stanley's letter by return mail, to confirm that their values were in agreement, offering the comment, "It will save our readers a certain amount of worry if we manage to give the impression that we are all telling the same story."[69] A few days later, on 17 November, Bawden and Pirie submitted to *Nature* a note coauthored with crystallographers J. D. Bernal and Isidor Fankuchen on their analysis of the "liquid crystalline substances from virus-infected plants." Pirie promptly sent Stanley a copy.[70]

Stanley had thus assented to Bawden and Pirie's corrections prior to the publication of Bawden and Pirie's critique—which was their first paper on TMV. This led Pirie to remark cynically, "Gradually he incorporated most of our description into his own. This hastened progress in virus research—it leaves unanswered the puzzle of what he made originally."[71] Behind the initial discrepancy between the publications by Stanley and the British pair over the chemical nature of the virus, a disagreement now reiterated in textbook histories of virology, was a complex set of negotiations, in which the competition was used by both teams as a form of confirmation of the chemical approach. As Bawden wrote Stanley in September 1937,

> The medical people here are extremely slow: most of them think we have imagined our results, and the few who do not credit us with such vivid imaginations don't really believe the stuff is virus. It is always a mixture of virus and protein [in their view]: why not apple sauce as well is my invariable reply now. These obstructionists will not realise that it is they and not us who are making the problem really complicated.[72]

The sense of shared adversaries did not attenuate the dissension over details, as recounted to Stanley by Bawden: "We think the virus is a nucleoprotein, and that it is not crystalline in the sense that it can be crystallized, or that real crystals of it can be produced. Not unnaturally, I think we are right and you are wrong."[73]

Nonetheless, scientific credit continued to accrue to Stanley. In March 1936, a writer for *Nature* highlighted the significance of Stanley's work.[74]

69. Pirie to Stanley, undated [between 9 Nov and 18 Nov 1936], Stanley papers, carton 11, folder Pf–Pla misc.

70. Pirie to Stanley, 19 Nov 1936, Stanley papers, carton 11, folder Pf–Pla misc.; Bawden et al., "Liquid Crystalline Substances."

71. Pirie, "Recurrent Luck in Research," 503–4.

72. Bawden to Stanley, 27 Sep 1937, Stanley papers, carton 6, folder Bawden, Sir Frederick Charles.

73. Ibid.

74. "Mutations in Tobacco Mosaic Virus," *Nature*.

Later that year, the same issue of *Nature* containing the letter from Bawden, Pirie, Bernal, and Fankuchen contesting Stanley's claim also included an address from John Caldwell praising Stanley's isolation of a pure virus protein as "the most outstanding recent development" in the field.[75] Given that Stanley himself disclaimed some of the results he presented in his 1935 announcement in *Science,* one may well question its enduring authority.[76] Beyond the media attention given Stanley's work, two other factors explain the continuing recognition he received.

First, Bawden and Pirie's papers never presented a clear claim to the virus themselves.[77] Instead, they built upon Stanley's initial work, opening their letter to *Nature* with a statement of Stanley's findings, and offering their observation as a confirmation as well as a correction: "We have confirmed these results, but have found that by further purification the protein in neutral aqueous solution can be obtained in liquid crystalline states."[78] Similarly, their longer 1937 paper stresses the agreement between the laboratories:

> We have now exchanged material with Dr. Stanley and find no gross differences in the activities of our respective products. We have found, however, that by further purification the protein in neutral solution can be obtained in liquid crystalline states. Also, as will be shown later, there are considerable differences in the chemical descriptions given of the virus protein; some of these differences have already been resolved, and others presumably will be by future work.[79]

If Bawden and Pirie never displaced Stanley as an author of the chemical virus (as Stanley had Vinson), this was partly because they corroborated Stanley even as they invalidated some of his results.

Second, the productivity and resilience of Stanley's experimental system reinforced the early appreciation of his discovery. Stanley's chemical rendition of TMV proved to be immensely productive for further research. By 1936, Stanley and his coworkers had obtained over a kilogram of purified virus to work with. Before the end of 1937, when Bawden and Pirie's longer paper came out (following their letter in *Nature*), Stanley

75. Caldwell, "Agent of Virus Disease in Plants." The address, published on 19 December 1938, had been delivered before Section K (Botany) of the British Association on 11 September.

76. Indeed, this is Kay's point of departure in "Stanley's Crystallization."

77. I am indebted to Judy Swan on this point; see Creager and Swan, "Fashioning the Virus."

78. Bawden et al., "Liquid Crystalline Substances," 1051.

79. Bawden and Pirie, "Isolation and Some Properties," 274–75.

had published three more announcements in *Science*, and had amassed fifteen other experimental reports on plant viruses (all published or in press).

The abundant skills and resources in the Princeton laboratories translated into formidable competition for the English team. In 1936, Stanley and Hubert Loring reported that they had been able to isolate and crystallize tobacco mosaic virus from a different plant host, tomato, and that the protein was chemically indistinguishable from the virus purified from tobacco.[80] They were also able to isolate the same protein, or a closely related one, from two other TMV-infected plants that do not belong to the tobacco family, spinach and phlox, and they demonstrated that these isolates displayed cross-reactivity to serum raised against TMV from tobacco.[81] By contrast, the crystalline material obtained from tobacco with a different viral disease, tobacco ring spot, exhibited strikingly different characteristics compared with TMV.[82]

Extending this comparative approach further, Stanley began working with the different strains of TMV that were being isolated by James Jensen in the Kunkel laboratory. A yellow-mottled variant of tobacco mosaic disease (aucuba mosaic) was already available, and Jensen isolated other strains from unusual patches of infected tobacco tissue. As the Institute's yearly overview mentioned, "These spots have proven to be veritable Pandora boxes out of which a varied assortment of diseases is ready to escape if given a little assistance. From such spots Doctor Jensen has isolated, by means of needle-puncture transfers, about fifty different viruses, and there is no reason to believe that the supply has been exhausted."[83] These diseases were taken to be new strains of the common TMV, varying from the generic version in host specificity and severity.

Stanley was interested in whether differences between these genetic variants could also be observed in terms of the isolated virus proteins, and Jensen's collection provided a wide assortment of candidates for chemical study. Stanley and Loring found that the protein isolated from the yellow strain, aucuba mosaic virus, exhibited a greater molecular weight, the formation of larger crystals, and a more alkaline isoelectric point. In other respects, including chemical composition and serological properties,

80. Stanley and Loring, "Isolation of Crystalline Tobacco Mosaic Virus Protein."

81. "The crystalline protein precipitates rabbit serum immune to tobacco-mosaic virus, and tobacco-mosaic virus precipitates serum from rabbits injected with the protein" (Director's Report to the Corporation, "Plant Pathology: Dr. Kunkel and Associates," RAC RU 439, box 6, vol. 24, 1935–36, 351).

82. Stanley, "Isolation and Properties of Tobacco Ring Spot Virus."

83. Director's Report to the Corporation, "Plant Pathology: Dr. Kunkel and Associates," RAC RU 439, box 6, vol. 24, 1935–36, 347–48.

aucuba mosaic was indistinguishable from ordinary TMV.[84] The pattern of similarities and differences led Stanley to speculate on how the variant TMV strains might allow for chemical characterization of genetic differences.[85] Moreover, the system's combination of improved methods, different hosts, and variant strains created a matrix of experiments and results that made visible the activity of the virus in many different registers, growing out of Stanley's initial experiments on TMV with enzymes and chemicals.

By 1937, Stanley was promoted to associate member of the Rockefeller Institute, with two fellows under him, Hubert Loring and Max Lauffer. Stanley set Lauffer, trained as a physical chemist, on the task of accounting for the crystallographic irregularities Bernal and Fankuchen reported in samples of TMV prepared as Stanley's "crystalline" virus had been.[86] Lauffer reproduced Bernal and Fankuchen's complex results, simplifying and re-presenting them in such a way as to reclaim part of their significance for Stanley's laboratory.[87] In effect, Stanley's experimental system subsumed the British researchers' corrections, and the sheer productivity of his research on TMV during these years helped substantiate his 1935 claim.

The Symbolic Uses of Crystalline TMV

Stanley's presentation of a crystalline virus left open the question of whether the substance should be regarded principally as an inert chemical or as a minuscule organism. His scientific papers on TMV left little doubt, either experimentally or rhetorically, that the virus was a chemical object, and none of his studies were aimed at understanding the biological agency of the substance—with the exception of his bioassay for infectivity, all of Stanley's determinations of the virus were *in vitro*, not *in vivo*. Nonetheless, fascination with Stanley's work derived from the impression that the virus was a fundamental element of life, and Stanley himself drew attention to its ambiguous status: "Whether this unusual, high molecular weight, crystalline protein is regarded as living, as non-living, as a gene, as a super-catalyst, as an organizer, or as a pathological protein, a complete study of its basic properties should prove of importance."[88]

Stanley's claim to have isolated a crystalline virus, particularly one viewed as an autocatalytic enzyme, had particular currency in the 1930s

84. Scientific Reports, RAC RU 439, box 6, vol. 25, 1936–37, 263.
85. This continuing line of research in Stanley's laboratory is taken up in chapter 6.
86. Bernal and Fankuchen extended their criticisms in the 1936 *Nature* article (coauthored with Bawden and Pirie) in a lengthy 1941 publication, "X-Ray and Crystallographic Studies."
87. Lauffer and Stanley, "Stream Double Refraction of Virus Proteins."
88. Stanley, "Dr. Stanley's Prize Research Described in His Own Words," 21.

by virtue of the active scientific debates about the origin of life. As Harmke Kamminga has observed, theories about the origin of life in the twentieth century can be roughly segregated according to whether the fundamental living unit was considered to be the protoplasm or the gene.[89] Biochemists regarded metabolism and self-regulation as key to vital functioning, and so tended to emphasize cellular systems of the cytoplasm as constitutive of life. By contrast, geneticists tended to regard self-duplication as the essential property of life, and thus the nucleus as necessary, and perhaps even sufficient, for life. As Scott Podolsky has noted, viruses could be mobilized as evidence for this second, "nucleocentric" view of the origin of life, positing the virus as the primordial bridge to cellular (and multi-cellular) life.[90] J. B. S. Haldane, an advocate of this view, asserted in 1929, "Life may have remained in the virus stage for many millions of years before a suitable assemblage of elementary units were brought together in the first cell."[91]

Podolsky's analysis of the "role of the virus in origin-of-life theorizing" helps account for the great significance ascribed to Stanley's crystallization of TMV. By the mid-1930s there was general consensus that viruses were obligate parasites, dependent on cellular machinery for their metabolic needs. This emphasis on parasitism "created the potential for viewing the virus as an unencumbered hereditary unit, able to turn the efficient machinery of the host cell to its own formal self-copying devices. The virus could become synonymous with self-duplication, at the expense of any mention of metabolism or self-maintenance."[92] In particular, this conception of the virus seemed to exemplify Troland's depiction of the gene as an autocatalytic enzyme, capable of self-reproduction.[93]

In 1929, colloidal chemist Jerome Alexander and geneticist Calvin Bridges proposed just such an explanation, according to which "life began . . . in the molecular order of complexity with an autocatalytic molecule of definite structure and less definite constituents."[94] During the same year, Haldane suggested that the relationship between bacteriophage and its host cell might be analogous to the relationship between the first "living or half-living things" and the "hot dilute soup" in which life first arose.[95] By

89. Kamminga, "Protoplasm and the Gene."
90. Podolsky, "Role of the Virus." The alignment of viruses with the nucleocentric view of the origins of life was based mostly on the observations that viruses could mutate like genes but lacked metabolism, not on the comparison of viruses to crystals, whose self-organizing properties might be more compatible with the biochemical view. I am indebted to Evelyn Fox Keller for discussions on this point.
91. Haldane, "Origin of Life," 247; also quoted in Farley, *Spontaneous Generation Controversy*, 164.
92. Podolsky, "Role of the Virus," 86–87.
93. Troland, "Biological Enigmas."
94. Alexander and Bridges, "Some Physicochemical Aspects of Life."
95. Haldane, "Origin of Life"; Podolsky, "Role of the Virus," 93–94.

presenting his crystalline virus as an autocatalytic enzyme, Stanley offered origin-of-life theorists tantalizing evidence that complex chemicals might self-duplicate in the right medium. Viewed in this light, Stanley's crystallization of TMV was momentous. Haldane sought to publicize Stanley's work for readers of the *Sunday Chronicle:*

> In 1935 Dr. W. M. Stanley, of the Rockefeller Institute for Medical Research in New York [*sic*], made a discovery which marks an epoch in science comparable with Pasteur's discovery of disease bacteria, or Schleiden and Schwann's that animals and plants are made up of cells. . . . Here . . . is a chemical substance which may be kept in a bottle and shows no signs of life. But given the right food, it can reproduce itself. Clearly, the gap between chemistry and life has been very much narrowed.[96]

Alexander, Bridges, and Haldane thus portrayed filterable viruses and bacteriophages as "functionally primitive parasites cycling through more advanced, foreign life forms," as Podolsky has put it.[97] The use of Stanley's TMV as a cornerstone for such a nucleocentric origin of life reached its apogee in a 1938 book by Richard Beutner, professor of pharmacology at Hahnemann Medical College, entitled *Life's Beginnings on the Earth*. In asking whether scientists might be able to create life in the laboratory, Beutner reflected:

> We have of course not yet learned to synthesize self-regenerating enzymes; this task would take perhaps five hundred instead of five years; but Dr. W. M. Stanley of The Rockefeller Institute at Princeton, New Jersey, has discovered where such substances occur in nature: a virus which causes infectious diseases, and hence constitutes a living entity, is at the same time a substance which can be obtained in pure form as crystals. Here is indeed a substance which is an enzyme capable of regenerating itself in a natural environment. Through Stanley's epoch-making discovery, more definite assumptions on the origin of life can now be made than ever before.[98]

However, at just this apparent high point for the representation of TMV as a progenitor of life, the compelling theory of a nucleocentric origin of life from the virus as autocatalytic enzyme was undermined by three developments. Alexander Oparin's monumental treatise *The Origin of Life,* also published in 1938, offered a comprehensive theory centered on metabolism, marginalizing self-duplication as the key original

96. Haldane, "Can We Make Life?" in *Keeping Cool,* 25–26.
97. Podolsky, "Role of the Virus," 86–87.
98. Beutner, *Life's Beginnings on the Earth,* v–vi.

manifestation of life.[99] The phylogenetic significance of the virus was also challenged. In 1935, Robert Green proposed that viruses were the product of retrograde evolution from their hosts, such that viruses followed unicellular organisms, not the other way around.[100] Sir Patrick Laidlaw presented further evidence that viruses as parasites were "lazy bacteria,"[101] and the retrograde theory of virus origin began to be called the "Green-Laidlaw hypothesis." Finally, no evidence could be found to support the popular notion that viruses replicate through enzymatic autocatalysis. As Stanley noted in 1937, "Since the normal plant proteins do not possess high molecular weights and do not possess properties similar to those of the high molecular weight virus proteins, the latter proteins appear to be produced, not by a simple rearrangement of a molecule of normal protein, but rather by polymerization or, more likely, by direct synthesis."[102] Within sixteen months of the initial *Science* report, Stanley dropped autocatalysis, and he became an early supporter of the Green-Laidlaw hypothesis.[103] By the end of the 1930s, the prospect that crystalline TMV might provide a model for the origin of life had been extinguished.

Beyond the significance of Stanley's isolation of TMV for the origin-of-life debates, the scientific reception of his chemical work on viruses was diverse. Bacteriologists and pathologists were, according to Stanley, his most skeptical critics, expressing doubts that a viral disease could be attributed to a chemical.[104] Pirie deplored the kind of sensationalism that attended media references to the crystallization, as he demonstrated with his usual caustic wit in his essay, "The Meaninglessness of the Terms Life and Living."[105]

For biochemists less vexed by the philosophical abuses of Stanley's research findings, the isolation of TMV was simply the most spectacular

99. Oparin, *Origin of Life*. See Podolsky, "Role of the Virus," 97–98; and Kamminga, "Studies in the History of Ideas."

100. Green, "On the Nature of Filterable Viruses." As Podolsky points out, his argument presumably drew on Haldane's recent book, in which Haldane contended that "most evolutionary change has been degenerative" (Haldane, *Causes of Evolution,* 139, quoted in Podolsky, "Role of the Virus," 100).

101. Laidlaw, *Virus Diseases and Viruses,* 47, quoted in Podolsky, "Role of the Virus," 101.

102. Stanley, "Chemical Studies on the Virus of Tobacco Mosaic. Part 8," 339.

103. Podolsky, "Role of the Virus," 103.

104. "The pill has been too much for the bacteriologists to swallow. They tend to resent the very idea that these proteins which multiply and undergo mutations like living things are also in a sense non-living, yet cause disease. Stanley had little to say about this beyond suggesting that bacteriology may soon have to cede to chemistry priority in the struggle against disease" (Harding, "What Is Life?" 236).

105. Pirie, "Meaninglessness of the Terms."

example in a series of successful chemical purifications of enzymes, hormones, and other proteins. J. B. Sumner's crystallization of urease in 1926
had been met with disbelief by many chemists and enzymologists, who
held (as most forcefully argued by Richard Willstätter) that enzymes were
composed of small catalysts adsorbed onto larger inert substances. By arguing that the crystalline protein was the enzyme itself rather than a carrier, Sumner rejected colloidal theories of enzyme action.[106] He was soon
vindicated: Northrop's purification of crystalline trypsin, chymotrypsin,
and pepsin during the early 1930s provided further proof that enzymes
were proteins. Hormones also proved amenable to isolation as proteins.
In 1932 Adolf Butenandt of Göttingen University purified 15 milligrams
of androsterone from a staggering 15,000 liters of male urine, and two
years later he obtained 20 milligrams of progesterone from the ovaries of
50,000 sows. Similarly heroic attempts were being made by chemists to
isolate adrenocorticol hormones: at the Mayo Clinic, Edward Kendall
used 1,250,000 cattle carcasses in his chemical fractionation of eight
crystalline corticol compounds.[107] Stanley was representative of a group
of organic chemists who used massive amounts of starting material, labor, and ingenuity to isolate increasingly complex natural products as
crystalline proteins.[108] By isolating a virus as a crystalline (or, more precisely, paracrystalline) form, Stanley extended the range of biochemical
fractionation up to the largest and most lifelike of subcellular entities,
reinforcing Beutner's declaration that "chemistry holds the key to the
secrets of life."[109]

This triumphalism was not lost on scientists in other fields promoting
the application of physical sciences to problems of life. In 1936, the geneticist H. J. Muller lectured before the Academy of Sciences in Moscow on
"Physics in the Attack on the Fundamental Problems of Genetics," and
subsequently published his address in *Scientific Monthly*. Muller argued
that Stanley's crystalline virus "represents a certain kind of gene," based
on its apparent self-reproduction in the cell.[110] As Muller had earlier
declared, this genetic action of "self-propagation . . . fulfills the chemist's
definition of 'autocatalysis'; it is what the physiologist would call 'growth';
and when it passes through more than one generation, it becomes 'heredity.' "[111] Thus virus research potentially offered wide-reaching biological

106. See Sumner, "Chemical Nature of Enzymes." On Willstätter, see Fruton, *Molecules
and Life*.
107. Slater, "Industry and Academy," 450–51.
108. Note, for instance, the publication by Allen of "Isolation of Crystalline Progestin"
in *Science* less than two months after Stanley's announcement.
109. Beutner, *Life's Beginnings on the Earth,* 49.
110. Muller, "Physics in the Attack," 213. I take up the analogy between viruses and
genes in chapter 6.
111. Muller, "Variation Due to Change," 33.

insights. At another level, Stanley's success in using chemistry to isolate a virus resolved decades-old debates about the character of the infectious entity, as well as validating the current Rockefeller Foundation program in the Natural Sciences, which promoted the use of the tools and methods of the physical sciences in biology.[112]

Conclusions

When Stanley announced that he had crystallized a protein with the properties of tobacco mosaic virus, he recast the virus from a biological pathogen to a chemical object. The definitiveness of his claim and the highly visual evidence of his experimental technique—the needle-shaped crystals—made the previous attempts of Olitsky, Vinson and Petre, and Barton-Wright and McBain seem obsolete and artifactual. Subsequent corrections to Stanley's characterization of the virus did not have similarly corrosive effects on his priority. Rather, Stanley's 1935 paper in *Science* quickly attained canonical status in virus research. Its author was recognized with numerous accolades, including a prize from the American Association for the Advancement of Science, the Isaac Adler Prize of Harvard Medical School, the Rosenberger Medal of the University of Chicago, the John Scott Medal of the City of Philadelphia, and honorary degrees from Harvard, Yale, and Earlham (his alma mater). In 1940 Stanley was elected to the American Philosophical Society, and the following year to the National Academy of Sciences. He was only thirty-seven.

What difference did it make that TMV—and, by extension, other viruses—were now treated as chemical objects? An immediate and sensational consequence was TMV's adoption as an emblem of primitive and elemental life, especially in discussions of the origin of life. As a writer for *Scientific American* reflected on Stanley's work in 1937,

> Hitherto, crystalline substances had been regarded, and for good reason, as inanimate—perhaps portions of animate beings, but not themselves alive. Yet there is no doubt that an almost infinitesimal bit of this dead-or-alive material, when placed in the living tobacco plant, soon shows its activity and "life," for the leaves wilt and die— while the virus feeds upon the living tissues and reproduces itself indefinitely.... Decidedly, it must be thought of as a being from the borderland of animate existence, an organization just at the threshold of life. In it we discover how the stages of increasing complexity of atomic combination have at last scaled up to the realm of life.... Thus, again we discover that Nature does not do things by big jumps. There is today a smooth slide up to life.[113]

112. Kohler, "Management of Science," 298; Abir-Am, "Discourse of Physical Power."
113. Newman, "Smooth Slide Up to Life," 305–6.

In the late 1930s, though, viruses fell out of favor in speculations on the origin of life, and awe over TMV's "alive but dead" status abated.

A more lasting and pragmatic consequence of Stanley's work was that, for many researchers, viruses came to be defined in terms of isolated proteins rather than disease symptoms. F. C. Bawden asserted in his 1939 monograph *Plant Viruses and Virus Diseases,* "at the present time it seems most reasonable and probable that these nucleoproteins are the viruses themselves, and in this book this view will be adopted as a working hypothesis."[114] The characterization of TMV in terms of specific composition and dimension further stabilized the attribution of viral diseases to chemical entities.

Stanley's depiction of TMV as a chemical entity also reinforced the analogies already being drawn between viruses, enzymes, and genes. At the Seventh International Genetical Congress in 1939 there was a double "Session on Protein and Virus Studies in Relation to the Problem of the Gene," chaired by J. B. S. Haldane. Historians have paid substantial attention to this symbolic consequence of the 1935 crystallization, arguing that these conceptual links played a seminal role in the early formation of molecular biology.[115] Yet there developed a striking disjunction between this view of TMV's significance and the ongoing experimentation in Stanley's laboratory. For one thing, by 1937 it was clear that TMV was not an autocatalytic enzyme, and that the crystallization of the virus shed no light on the mode of its self-multiplication. Even more striking was that behind the perceived ontological stability of the needle-shaped crystals was a complete reworking of Stanley's experimental system. The highly evocative method of crystallization was entirely superseded during the same years that Stanley's TMV attained a secure public status. By 1940, the ultracentrifuge came to replace techniques of chemical purification in the production of pure virus, and TMV was recast from an autocatalytic enzyme to a macromolecule. This critical transition at the level of experimental practice in Stanley's laboratory is not part of the received story about the chemical purification of TMV, and, as the next chapter details, it was this development that solidified TMV's role as an exemplar for other viruses.

114. Bawden, *Plant Viruses and Virus Diseases,* 9.
115. Olby, *Path to the Double Helix;* Judson, *Eighth Day of Creation.* See also Stent, *Molecular Genetics.*

4

That "Whirligig of Science": The Ultracentrifuge in Virus Research

The new whirligig of science, the ultracentrifuge,...extracts into plain view disease viruses which heretofore could not be seen even with microscopes.

New York Times, 24 June 1937

Stanley's demonstration in 1935 that TMV could be crystallized as a protein opened virus research to a range of new physical methods and machines for characterizing macromolecules. This new direction of research took Stanley away from his base of skills as an organic chemist and toward collaborative projects with other physical scientists. Not only did their physicochemical instruments yield new representations of TMV, but the means of producing the virus in the laboratory changed. One particular machine came to be dominant in both the shifting depictions of and means of producing pure TMV: the ultracentrifuge.

Stanley's adoption of the ultracentrifuge had profound consequences for biochemical research on viruses, but it also had implications for the design and production of the machine, which became standardized for molecular biological research in the 1940s. In response to the demands of virus research—and contrary to the intentions of the instrument's exalted inventor—the ultracentrifuge became a tool for *preparing* biological objects as well as *analyzing* them. World War II played a role in displacing the ultracentrifuge from its function as an analytical tool to that of a production tool, as Stanley extended the use of high-speed centrifugation to produce vaccines for the war effort.

This chapter recounts the role of the ultracentrifuge in rendering TMV as a macromolecular system, emphasizing the work required in order to reconcile results from the ultracentrifuge with the representations being generated by other new tools, such as the electron microscope and the double-refraction-of-flow apparatus. The ongoing debate as to whether proteins were best understood as colloidal aggregates or as simple molecules framed scientific discussions about the structure of viruses.[1] By 1940, most researchers had adopted the view that viruses were macromolecules, homogeneous but composed of protein and nucleic acid components. This conception did not prevail on the basis of one unequivocal experiment, but ensued from a long process of tinkering with the instruments, optimizing the experimental systems, and marginalizing alternative colloidal explanations. As TMV was at the vanguard of experiments on biological macromolecules, it served as an important—and contested—reference system for the emerging consensus that proteins and viruses were homogeneous structural units (even though they might be composed of discrete subunits). So while Stanley's initial crystallization of TMV may have provided evidence that viruses were "simply" chemical, several years of physical-chemical experiments and interpretation were required to sort out what this claim meant.

The use of the ultracentrifuge to investigate TMV structure also sheds light on how a new generation of physicochemical tools for biological research emerged in the mid-twentieth century. During the 1930s, the Rockefeller Foundation nurtured collaborations between physical scientists and life scientists in an effort to ground biological and medical research in more rigorous and quantitative theories and methods. As Robert Kohler has argued, perhaps the most important legacy of this investment in interdisciplinary research was the development of a set of new instruments and methods that were taken up widely in biomedical research in the 1940s and 1950s.[2] However, I stress that the organization of this kind of research changed dramatically after World War II, as the role of physical scientists as innovators in biological work was replaced in many instances by commercial instruments. To assemble and use a Tiselius electrophoresis apparatus during the 1930s, as Lily Kay has observed, required the assistance of physical scientists; electrophoretic studies of biological substances involved a team of scientists organized around a complicated and often massive apparatus. In the postwar period, by contrast, such collaborative efforts were "black-boxed" into standardized, industrially produced machines such as the Perkin-Elmer electrophoresis setup, available for $3,000 in 1948.[3]

Stanley's use of the ultracentrifuge coincided with—and exemplifies—this shift from collaboration to commercialization. Although Stanley was

1. On the debates over whether proteins were colloidal aggregates or true macromolecules, see Edsall, "Proteins as Macromolecules"; Olby, "Macromolecular Concept"; and Fruton, *Molecules and Life*. This debate was itself shaped by the growth of physical chemistry as a specialty; see Servos, *Physical Chemistry from Ostwald to Pauling*.

2. See chapter 13, "Instruments of Science," in Kohler, *Partners in Science*, 358–91. On the development of spectroscopy, see Zallen, "Rockefeller Foundation and Spectroscopy Research."

3. Kay, "Laboratory Technology and Biological Knowledge."

a chemist, he did not possess expertise in building and using biophysical instruments. Consequently, as soon as he obtained purified TMV late in 1935, he established a web of collaborations with instrument specialists, the majority of which were being funded by the Rockefeller Foundation. Three ultracentrifuge experts and several other physical chemists with specialized equipment began to characterize Stanley's TMV preparations. However, the ultracentrifuge came to play such an important role in virus research that Stanley soon sought to domesticate the instrument, so that his Princeton laboratory could make its own structural determinations of TMV.[4]

The impact of routine use of the ultracentrifuge in Stanley's laboratory was extensive. Viruses soon came to be viewed in terms of the sedimentation boundaries they exhibited through the instrument's optical system. Ultracentrifuge experiments provided a new criterion for molecular purity other than that of crystallinity, whose invocation for TMV by Stanley had been strongly contested by X-ray crystallographers such as J. D. Bernal. Through use of the ultracentrifuge to visualize TMV, researchers at the Rockefeller Institute soon realized that the machine could itself be used to purify the virus by physically isolating it from the other constituents of the infected cell. The purification of TMV in this manner demonstrated the potential utility of the instrument for the preparation of other viruses, including many which could not be isolated chemically. The consequent demand among virologists for preparative ultracentrifuges changed the meaning and the design of the instrument in two respects. The demarcation between "preparative" and "analytical" machines, which was blurred in the early isolation of TMV in an ultracentrifuge, was redrawn, with "quantity ultracentrifuges" being customized for virus research by the late 1930s. In turn, the demand for air-driven ultracentrifuges, which could be made in both "quantity" and "analytical" models, soon rivaled that for The Svedberg's expensive oil-driven version. After the war, the availability of the commercial Spinco Model E machine, an analytical machine which combined the air-driven design with electrical power, dramatically changed the scale of use of the analytical ultracentrifuge. Whereas only about sixteen custom-built machines (half of them oil-turbine and half air-turbine design) were in operation in 1947, there were more than three hundred additional, electrically driven ultracentrifuges in 1959, mostly situated in biophysical and biochemical laboratories.[5] Stanley was in fact the first customer

4. On the work involved in domesticating a new material or instrument in the laboratory, see Burian, "How the Choice of Experimental Organism Matters"; on the possibility of failure in domesticating a new system, see Creager and Gaudillière, "Meanings in Search of Experiments."

5. Schachman, *Ultracentrifugation in Biochemistry,* 1.

to purchase a Model E, for the Virus Laboratory he was equipping at Berkeley.

As ultracentrifuges proliferated, TMV remained a key example for the use of the machine in visualizing and purifying viruses as macromolecules. The role of TMV as a model virus illuminates the material synergy between the growth of virus research and the increased use of laboratory ultracentrifuges (particularly the air-driven machines). Researchers in the postwar period relied on the quantity ultracentrifuge in order to obtain viruses in pure form, giving this instrument a special place in the repertoire of physicochemical techniques for virology. As Joseph Beard stated in his 1948 review of "Purified Animal Viruses," "The efficiency and utility of the ultracentrifuge in the purification of viruses of small size were demonstrated in 1936 by Wyckoff and Corey in studies of [Stanley's] tobacco-mosaic virus. Since that time, ultracentrifugation has been the principal procedure for the purification and physical study of viruses."[6]

The Development of Analytical Ultracentrifuges

Historian of technology Boelie Elzen has offered a fine-grained analysis of the development of biochemical ultracentrifuges.[7] I focus on the consequential interactions of virus researchers with the inventors of these high-speed sedimentation machines. Early on, the Rockefeller Foundation supported the development of ultracentrifuges through grants to physicists The Svedberg and Jesse Beams and physical chemist James McBain. By the mid-1930s, there were several different high-speed centrifuges that could spin and sediment microscopic particles; the machines could be roughly classified according to turbine design into two groups, oil-driven or air-driven. Svedberg had designed a very precise optical machine that required an expensive oil-driven turbine, and Beams and Pickels offered a less expensive, air-driven machine that could be utilized to prepare as well as analyze macromolecules. The Rockefeller Foundation prized Svedberg's high-precision apparatus, but the prohibitive cost of the oil-driven machine slowed its wider dissemination.[8] By the 1950s, the air-driven ultracentrifuge had become the standard piece of laboratory equipment, partly because of its suitability for virus purification. Research

6. Beard, "Review: Purified Animal Viruses," 51.

7. Elzen, *Scientists and Rotors*.

8. In 1936, Warren Weaver indicated that the Foundation would fund the construction of only one such instrument in an academic setting in the United States (Weaver to The Svedberg, telegram 21 May 1936. Similarly, in "Staff Conference," 4 Jun 1936, there is the following memo: "Wisconsin centrifuges—question raised as to the number of centrifuges RF would be interested in establishing. Answer: Probably no other." RAC RF 1.1, 200, box 164, folder 2013).

on "filterable viruses," particularly at the Rockefeller Institute, favored the development of "quantity" ultracentrifuges as inventors competed for market share.

The genealogy of the ultracentrifuge is conventionally traced to the Swedish chemist The (Theodor) Svedberg, whose laboratory at Uppsala was the preeminent site for ultracentrifuge development through the 1930s.[9] His original instrument was assembled not in Sweden, but in the United States, at the University of Wisconsin with graduate student James Burton Nichols. Svedberg spent eight months in Madison in 1923 as a visiting professor of colloid chemistry and sought to contribute to current understanding of the Brownian motion of colloids by ascertaining the distribution of particle sizes in colloidal suspensions. This task proved difficult to accomplish with available instruments. Most experimental research on colloidal particles relied on the ultramicroscope, which German chemist Richard Adolf Zsigmondy developed in 1903 with H. F. Siedentopf, an optician at Zeiss.[10] Svedberg recognized that sedimentation provided an alternative way to segregate particles by weight.[11] The machine he and Nichols designed, the "optical centrifuge" (see fig. 4.1) allowed the researcher to follow sedimentation visually, as demonstrated through their study of gold sols, a model material for colloidal chemistry.[12]

Centrifuges were already commonly used in dairies and other industrial settings for separating materials of different weights. Upon returning to Uppsala, Svedberg and his student Herman Rinde adapted a cream separator for their research work by installing equipment to watch and photograph materials as they sedimented (see fig. 4.2). They found that by using sector-shaped cells and spinning the rotor in hydrogen at atmospheric pressure, they could eliminate convection currents to attain "faultless sedimentation."[13] Svedberg and Rinde christened their modified milk separator an "ultra-centrifuge."[14]

By 1930 Svedberg and his coworkers had developed two more specialized analytical machines. One allowed visual measurements of the *sedimentation velocity* of materials as they sedimented in solution at high speeds and experienced forces from 15,000 to 75,000 times that of gravity.

9. Although I am using the current spelling of Uppsala, Elzen notes that in the 1920s and 1930s the city was spelled Upsala (Elzen, *Scientists and Rotors*).

10. Elzen, "Two Ultracentrifuges," 628; Siedentopf and Zsigmondy, "Über die Sichtbarmachung und Grössenbestimmung ultramikroskopischer Teilchen."

11. Elzen, "Two Ultracentrifuges," 630; Svedberg and Rinde, "Determination of the Distribution of Size of Particles."

12. Gray, "Ultracentrifuge," 43.

13. Svedberg, "Ultracentrifuge and Its Field of Research," 114.

14. Elzen, *Scientists and Rotors,* 28; Svedberg and Rinde, "Ultra-Centrifuge, a New Instrument."

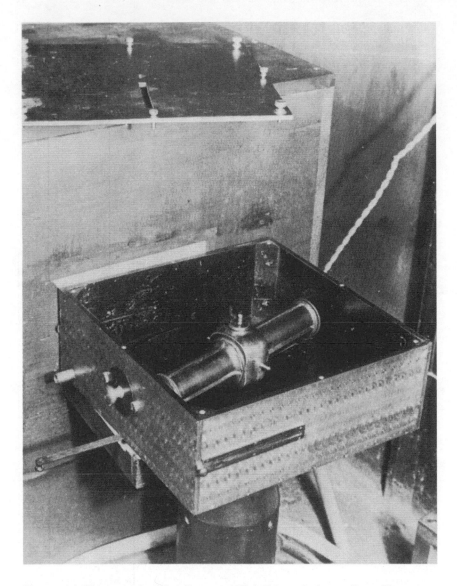

FIGURE 4.1 Photograph of Svedberg and Nichols's optical centrifuge. Picture taken in Madison, Wisconsin, 1923. (From Pedersen, "The Development of Svedberg's Ultracentrifuge," 4.)

The other enabled studies of *sedimentation equilibrium,* gauging the position of materials spun to equilibrium at lower speeds, using forces from 500 to 15,000 times that of gravity.[15] The first (velocity) apparatus could reveal whether the sedimenting species was homogeneous, as indicated by a sharp boundary in the ultracentrifuge cell. The position of the particle in the second (equilibrium) apparatus could be used to determine its molecular weight.[16] As the second apparatus could be run solely with an electric motor, Svedberg's "oil-turbine ultracentrifuge" (see fig. 4.3) usually referred to the sedimentation velocity machine.[17]

Since Svedberg's instrument enabled researchers to visualize colloids that could not be seen in the ultramicroscope, medical researchers and biochemists began collaborating with him in 1924 to look at proteins, a category of colloidal materials attracting particular attention from both physical and life scientists. Svedberg firmly supported the existence of molecules (as indicated by the title of his 1912 monograph *Die Existenz der Moleküle*),[18] but he shared with most chemists the belief that molecular structure was limited to small particles. In his view, complex and larger entities, including proteins, soaps, and industrial polymers, must be composed of aggregated molecules. Svedberg's early experiments with the proteins egg albumin and casein confirmed his assumption that proteins

15. See Svedberg, "Ultracentrifuge and Its Field of Research," 115; Elzen, *Scientists and Rotors,* 79–84.

16. By the late 1930s, the high-speed sedimentation velocity machine overshadowed the lower-speed sedimentation equilibrium instrument; it was the high-speed version that most researchers referred to in discussing Svedberg's innovations. Because that machine was so costly, the perception of its superiority supported Svedberg's scientific monopoly. As Elmer Kraemer commented in 1938, "As is well known, the oil-turbine ultracentrifuge is a rather formidable machine, and very few institutions can afford to install it. Owing to the publicity that the high-speed machine has received, there is a widespread impression that ultracentrifuge research is out of the question for the average chemical laboratory. This is by no means the case, for a wide variety of problems can be successfully attacked with the low-speed ultracentrifuge" (Kraemer, "Discussion," 128).

17. In addition, within a few years the sedimentation equilibrium machine became obsolete because, with the development of a new cell, one could determine molecular weight just as easily with a sedimentation velocity apparatus. As Frank Horsfall observed upon arriving in Svedberg's laboratory early in 1938, "The centrifuges here run about as constantly and as hard as the N.Y. subways. They don't ever seem to stop, and I've heard them as early as eight in the morning and as late as ten in the evening. At the end of one run the cell is taken out, cleaned, refilled and put back for another run all in no more than ten minutes. It is a regular sedimentation constant factory and they have a huge staff of technicians who do practically all the work. . . . The equilibrium centrifuges, of which there are four, simply don't run at all any more. The diffusion cell with its scale has taken over this job completely and the curves that are obtained with it are beautiful." Horsfall to Johannes Bauer, 21 Jan 1938, RAC RF 5, 4, box 14, folder 155.

18. Svedberg, *Die Existenz der Moleküle;* Kerker, "The Svedberg and Molecular Reality"; Nye, *Molecular Reality,* esp. 122–27.

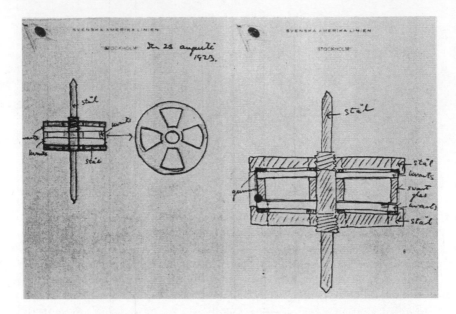

FIGURE 4.2 Svedberg's first ultracentrifuge in his Uppsala laboratory. *Above:* "The first drawing of an ultracentrifuge rotor made by Svedberg on the Atlantic on his way home from the United States. Notice the sector-shaped cells." *Facing page:* "Photograph of Svedberg and Rinde's ultracentrifuge." (Images and description reprinted from Pedersen, "The Development of Svedberg's Ultracentrifuge," *Biophysical Chemistry* 5 [1976]: 4 and 6 with permission from Elsevier Science.)

were colloidal suspensions of large and heterogeneous aggregates.[19] Much more surprising was the 1924 result of Svedberg's and Robin Fåhraeus's experiment with hemoglobin extracted from horse blood; they observed a sharp boundary separating the colorless solvent from the red blood protein. All of the hemoglobin particles were settling at the same speed, indicating that the protein possessed a singular molecular weight, which they determined (using the sedimentation equilibrium method) to be about 66,800 daltons (one dalton being the weight of a hydrogen atom).[20] Similarly, hemocyanin from snails gave a sharp boundary, this time yielding a uniform particle weight in the millions of daltons.

19. Elzen, *Scientists and Rotors,* 43. Svedberg's initial experiments centrifuging egg albumin were done with J. W. Nichols, who came over to Uppsala from Wisconsin on a grant from the "American Scandinavian Foundation." The experiments with casein were conducted with Robin Fåhraeus, assistant professor of pathology at the Caroline Institute in Stockholm.

20. Svedberg and Fåhraeus, "New Method for the Determination." See also Elzen, *Scientists and Rotors,* 43; Pedersen, "The Svedberg and Arne Tiselius," 240–41.

FIGURE 4.2 (*continued*)

FIGURE 4.3 Oil-turbine ultracentrifuge in Uppsala with adjoining equipment. *A:* Lamp-house; *B:* Centrifuge; *C:* Camera; *D:* Turbine oil inlet; *E:* Turbine oil outlet; *F:* Bearing oil outlet and oil drain; *G:* Main oil container; *H:* Oil coolers; *I:* Steps to compressor pit. (From Svedberg and Pedersen, *The Ultracentrifuge,* 189. Copyright 1940 Oxford University Press. Reprinted by permission.)

Just as Svedberg's experiments with proteins began to draw substantial interest from biologists and medical researchers, a physicist at the University of Virginia was designing a different high-speed centrifuge, based on a much simpler drive mechanism. Jesse Wakefield Beams reported his design in a 1931 article in *Science* entitled, "A Simple Ultra-Centrifuge."[21] Beams and his machine-shop engineer, A. J. Weed, adapted Henriot and Huguenard's "spinning top" design to make a centrifuge rotor supported

21. Beams and Weed, "A Simple Ultra-Centrifuge."

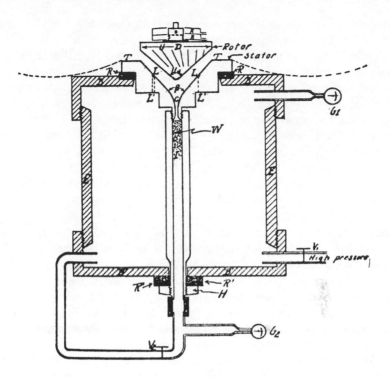

FIGURE 4.4 Diagram of a spinning top apparatus adapted to be an ultracentrifuge. (Reprinted, with permission, from Beams and Weed, "A Simple Ultra-Centrifuge," *Science* 74 [1931]: 45. Copyright 1931 American Association for the Advancement of Science.)

and driven by jets of compressed air (see fig. 4.4). The new machine maintained high rotor speed while maximizing the radius of the rotor, allowing for greater separation of sedimenting components. Beams added a viewing apparatus—"a simple arrangement... by which the light absorption and approximate index of refraction of the materials being centrifuged [could] be measured"—but his principal objective was separating rather than visualizing materials as they sedimented.[22]

One of Beams's graduate students, Edward G. Pickels, began developing the spinning-top rotor further so that he could use it to measure molecular weights.[23] Molecular weight determinations relied on integrating a good optical system that would allow the researcher to view the pattern of sedimentation in the spinning sample. Pickels, aware of Svedberg's achievements, sought to emulate key features of the Swede's oil-turbine

22. Beams, "A Simple Ultracentrifuge."
23. Elzen, *Scientists and Rotors,* 156.

machine in his air-driven spinning tops.[24] He built a rotor with sector-shaped cells and two glass windows, so that, with a light source, the rate of sedimentation could be obtained through measuring either the light absorption or index of refraction in the cell. Even with these changes, considerable technical problems remained. Convection currents disturbed the particle sedimentation in Pickels's early design of the air-driven apparatus, rendering it unreliable for molecular weight determinations. Nonetheless, Beams's and Pickels's efforts attracted the attention of Johannes Bauer, a medical researcher at the Rockefeller Foundation's International Health Division laboratory in New York.[25] Bauer was interested in adapting the air-driven ultracentrifuge to use for his isolation and study of yellow fever virus, and he arranged for Pickels to work with him during the summer of 1934.[26] While Beams continued to seek higher and higher rotational speeds, setting a world's record in 1934 of 1,200,000 rpm,[27] Pickels's vision of the ultracentrifuge was shaped by its potential uses in biomedical laboratories, particularly for virus research.

Pickels redesigned the air-driven ultracentrifuge early in 1935. Rather than resting on the stator and being driven directly by air jets, the rotor was hung in a vacuum chamber from a piano wire in a thin shaft below the air-driven turbine (see fig. 4.5). The new design of the vacuum chamber helped overcome earlier problems associated with convection currents, and the chamber could be water-cooled to prevent heat-induced denaturation of biological materials.[28] As Warren Weaver noted upon seeing a demonstration of this design, the "present technique, when and if perfected, will serve in many instances as a ridiculously cheap and simple substitute for the Svedberg technique."[29] Pickels was keenly aware of its

24. Beams, Weed, and Pickels, "Ultracentrifuge"; Elzen, *Scientists and Rotors,* 156–58. On Pickels's opinion of Svedberg, Elzen includes this quotation from Pickels's Ph.D. thesis: "Recent years have witnessed a rapidly increasing interest in the production of high rotational speeds and large centrifugal forces. Perhaps the most outstanding example of practical application is the work of T. Svedberg and his collaborators, who have used high-speed centrifuges of the electrically driven and oil-turbine types to determine the molecular weights of a great many organic materials, particularly proteins. The fixing of the molecular weight of haemoglobin as approximately 69,000 is indeed one of the classical experiments of modern times. The particular disadvantages of their higher-speed ultracentrifuges are the excessively high costs of construction and operation, and the fact that the design is not so easily varied to meet the needs of different investigators" (Pickels, *Air-Driven Ultra-centrifuge,* 1–2).

25. This laboratory, initially focused just on yellow fever research, was established in 1928 with an orientation toward "fundamental problems of disease rather than [the] applied field problems" with which the Rockefeller Foundation's International Health Organization was largely concerned (Farley, "International Health Division," 204).

26. Elzen, *Scientists and Rotors,* 161.

27. Ibid., 162; "1,200,000 R.P.M."

28. Elzen, *Scientists and Rotors,* 167–68.

29. Warren Weaver diary excerpt, 17 Oct 1935, RAC RF 1.1, 252, box 1, folder 8.

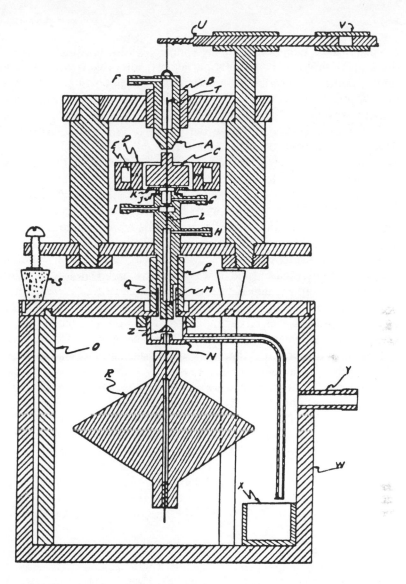

FIGURE 4.5 Cross-sectional drawing of Beams and Pickels's ultracentrifuge. "This machine includes an apparatus for spinning rotors in a vacuum. The rotor (R) hangs inside a vacuum chamber. Compressed air enters the annular space E, inside D, through a flexible rubber tube connection not shown. The turbine C rides upon an air film formed between its lower surface and the upper surface of the Bakelite collar K." (Diagram and description used with permission, from Beams and Pickels, "The Production of High Rotational Speeds," *Review of Scientific Instruments* 6 [1935]: 305, © American Institute of Physics.)

possibilities for biomedical research and concluded his dissertation, "For the study of biological materials, particularly filterable viruses and bacteria, the contributions may be a definite step in opening an entirely new field of research."[30] After filing his Ph.D. in Virginia in 1935, he joined Bauer to resume the work on viruses.[31]

Bauer's motivation in hiring the young physicist was technical—to develop a new tool for virus isolation—yet Pickels's work was seen to encompass a more scientific issue, the study of the nature of viruses with physical tools.[32] The Rockefeller Foundation's International Health Division Virus Laboratory was housed at the Rockefeller Institute; hence the presence of Pickels brought a diverse group of medical researchers into contact with the possibilities of the ultracentrifuge. The head of biophysics at the Rockefeller Institute, Ralph W. G. Wyckoff, had been developing techniques for using X-ray diffraction, ultraviolet microscopy, and radiation for medical research, and he was eager to expand his armamentarium of instruments to include the air-driven ultracentrifuge.[33] Wyckoff had the best machine shop at the Institute, so he suggested a collaborative effort with Pickels. Pickels declined to share his project, but Wyckoff did work for several years on ultracentrifuge development. He and Pickels published two papers together in 1936.[34] Bauer and Pickels continued to optimize the instrument for virus research.[35] Other American physicists, most notably James McBain, were publishing other designs and imagining diverse applications for air-driven spinning tops.[36]

As other historians have noted, the Rockefeller Foundation nurtured the development of the ultracentrifuge from the outset of its promise as a laboratory tool.[37] In 1927, a year after receiving his Nobel Prize in Chemistry, Svedberg wrote to the Rockefeller Foundation for financial support as he sought to expand his laboratory into a fully equipped Institute of Physical Chemistry. The International Education Board awarded Svedberg's laboratory $50,000 in 1928, and the Rockefeller Foundation

30. Elzen, *Scientists and Rotors,* 173, quoting Pickels, *Air-Driven Ultra-centrifuge,* 56.

31. "Mr. Pickels—Appointment," 19 Sep 1936, RAC RU 5, 4, box 22, folder 260.

32. Annual Reports of the International Health Division Virus Laboratory, 1935–42, RAC RF 5, 3. The reports from this group in 1936, 1937, and 1938 included a section under "Studies on the Nature of Viruses."

33. In 1939, Wyckoff claimed that he had "suggested about ten years ago to Dr. Rivers that if the pathologists at the Institute would utilize [the ultracentrifuge's] possibilities, my department of biophysics would undertake to make one. We received no encouragement" (Memorandum to Dr. Arthur F. Coca, 2 May 1939, Francis Peyton Rous papers, folder Wyckoff, Ralph #2, APS BR77).

34. Biscoe, Pickels, and Wyckoff, "Air-Driven Ultracentrifuge," and "Light Metal Rotors."

35. Bauer and Pickels, "High Speed Vacuum Centrifuge."

36. McBain, "Some Uses of the Air-Driven Spinning Top."

37. Elzen, *Scientists and Rotors;* Kohler, *Partners in Science.*

continuously supported Svedberg for the next twenty-three years, with a total investment of $242,000.[38] The initial grant was made to Svedberg on the basis of his achievements as a physical chemist. However, it was Svedberg's interest in using his apparatus to analyze protein structure that made his research so attractive to this patron in the 1930s, when Warren Weaver became director of the Natural Sciences Division of the Rockefeller Foundation. Svedberg's development of the analytical ultracentrifuge to characterize biological materials ideally suited Weaver's interdisciplinary mission of bringing the quantitative methods of the physical sciences to bear on questions in the life sciences. The project may have had added appeal to Weaver insofar as the technology actually drew on work which he had done as a physicist before coming to the Rockefeller Foundation.[39]

Weaver also awarded $13,000 of Rockefeller Foundation support to Jesse Beams from 1934 to 1939 and $7,200 to James McBain at Stanford from 1934 to 1937 for development of simpler, air-driven ultracentrifuges.[40] Thus, while the Rockefeller Foundation privileged Svedberg's status as creator and leader of the field, the Foundation also invested in the development of alternative ultracentrifuges. Of course, since the air-driven machines were quite cheap to build in comparison with Svedberg's apparatus, Beams's and McBain's grants were small in comparison to the level of support Svedberg received.

During the 1930s, the few operating ultracentrifuges were largely in the hands of the physicists and physical chemists who could construct them. The complexity and cost of Svedberg's machine limited dissemination of the technology. All eight of the oil-turbine analytical ultracentrifuges in the world in use by 1950 were built in Sweden, either by Svedberg's machine shop or by the company associated with Svedberg, LKB-Produkter.[41] The Rockefeller Foundation paid for the purchase of

38. "Machines That Sort and Weigh Molecules," Excerpt from Trustees Bulletin, RAC RF 1.1, 252, box 1, folder 13.

39. Mason and Weaver, "Settling of Small Particles." See Kay, *Molecular Vision of Life*, 43.

40. "Machines That Sort and Weigh Molecules" (see note 38 above); on Beams's support, see correspondence in RAC RF 1.1, 252, box 1, folder 10; on McBain's support, see RAC RF 1.1, 205D, box 9, folder 119. McBain had also been sponsored by the RF (with a $1,500 fellowship) as a visiting professor in Svedberg's laboratory in 1931.

41. According to Elzen, whose Ph.D. dissertation remains the best source of information on the dissemination of ultracentrifuges, LKB produced five oil-turbine ultracentrifuges in 1947, the last of which was sold in 1956, as well as selling one or two Svedberg equilibrium ultracentrifuges per year until 1954. The company also offered a "Beams centrifuge," which was preparative (lacking an optical system), and this machine sold more briskly at five per year. The prices of the instruments in 1946 were about $17,000 for a Svedberg sedimentation velocity machine, about $6,000 for a Svedberg equilibrium ultracentrifuge, and $4,000 for a Beams preparative centrifuge. In 1951, the company had never been profitable. Elzen, *Scientists and Rotors*, 338–40.

two ultracentrifuges from Svedberg for the University of Wisconsin in 1937 and one for the Lister Institute in 1936. Two other analytical ultracentrifuges had been bought from Svedberg through other funding sources: one went to the Biochemical Laboratories of Oxford in 1937, and the other was installed at DuPont Experimental Station in 1934, where Svedberg's former student Nichols conducted industrial research in polymer chemistry.[42]

In the mid-1930s, Pickels was finishing the adaptation of an air-driven ultracentrifuge for virus research, which was used as the prototype for analytical ultracentrifuges built at Harvard Medical School and at the Rockefeller Institute laboratories in Princeton (for which three machines were constructed). Pickels built ultracentrifuges for Harold Cox and Thomas Francis in 1940 as they moved from positions at the Rockefeller Institute to other laboratories, but the director of the Rockefeller Foundation discouraged Bauer and Pickels from manufacturing machines for outside researchers.[43] Jesse Beams, however, had no policy against filling outside orders for ultracentrifuges, and kept his department's machine shop in Virginia busy making customized machines as well as sending blueprints to inquirers.[44] "Customer" problems with the apparatus were numerous, requiring ongoing consultation by Beams and his workers (or Pickels).[45] Although Beams's basic design was established, each apparatus was individually built and adapted to particular purposes. The ultracentrifuge was far from standardized.

Given the variability of available designs, what distinguished an *ultracentrifuge* from an ordinary laboratory or industrial centrifuge? The contending instrument inventors disagreed, emphasizing different features as salient. Svedberg's coining of the term in 1924 drew on the analogy with other *ultra*-instruments: "The new centrifuge constructed by us allows the determination of particles that cannot be made visible in the ultra-microscope. In analogy with the naming of the ultra-microscope

42. See "Machines That Sort and Weigh Molecules" (see note 38 above); and Elzen, *Scientists and Rotors,* 311–21, 324–35. On the construction of Williams's ultracentrifuges at Madison in 1936 (installed in 1937), which included a Svedberg velocity machine, a Svedberg equilibrium machine, and also an air-driven machine, see correspondence in RAC RF 1.1, 200, box 164, folder 2013.

43. Thomas Francis Jr. to Johannes Bauer, 11 Aug 1941; Bauer to Francis, 10 Oct 1941, RAC RF 5, 4, box 12, folder 122; Bauer to W. A. Sawyer, 24 Apr 1941 and 21 May 1941, RAC RF 5, 4, box 26, folder 295. In 1936, a lawyer advising the Rockefeller Foundation recommended that Pickels not "make for others, either without charge or at cost, any patented device or any device containing any patented principle or process unless permission of the owner of the patent has first been obtained" (Thomas M. Debevoise to Thomas B. Appleget, 5 Nov 1936, RAC RF 1, 100, box 10, folder 80).

44. See correspondence with Beams in RAC RF 1.1, 252, box 1, folder 10.

45. See, for example, Bauer to Sawyer, 24 Apr 1941, RAC RF 5, 4, box 26, folder 295.

and ultra-filtration apparatus, we propose the name *ultra-centrifuge* for this apparatus."[46] In his view, what differentiated an ultracentrifuge from an ordinary centrifuge was the integration of an optical system, which made the instrument, like the ultramicroscope, capable of visualizing particles.[47] Svedberg built an apparatus for high-precision sedimentation measurements, with small cells designed to be free of convection currents. Because his rotor held only small amounts of material, it was intended for analyzing materials, not preparing them; Svedberg argued that any preparative devices should be called "supercentrifuges," reserving the name ultracentrifuge for analytical instruments.[48] McBain referred to his instrument simply as an "air-driven spinning top," whereas Beams and Pickels referred to any high-speed rotational machine as an ultracentrifuge, regardless of whether it included optical equipment. In a 1938 review of the field, Beams stated, "All of the centrifuges described in this paper are entitled to the name ultracentrifuge."[49] In the end, the term *ultracentrifuge* was applied to both air-driven and oil-driven turbines, and referred to both producing and visualizing substances. By 1940, these purposes were differentiated by use of adjectives: researchers spoke of the preparative ultracentrifuge and the analytical ultracentrifuge.

Beyond the attempts by various physicists to "trademark" the term *ultracentrifuge* according to a particular design feature, actual patent issues complicated the definition of the ultracentrifuge as an instrument. Svedberg and Nichols had patented their optical arrangement for the ultracentrifuge in 1927, but not the drive mechanism. Beams's and Pickels's publications included drawings and measurements of their designs that instrument makers at universities could use to construct customized ultracentrifuges.[50] They emphasized that not only was their machine much less expensive than Svedberg's apparatus, but their design was more adaptable. At the same time, the Sharples Specialty Company, a manufacturer of cream separators, became interested in producing ultracentrifuges for laboratory use. They already produced a Laboratory Super

46. Svedberg and Rinde, "Ultra-Centrifuge, a New Instrument," 2677–78; emphasis in original.

47. Gray, "Ultracentrifuge," 50.

48. Svedberg to Stanley, 12 Apr 1937, as cited from Svedberg papers in Uppsala by Elzen, *Scientists and Rotors,* 121. This letter is apparently missing from the Stanley papers at the Bancroft Library in Berkeley.

49. Beams, "High Speed Centrifuging," 246n; Elzen, *Scientists and Rotors,* 210.

50. Beams and Pickels, "Production of High Rotational Speeds"; Biscoe, Pickels, and Wyckoff, "Light Metal Rotors"; Pickels, "Practical Speed-Measuring Devices for High Speed Centrifuges," and "Improved Type of Electrically Driven, High Speed Laboratory Centrifuge." The specifications given in these articles were probably not sufficient to enable duplication of the machine; many researchers wanting to build their own instruments contacted Beams and Pickels for information, advice, and blueprints.

Centrifuge, whose top speed was increased to 50,000 rpm in 1935. Because Svedberg's oil-driven machine was so expensive to build, they decided to commercialize Beams's and Pickels's vacuum, air-driven ultracentrifuge.[51]

Philip Sharples tried to persuade Beams to take out a patent on his design, and, failing that, he offered Pickels a job at the company about the time that Bauer invited Pickels to the Rockefeller Institute.[52] Although Pickels declined the job with Sharples, he did consult with the company on the design and development of their commercial apparatus. Beams continued to resist Sharples's pleas to take out patents on some of his innovations, instead encouraging interested researchers to save money by having their university machine shop build a machine rather than buying a commercial apparatus.[53] Failing to recruit Pickels, Sharples took out an exclusive license on the Svedberg-Nichols patent. Since Bauer and Pickels continued furnishing blueprints for the air-driven ultracentrifuge to interested individuals and institutions, Sharples wrote an angry letter to Pickels, pointing out "that it is not legal for anyone to build or use an ultracentrifuge that infringes claims of that patent."[54] After consulting with the Institute's lawyer, Bauer responded curtly, "Our counsel sees no reason to change the advice already given to the effect that the Foundation has a right to use patented devices, principles, or processes when such use is not prompted by a profit motive and when its purpose is scientific."[55] Beams was at the same time building ultracentrifuges at cost for other laboratories for several hundred dollars per machine, arguing that his were better machines than the Sharples ones, which cost over $3,000.[56] To further complicate the commercial claims, a Swedish

51. Ayres, "80,000 Revs per Minute?"

52. Beams to Weaver, 21 Sep 1937, RAC RF 1.1, 252, box 1, folder 10: "Two or three different times during the last few years Mr. Sharples has written me urging that his company be allowed to take out patents in my collaborators' and my name on the air driven ultracentrifuge. In return for this he was willing to turn over to [sic] a certain per cent of the returns from sales." See also Elzen, *Scientists and Rotors,* 349–51.

53. Elzen, *Scientists and Rotors,* 352.

54. Warren Weaver to Jesse Beams, 13 Sep 1937, RAC RF 1.1, 252, box 1, folder 10. Sharples was particularly angry because he was under the impression that not only educational institutions, but American Cyanamid, had access to the ultracentrifuge blueprints. In response to Sharples's complaint, Bauer wrote Svedberg to ask if he and Pickels should obtain a license from Svedberg and Nichols for the use of their air-driven ultracentrifuge (Bauer to Svedberg, 11 Jan 1938, RAC RF 1, 100, box 10, folder 81).

55. Bauer to Sharples, 15 Feb 1938, RAC RF 1, 100, box 10, folder 81. Sharples agreed "to let the patent matter drop for the time being" (Bauer to J. B. Nichols, 7 Mar 1938, RAC RF 1, box 10, folder 81; see also Svedberg to Weaver, Nichols, Bauer, and Sharples, all from 26 Jan 1938, same folder).

56. As Beams wrote, "I have furnished anyone with drawings and showed them all we knew about centrifuges whenever they came to Virginia. I encourage everyone to make their

ultracentrifuge company approached both Beams and Pickels in 1937 claiming that they held a patent over all angular centrifuges.[57] The following year the Research Corporation, which held patents on behalf of university researchers, wrote Beams and Weaver about their interest in licensing patents for the air-driven ultracentrifuge.[58]

The Rockefeller Foundation was keen to avoid the appearance that they had enabled commercial profit. As Warren Weaver declared to Beams, "We have no interest in the development of the 'ultracentrifuge industry' and only wish that developments in this field be made freely and economically available."[59] Beams took his cue from Weaver that if he were to patent his technology, it would be viewed dimly by Rockefeller Foundation officers.[60] (Beams offered to sign his innovations over for the Rockefeller Foundation to patent, but Weaver declined.)[61] At the same time, the Foundation was funding physicists such as Beams and McBain with the expectation that their new designs would "make this technique available to a large number of biologists and medical men who would not be prepared to undertake the maintenance of the more complicated [Svedberg-type] machine."[62] The question of how such diffusion would

own machines when possible" (Beams to Weaver, 21 Sep 1937, RAC RF 1.1, 252, box 1, folder 10; see also Frank Blair Hanson diary excerpt regarding Beams and patents, 28–30 Dec 1937, same folder).

57. See Beams to Weaver, 21 Sep 1937, RAC RF 1.1, 252, box 1, folder 10. The company may have been LKB-Produkter. Correspondence in RAC RF 1, 100, box 10, folder 80 documents a similar dispute in the fall of 1936 with the Ivan Sorvall Company, which argued that 50°-angle centrifuges were protected under their patent.

58. Howard Poillon, president of Research Corporation, to Weaver, 10 Feb 1938, RAC RF 1.1, 252, box 1, folder 11.

59. Weaver to Beams, 13 Sep 1937, RAC RF 1.1, 252, box 1, folder 10. Weaver was also unhappy with Svedberg about the patent situation. As he wrote Jack Williams, "I had a very pleasant evening with Svedberg and you. In a way I was a little relieved that the time of my train made it necessary for me to run when I did, because I thought the conversation was about to enter possibly embarrassing phases relative to Svedberg's patent relations with Sharples. I would personally feel a good deal better about the whole thing if Svedberg and Nichols had never taken out the original patent, and had no present interest in that side of things" (16 Nov 1937, RAC RF 1.1, 200, box 164, folder 2013).

60. Beams to Poillon, 5 Feb 1938. In his own letter to the head of Research Corporation, Weaver stated, "It is usually possible to phrase this answer [that the RF has no firm patent policy] in such a way as to make it reasonably clear that a person's future relations with The Rockefeller Foundation would necessarily be influenced, should his own judgment lead him to make a decision that was personally selfish or scientifically unsound" (Weaver to Poillon, 25 Feb 1938; both letters from RAC RF 1.1, 252, box 1, folder 11).

61. Weaver to Poillon, 25 Feb 1938, RAC RF 1.1, 252, box 1, folder 11, and Weaver to Beams, 3 Dec 1937, RAC RF 1.1, 252, box 1, folder 10.

62. Grant action to Jesse Beams, University of Virginia—Ultracentrifuge Development, 15 Jan 1937, RAC RF 1.1, 252, box 1, folder 10. The RF had awarded $2,500 to Beams in 1934 and authorized $4,000 a year from 1937 to 1939.

be effected remained unclear. The issue went beyond patents; the inventors did not yet feel the technology was advanced enough to create a general-purpose, standardized instrument. As Beams commented in 1937, "The ultracentrifuge is primarily a research apparatus and cannot be completely standardized."[63]

Beyond the trade-offs between standardization and continued optimization, and the ethical and legal difficulties connected with patents, the transfer of ultracentrifugal technology to commercial development threatened to disrupt the collaborative relationships between physical chemists and biomedical researchers nurtured by the Rockefeller Foundation's program in experimental biology. In the late 1930s, an ambitious protein chemist or virus researcher could construct an ultracentrifuge, but relied on direct interaction with Svedberg, Beams, or Pickels to draw up blueprints, get the machine running, and consult on maintenance.[64] The Sharples Company, in seeking to commercialize the same instrument, emphasized the advantages of having company representatives take over the roles of design and maintenance.[65] Few laboratories could afford a $3,000 ultracentrifuge, particularly if the Rockefeller Foundation did not support such a purchase. As Elzen has surmised, it is likely that Sharples sold only one ultracentrifuge before World War II.[66] Thus, despite the prominence of the ultracentrifuge investigations of proteins and viruses at these research centers, the analytical ultracentrifuge was still an uncommon piece of equipment. Many investigators wishing to obtain sedimentation analysis of their materials brought them or mailed them to Uppsala, where Svedberg's three machines ran almost continuously.[67]

Svedberg's position in biological ultracentrifugation during this time has been described as a "kind of scientific monopoly."[68] Svedberg's analytical ultracentrifuges were superior in the precision of their measurements, and the resources and credibility of the Rockefeller Foundation enhanced his status in the community of physical biochemists. At the same time, a close analysis of how the ultracentrifuge became used in Stanley's studies of TMV reveals the limits of the Swede's dominance in two

63. Beams to Weaver, 21 Sep 1937, RAC RF 1.1, box 1, folder 10, 3.

64. For example, Pierre Lépine of the Pasteur Institute requested blueprints, photographs, etc. of Pickels's ultracentrifuge so that he could build one in France, with the assistance of E. Huguenard, one of the inventors of the spinning top; see memo from Lépine to Rivers with letter from Johannes Bauer to Sawyer, 26 May 1941, RAC RF 5, 4, box 26, folder 295.

65. Elzen, *Scientists and Rotors*, 359.   67. Ibid.

66. Ibid., 364.

68. Kay, "Stanley's Crystallization," 462. Kohler similarly speaks of Svedberg's "virtual monopoly in ultracentrifuge technology" in *Partners in Science,* 337.

respects. First, because viruses were generally larger than proteins, one could obtain sharp boundaries and definite molecular weights with instruments less sophisticated than Svedberg's. In this respect, virus research, particularly as conducted at the Rockefeller Institute, helped unsettle the superiority of the costly oil-driven instrument over the air-driven machine. Second, the role of the ultracentrifuge was radically transformed (and, to Svedberg's mind, perhaps subverted) by its adaptations for viruses. In 1936 it became clear that the ultracentrifuge would be the tool of choice for producing chemically pure viruses.

Placing Tobacco Mosaic Virus in the Analytical Ultracentrifuge

As soon as Stanley crystallized TMV, he sought opportunities to have the material analyzed in an ultracentrifuge. Stanley's crystals, which seemed to be composed of a high molecular weight protein, provided, in the words of James McBain, "a clear case for test with the ultracentrifuge."[69] Stanley was already acquainted with Svedberg, who had visited the Princeton laboratories on a trip sponsored by the Rockefeller Foundation a year earlier. Stanley referred to this previous point of contact when he wrote Svedberg on 10 July 1935 with an offer to send some purified virus to Sweden for analysis.[70] This letter was likely prompted by the receipt of a letter dated 6 July from McBain, who had seen Stanley's *Science* announcement the week before and offered to determine the sedimentation velocity of crystalline TMV in his ultracentrifuge.[71] Stanley sent samples of TMV to both Svedberg and McBain, but requested that McBain defer publishing his results until after Svedberg, the recognized expert, completed his study.[72] Closer to home, Ralph Wyckoff, who was working in concert with Bauer and Pickels on ultracentrifugation of proteins and viruses, also offered to collaborate with Stanley.[73] Stanley had already sent him some crystals of TMV to be used for X-ray diffraction studies, but Wyckoff was eager to determine the molecular weight of TMV in the ultracentrifuge. He wrote Stanley on 22 January 1936,

69. James McBain to Warren Weaver, 23 May 1936, RAC RF 1.1, 205D, box 9, folder 121.
70. Stanley to Svedberg, 10 Jul 1935, Stanley papers, carton 12, folder Svedberg, Theodor.
71. McBain to Stanley, 6 Jul 1935, Stanley papers, carton 10, folder McBain, James William.
72. Stanley to Svedberg, 10 Jul 1935, Stanley papers, carton 12, folder Svedberg, Theodor; Stanley to McBain, 10 Jul 1935, Stanley papers, carton 1, folder Jul 1935; McBain to Stanley, 20 Jul 1935, Stanley papers, carton 10, folder McBain, James William; Svedberg to Stanley, 29 Jul 1935, Stanley papers, carton 12, folder Svedberg, Theodor; and Stanley to McBain, 10 Aug 1935, carton 1, folder Aug 1935.
73. Corner, *History of the Rockefeller Institute,* 183–84.

We are now ready to run ultracentrifuge sedimentation constants by
the absorption method. What are your relations with Svedberg in
the matter? Would you prefer to wait till he publishes on the material you sent him; or would you be willing to go in with us on some
sedimentations of your two strains and of normal globulin[?] We are
now finishing some check results (against Svedberg) with hemoglobin
and would like to do some bigger molecules. Dr. TenBroeck said you
were getting molecular weights by diffusion. The two methods would
check one another nicely. If you should want to do this, we might talk
the thing over soon; but if you'd prefer to leave Svedberg a clear field,
don't hesitate to say so.[74]

Stanley went ahead and sent Wyckoff some samples, beginning a steady
collaboration, and within the year Wyckoff relocated his laboratory to
Princeton, in part to collaborate more closely with virus researchers there.
Even as Stanley made clear that Svedberg would be given priority in
publication (as he wrote McBain of his work with Wyckoff, "We are giving [Svedberg] a clear field with respect to the tobacco protein"),[75] he
managed the competition and available resources so as to maximize his
potential results. Maintaining these multiple alliances was difficult; the
race to determine the size and shape of TMV in the analytical ultracentrifuge was likely to be a winner-take-all venture.

Ultracentrifugation was only one of the recently developed techniques
that Stanley wished to apply to his pure virus, but other new methods
also relied on collaborating with a specialist. As just mentioned, Wyckoff
began X-ray diffraction studies of TMV early in 1936. At about the
same time, Stanley established a collaboration with George Lavin in the
Rockefeller Institute's Chemical Pharmacology Laboratory; Lavin used
his ultraviolet spectroscopic equipment to study both the absorption
spectrum of the crystalline material and to attempt ultraviolet light inactivation of the virus.[76] Stanley also made arrangements with John Edsall,
in Edwin Cohn's laboratory at the Department of Physical Chemistry
at Harvard Medical School, to measure the double refraction of flow

74. Ralph Wyckoff to Stanley, 22 Jan 1936, Stanley papers, carton 13, folder Wyckoff,
Ralph Walter Graystone. Wyckoff went ahead with some runs of TMV, and he was at the
same time trying to get X-ray diffraction patterns from the material, to no avail (Wyckoff to
Stanley, 13 Feb 1936, Stanley papers, carton 13, folder Wyckoff, Ralph Walter Graystone).
75. Stanley to McBain, 23 Mar 1936, Stanley papers, carton 1, folder Mar 1936.
76. Stanley to George I. Lavin, 9 Jan 1936, Stanley papers, carton 1, folder Jan 1936;
Stanley to Lavin, 27 Feb 1936, Stanley papers, carton 1, folder Feb 1936; Corner, *History of
the Rockefeller Institute,* 452. The collaboration resulted in two publications: Lavin and
Stanley, "Spectrum of Crystalline Tobacco Mosaic Virus Protein"; Lavin, Loring, and
Stanley, "Spectra of Latent Mosaic and Ring Spot Viruses."

and relaxation time of the virus protein using their special equipment.[77] Arne Tiselius in Svedberg's laboratory used the samples Stanley sent Svedberg to run TMV on his electrophoresis apparatus and to determine the isoelectric point. Credit was not always divided equally; whereas Stanley and Lavin coauthored their papers on TMV, Stanley was not an author on Svedberg's study of TMV.[78] At the same time, Stanley did report Svedberg's results in his lectures.[79] Beyond these specific collaborations, many scientists were writing Stanley for samples of his crystalline virus for various experiments, and he responded favorably to most requests. But he was always keen to avoid direct competition: he asked his senior colleague at the Institute, Max Bergmann, to specify his plans for using TMV before he sent a sample to New York City.[80]

Results from the analytical ultracentrifuge held particular promise for demonstrating that the virus was a discrete protein, but Stanley's hopes were soon disappointed. Svedberg's initial runs indicated that the material was not homogeneous in terms of its size, and that the heterogeneity increased when the pH was increased or lowered. Communicating his preliminary results in a letter to Stanley, Svedberg wrote, "It is not composed of a small number of well-defined compounds either, but has molecules varying almost continuously over a certain range."[81] This result was a setback for Svedberg as well as for Stanley, for Svedberg was at this time collecting evidence to support his theory that all proteins are discrete aggregates composed of well-defined components of molecular weight 17,000.[82] In striking contrast, when Stanley's sample was run on Tiselius's electrophoresis apparatus, it appeared as a single homogeneous species.

In his publication of these equivocal results for TMV, Svedberg pointed out that the experiments left unresolved whether Stanley's crystalline virus had been chemically degraded through its isolation or whether the virus was not a unitary molecular species *in vivo* (see fig. 4.6). Stanley's

77. Stanley to E. J. Cohn, 18 Dec 1935, and Stanley to J. T. Edsall, 18 Dec 1935, carton 1, folder Dec 1935; letter from John T. Edsall to Stanley, 23 Dec 1935, carton 8, folder Edsall, John Tileston; letter from E. J. Cohn to Stanley, 23 Dec 1935, carton 7, folder Cohn, Edwin Joseph, all from Stanley papers.

78. Lavin and Stanley, "Spectrum of Crystalline Tobacco Mosaic Virus Protein"; Lavin, Loring, and Stanley, "Spectra of Latent Mosaic and Ring Spot Viruses"; Eriksson-Quensel and Svedberg, "Sedimentation and Electrophoresis."

79. Stanley to Svedberg, 24 Mar 1936, and Svedberg to Stanley, 1 Jun 1936, Stanley papers, carton 12, folder Svedberg, Theodor.

80. Kay, "Stanley's Crystallization," 467, citing Stanley to Max Bergmann, 28 Apr 1937, Stanley papers, carton 1, folder Apr 1937; along similar lines see Stanley to P. A. T. Levene, 9 Jan 1939, Stanley papers, carton 1, folder Jan 1939.

81. Svedberg to Stanley, 7 Mar 1936, Stanley papers, carton 12, folder Svedberg, Theodor.

82. See Svedberg, "Protein Molecules."

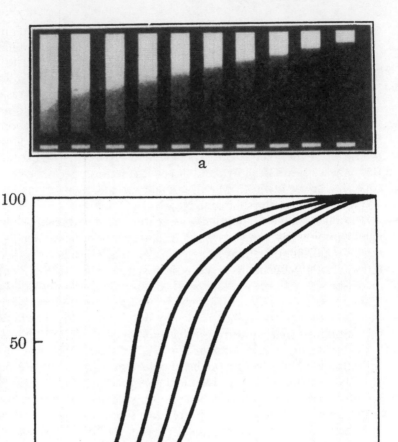

FIGURE 4.6 Sedimentation of TMV in Svedberg's oil-turbine ultracentrifuge. *a:* Sedimentation pictures of TMV protein obtained by means of the light absorption method. *b:* Concentration curves for the TMV protein at pH 6.8. Centrifugal force: 15,000 times gravity; time between the exposures: five minutes. The blurring of the boundary and the change in shape of the curves with time of sedimentation indicate the sample's polydispersity. (Reprinted, with permission, from Eriksson-Quensel and Svedberg, "Sedimentation and Electrophoresis of the Tobacco-Mosaic Virus Protein," *Journal of the American Chemical Society* 58 [1936]: 1864. Copyright 1936 American Chemical Society.)

TMV sedimented as a high molecular weight blur; Svedberg estimated that 65 percent of the material "falls in the molecular weight interval 15–20 millions."[83] (Assuming the virus particle to be roughly spherical, Svedberg assigned a provisional mean molecular weight of 19 million, soon revised down to 17 million.) While the results did not provide much information about the structure of the virus, Svedberg argued that they ruled out the possibility that TMV was a small bacterium. In his view, no living organism existed whose size would change depending on the pH and whose electrophoretic mobility would show molecular homogeneity.[84] Not wishing to abandon the promising notion that the virus was an authentic and chemically homogeneous protein species, Stanley and Svedberg chose to blame the preparation's heterogeneity on the lengthy extraction process, which had disrupted the intact virus structure.[85] Svedberg commented at the end of his paper, "The most likely interpretation of the facts revealed by us seems to be that the virus is a chemically well-defined protein, probably homogeneous with regard to molecular weight (17 millions) in the plant. It is very sensitive to deviations from neutral pH and is thus rendered inhomogeneous."[86] In support of this explanation, Svedberg found that recrystallization of the virus increased the molecular inhomogeneity of the species, so that greater chemical purity and greater molecular purity did not correlate.

By the time Svedberg's paper was published, Stanley had turned to other centrifugers in the hopes of getting better results. Wyckoff was his most available collaborator, because of both institutional affiliation and geographical proximity. Stanley was sending him samples of TMV from several hosts and different strains of TMV, and despite recurrent equipment problems,[87] Wyckoff's runs showed TMV to be more homogeneous than Svedberg's experiments had indicated (see fig. 4.7).[88] By mid-1936, Stanley and Wyckoff were performing experiments together in New York, as well as exchanging materials and results.[89] At the same time, McBain was also testing a sample of Stanley's preparation and reported, "Your virus is definitely monodisperse."[90]

83. Eriksson-Quensel and Svedberg, "Sedimentation and Electrophoresis," 1866.

84. Ibid., 1867.

85. Stanley to Svedberg, 9 Jul 1936, Stanley papers, carton 12, folder Svedberg, Theodor.

86. Eriksson-Quensel and Svedberg, "Sedimentation and Electrophoresis," 1867.

87. Wyckoff to Stanley, 6 May 1936, Stanley papers, carton 13, folder Wyckoff, Ralph Walter Graystone.

88. See Wyckoff, Biscoe, and Stanley, "Ultracentrifugal Analysis."

89. See Stanley to Wyckoff, 2 Sep 1936, Stanley papers, carton 1, folder Sep 1936; and Wyckoff to Stanley, 3 Sep 1936, Stanley papers, carton 13, folder Wyckoff, Ralph Walter Graystone.

90. McBain to Stanley, 23 May 1936, Stanley papers, carton 10, folder McBain, James William; see also Stanley to McBain, 23 Mar 1936, carton 1, folder Mar 1936.

FIGURE 4.7 Pictures of TMV sedimenting in Wyckoff's air-driven ultracentrifuge. The pictures compare homogeneous and nonhomogeneous samples of purified TMV. "*Panel 4:* A print of the series of photographs constituting an absorption run on a solution of the ordinary TMV protein at pH 9.3. The continued sharpness of the boundary down to the last picture is indicative of molecular homogeneity. *a,* air bubble; *b,* solution above sedimentation boundary. *Panel 5:* A similar run on a less homogeneous TMV protein sample. Boundaries in the later photographs are diffuse. The presence of unsedimentable material is shown by the greater transparency of the air bubble (*a′*) than of the solution (*b′*) above the sedimenting boundary. *Panel 6:* Photographs of the absorption run on the ordinary TMV protein taken from two-week-old plants. The double boundary is clearly evident in the later pictures." (Images and captions from Wyckoff, Biscoe, and Stanley, "An Ultracentrifugal Analysis of the Crystalline Virus Proteins" [1937]: 67. Reprinted by permission of the American Society for Biochemistry and Molecular Biology.)

McBain's news was cheering, for Stanley wanted to establish that TMV was a single entity, a homogeneous protein, despite its poor behavior in the ultracentrifuge compared to similarly large proteins such as hemocyanin. Wyckoff provided Stanley with good evidence that TMV was molecularly homogeneous in plant cells. He found that the sedimentation boundary for the particles of a single-lesion isolate of TMV in clarified tobacco sap was sharper than that of the chemically purified virus, despite the impurities from the tobacco leaf in the lesion sample. Stanley also took encouragement from Svedberg's observation that "the degree of homogeneity of the virus protein with regard to molecular weight depends on its pH-history."[91] This is not to say that Svedberg's early results with TMV could be altogether relegated to the status of artifact. Wyckoff and Stanley also reported several ultracentrifuge runs in which TMV exhibited more than one boundary or was heterogeneous.[92] Nonetheless, Stanley was so confident that the polydispersity of TMV in the ultracentrifuge was arti-factual that, in an update to Simon Flexner in April of 1936, he reported that "the protein is homogeneous, viz., the molecules are all of the same size."[93] His first publication on the "ultracentrifugal analysis of crystalline virus proteins," while not quite as audacious as his statement to Flexner, argued for the molecular homogeneity of undamaged virus protein.[94]

In July of 1936, Stanley reported Wyckoff's results to Svedberg. "I am afraid that in my efforts to prepare an especially 'pure' sample of protein for you, I defeated my own purpose. The repeated fractionation to which I subjected your sample, in an effort to obtain an especially homogeneous preparation, evidently caused it to become very inhomogeneous with respect to molecular weight.... Using our regular, so-called 'crude,' ma-terial, he [Wyckoff] obtains boundaries that are fully as sharp as those given in ... your paper for the hemocyanin."[95] Stanley sought to supplant Svedberg's original pronouncement on TMV's behavior in the ultracentri-fuge with his own experiments and explanation.

An unexpected observation by Wyckoff and his coworker Robert Corey offered a means by which to try to isolate TMV in a more pristine form. When they centrifuged clear juice from mosaic-diseased tobacco plants at 25,000 rpm, a pellet of fibrous material sedimented to the bottom of the cell. The pellet, when analyzed in a microscope, was seen to be composed of needle-shaped crystals.[96] In the *Science* announcement of their observation, "The Ultracentrifugal Crystalliza-tion of Tobacco Mosaic Virus Protein," Wyckoff and Corey compared the X-ray diffraction patterns of these centrifuge-produced crystals with the crystalline protein Stanley prepared by chemical means, and found them to be indistinguishable.[97]

This method of isolation by sedimentation held particular promise for purifying viruses too dilute or fragile for chemical extraction. Wyckoff was eager to extend his approach from TMV to other plant and ani-mal viruses. He and Stanley used the ultracentrifuge to isolate potato

91. Svedberg to Stanley, 1 Jun 1936, Stanley papers, carton 12, folder Svedberg, Theodor.
92. Wyckoff, Biscoe, and Stanley, "Ultracentrifugal Analysis," esp. table 1 (60–61) and figure 1 (62).
93. Stanley to Flexner, 7 Apr 1936, Stanley papers, carton 1, folder Apr 1936.
94. Wyckoff, Biscoe, and Stanley, "Ultracentrifugal Analysis."
95. Stanley to Svedberg, 9 Jul 1936, Stanley papers, carton 12, folder Svedberg, Theodor.
96. Wyckoff and Corey, "Ultracentrifugal Crystallization."
97. Ibid.

X virus (Bawden and Pirie's original subject of study) and tobacco ring spot virus. Wyckoff and Price (in Kunkel's group) isolated two virus strains of cucumber mosaic disease, and showed that they were chemically similar to TMV, whereas the tobacco necrosis virus they isolated in the ultracentrifuge proved quite distinct from TMV.[98] With Joseph Beard, from the animal pathology group, Wyckoff used the ultracentrifuge to isolate papilloma virus from rabbits.[99] The purification of an animal virus was a particularly significant achievement. By 1939 Wyckoff referred to this general method of isolation as ultracentrifugal fractionation.[100] In all of these studies, TMV provided a precedent for how the analytical ultracentrifuge could be integrated into various experimental systems, not only to visualize but actually to produce viruses as pure macromolecules.[101] Stanley was keen to point to the significance of extending the physical-chemical methods used to study plant viruses, as exemplified by TMV, to investigate animal viruses, from poliomyelitis to cancer agents.[102]

Wyckoff advocated the general preparative possibilities of the ultracentrifuge in 1937 in a *Science* article, "The Ultracentrifugal Purification and Study of Macromolecular Proteins." Ultracentrifugal isolation had already expanded the repertoire of biological macromolecules: "There are...many unstable substances with specific biological properties—such as protein-linked hormones and viruses—to which these chemical procedures [i.e., precipitation and crystallization] have not been successfully applied. Ultracentrifugation, in the sense of centrifuging at speeds sufficiently high so that molecules are actually sedimented, has already provided a number of these less stable proteins in purified form."[103] Even for proteins and viruses that could be isolated using conventional chemical means, the ultracentrifuge furnished a better alternative: "The molecular state of the ultracentrifuged sample is nearer that prevailing in

98. Stanley and Wyckoff, "Isolation of Tobacco Ring Spot"; Price and Wyckoff, "Ultracentrifugation of the Proteins," and "Ultracentrifugation of Juices." Bryan Harrison has pointed out (personal communication, 1 Mar 1999) that what Price and Wyckoff isolated was not cucumber mosaic in the regular sense, but rather cucumber green mottle mosaic virus.

99. Beard and Wyckoff, "Isolation of a Homogeneous Heavy Protein." For more on this development and its place in the history of tumor virus research, see Creager and Gaudillière, "Experimental Arrangements."

100. See Price and Wyckoff, "Ultracentrifugation of Juices."

101. For example, in Wyckoff's lecture for the American Philosophical Society in January 1937, his generalizations about the utility of quantity ultracentrifugation for virus research were all exemplified by his experiments with TMV ("Ultracentrifugal Study of Virus Proteins").

102. Stanley to Svedberg, 26 Mar 1937, Stanley papers, carton 12, folder Svedberg, Theodor.

103. Wyckoff, "Ultracentrifugal Purification," 92.

the animal or plant from which it is derived."[104] Wyckoff pointed out that the air-driven machine cost a small percentage of the price of Svedberg's oil-driven type, and one could interchange analytical and quantity rotors. The preparative (quantity) rotor could be spun 50–100 times faster than possible in ordinary centrifugation.[105]

The ultracentrifuge might be faster than ordinary laboratory centrifuges, but the use of the ultracentrifuge as a preparative device during these years was mirrored by new uses of high-speed laboratory centrifuges to fractionate cell components by size. At the Rockefeller Institute, Alfred Claude employed a preparative centrifuge at speeds up to 17,000 rpm to concentrate chicken tumor agent.[106] Even ordinary high-speed centrifuges were beginning to be called ultracentrifuges when used in this way.[107]

Svedberg had already opposed the adaptation of the ultracentrifuge to serve as a preparative device. In 1934, the Rockefeller Foundation gave the Lister Institute of Preventive Medicine in London funds to purchase an oil-turbine machine from Svedberg's own machine shop. The director, J. C. G. Ledingham, wrote Svedberg requesting customization of the instrument for their virus studies, including larger cells that could fractionate materials for biological assays.[108] Svedberg rebuked Ledingham for asking him to alter his analytical machine:

> I should like to point out, however, that my primary interest is to supply methods for accurate quantitative studies of high molecular [weight] compounds such as the proteins, and that I have always been a little suspicious of such qualitative and semi-quantitative methods as you suggest in your letter. . . . The types of ultracentrifuges which we have worked out here are not meant for the separation "in substance" but for optical observations.[109]

Increasing the size of the ultracentrifuge cell to allow for its preparative use threatened to introduce convection currents, disturbing sedimentation. For Svedberg, anything that compromised the precision of his instrument was out of the question; quantitative "ultracentrifugation" was an oxymoron.

104. Ibid., 92.
105. Ibid.
106. Claude, "Properties of the Causative Agent"; Rheinberger, "From Microsomes to Ribosomes"; Gaudillière, "Molecularization of Cancer Etiology," 143.
107. In this chapter, however, I will continue to follow the adaptation of analytical ultracentrifuges (oil-turbine or air-turbine) for quantity uses, rather than cover the history of preparative centrifuges generally.
108. Elzen, *Scientists and Rotors,* 325.
109. Svedberg to J. C. G. Ledingham, Nov 1934, as quoted by Pedersen, "The Svedberg and Arne Tiselius," 265.

Virus researchers had two reasons to resist Svedberg's stringent view of ultracentrifugation. As a class of macromolecules, viruses relaxed slightly the constraints on ultracentrifuge design necessitated by smaller particles such as mid-range proteins (e.g., hemoglobin). Ultrafiltration studies were showing most viruses to be generally larger than globulins (proteins). Because larger particles diffuse more slowly, it was easier to obtain sharp boundaries in the ultracentrifuge with viruses than with proteins (at least in theory), and the speeds necessary for sedimentation were not as great.[110] Although Svedberg felt that the air-driven machine was so inferior to his oil-turbine design as to make it an undesirable apparatus to use, many virus researchers found it perfectly adequate for sedimenting viruses. Second, the air-driven ultracentrifuge was readily adapted for quantity use. As Ralph Wyckoff observed in 1937, "The analytical observations [on the molecular behavior of purified TMV] just described, which were made with our ultracentrifuge, could equally well have been carried out on the classical machine of Svedberg. But as so often happens when a new kind of apparatus is made, the air ultracentrifuge quickly proved to have potentialities in addition to those of the machine which inspired it."[111]

In the end, use of the ultracentrifuge made obsolete the chemical crystallization of TMV, which had created such a sensation in 1935. As Stanley observed by 1939 in a private letter, "The question as to whether or not a virus is crystalline has no direct bearing on problems concerning the purity or the nature of the virus."[112] In addition, Stanley's adoption of the ultracentrifuge eased the move away from the "autocatalysis" analogy with Northrop's precursor enzymes. Experiments using the ultracentrifuge to visualize the contents of the tobacco cell failed to turn up any possible precursor proteins for TMV, which was the largest cellular constituent. By 1937 Stanley admitted that infected cells had to produce TMV from smaller constituents—by polymerization or direct synthesis.[113] Whereas crystalline TMV had been modeled on Northrop's proenzymes and bacteriophage—it was represented as an autocatalytic protein available through chemical isolation—now the analogies were flowing in the opposite direction. In August 1936, Stanley wrote Northrop of his recent visit with Max Schlesinger at the National Institute for Medical Research in London, offering Northrop information on Schlesinger's filtration method for isolating coli bacteriophage.

110. On the estimated sizes of bacteriophages in the early 1930s, see Elford and Andrewes, "Sizes of Different Bacteriophages."

111. Wyckoff, "Ultracentrifugal Study of Virus Proteins," 457.

112. Stanley to C. Levaditi, 23 Mar 1939, Stanley papers, carton 1, folder Mar 1939. Stanley is referring to his less concise statement of the same point in "Architecture of Viruses," 541.

113. Stanley, "Chemical Studies on the Virus of Tobacco Mosaic. Part 8."

Stanley informed Northrop that Schlesinger had an excellent method for preparing phage solutions, and was keen on employing the centrifuge in his phage work, although he was "not interested in working with the material as a protein."[114] Northrop, who had just published an account of his "Concentration and Partial Purification of Bacteriophage" in *Science,* began at the same time collaborating with Ralph Wyckoff on ultracentrifugation of his phage preparations.[115]

Stanley's TMV had changed in other respects. On the one hand, Bawden and Pirie's contention that the initial chemical analysis was faulty for its omission of nucleic acid had been thoroughly assimilated, so that Stanley spoke of viruses as nucleoproteins.[116] (This is not to imply that Stanley viewed the nucleic acid as having a particular role in virus multiplication—or that he was entirely consistent in referring to viruses as nucleoproteins rather than as proteins.) On the other hand, TMV as a product of the ultracentrifuge was materially distinct from the virus as a chemical precipitate. By virtue of Wyckoff and Corey's observation that TMV could be crystallized through sedimentation, the understanding of the virus as a crystalline chemical was successfully transferred to the ultracentrifuge, but only the sedimented TMV was a distinct macromolecular species, with a definite molecular weight.

Yet the ultracentrifuge, as the new means of obtaining pure virus, was not merely consigned to being a tool of "materials and methods." Rather, the scientific and technical uses of the ultracentrifuge continued to be intertwined in research. Wyckoff noted in 1938 that the analytical ultracentrifuge provided a useful visual "guide during the purification of ... macromolecular substances." He noted, "The number of sedimenting boundaries is an index of the number of macromolecular species; the continued presence of unsedimentable ultraviolet-absorbing material is an indication either that the macromolecules are steadily breaking down or that they have not yet been completely freed of small-molecular impurities. Observations on the sharpness of sedimenting boundaries are especially illuminating. ... If the boundary of the native substance is sharp, then observed diffuseness is a measure of the degree of molecular

114. Stanley to J. H. Northrup [*sic*], 26 Aug 1936, Stanley papers, carton 1, folder Aug 1936. Stanley sent Northrop an abstract of Schlesinger's results a week later (Stanley to Northrop, 31 Aug 1936, Stanley papers, carton 1, folder Aug 1936).

115. Northrop, "Concentration and Partial Purification"; Wyckoff, "Ultracentrifugal Analysis of Concentrated Staphylococcus Bacteriophage." Northrop's chemical methods had also lost out to the ultracentrifugal approach for isolating tumor viruses: Northrop and Beard attempted in January of 1937 to chemically precipitate Shope's rabbit papilloma virus, but their efforts failed, whereas Wyckoff was able to isolate the same virus in his preparative ultracentrifuge; see Beard and Wyckoff, "Isolation of a Homogeneous Heavy Protein."

116. Stanley conceded the nucleic acid in his paper on aucuba mosaic virus (Stanley, "Chemical Studies on the Virus of Tobacco Mosaic. Part 8," 329n).

damage undergone during purification."[117] Combining the preparative and analytical capabilities of the machine, Stanley used the purification method for TMV in the ultracentrifuge as a means to demonstrate more unequivocally than before the correlation of virus activity (as quantified through the local lesion assay) with the amount of virus protein.[118]

The growing reliance on the ultracentrifuge made Stanley quite dependent on his collaborators, particularly Wyckoff. At the end of 1937, only six months after Stanley hired physical chemist Max Lauffer to join him in the structural studies of TMV, Wyckoff abruptly left the Rockefeller Institute, taking all of his ultracentrifuge equipment with him.[119] Stanley, anxious to keep his research moving forward, contacted Johannes Bauer to see if Pickels could begin to collaborate with him on a project.[120] Stanley acquired two new, air-driven ultracentrifuge machines for the purification of viruses, although it took until 1939 for an analytical tool to be built (under the guidance of Pickels).[121] Thus at the time that Clarence Cook Little sent an urgent telegram to Stanley inquiring where he could purchase an ultracentrifuge, Stanley himself lacked complete equipment.[122]

Stanley looked to new collaborators to keep TMV at the forefront of protein chemistry. Wyckoff's departure (for the research laboratories at pharmaceutical house Lederle) not only interrupted Stanley's ultracentrifuge work, but also ended their joint work on the X-ray crystallographic determination of TMV's structure. After conversations with W. T. Astbury at the Cold Spring Harbor Symposium on Proteins in 1938, Stanley began sending him samples of TMV protein and nucleic acid.[123] He also sent samples of purified TMV for diffusion studies to Hans Neurath, a protein chemist at Duke.[124] Stanley sought to establish other

117. Wyckoff, "Ultracentrifugal Study of Macromolecules," 362.

118. Stanley, "Chemical Studies on the Virus of Tobacco Mosaic. Part 9."

119. Stanley to Walter J. Nungester, 31 Oct 1938, Stanley papers, carton 1, folder Oct 1938; Stanley to Bawden, 25 Feb 1938, Stanley papers, carton 1, folder Feb 1938.

120. Stanley to Bauer, 24 Jan 1938, Stanley papers, carton 1, folder Jan 1938.

121. Pickels to Stanley, 15 Mar 1938, Stanley papers, carton 11, folder Pickels, Edward Greydon; Stanley to W. J. Elford, 12 Oct 1938, Stanley papers, carton 1, folder Oct 1938; Stanley to J. F. McClendon, 19 Aug 1941, Stanley papers, carton 1, folder Aug 1941. Stanley wrote J. W. Williams, "We got our analytical ultracentrifuge working a few months ago and are now getting results from it" (10 Jan 1940, Stanley papers, carton 1, folder Jan 1940).

122. Western Union telegraph Clarence C. Little to Stanley, 22 Nov 1938, Stanley papers, carton 10, folder Little, Clarence Cook. Little's urgent request was likely prompted by John Bittner's recent identification of a mouse mammary tumor factor in milk, a possible tumor virus. On Bittner's work and the Bar Harbor Laboratories, see Gaudillière, "Circulating Mice and Viruses," 89–124; and Rader, *Making Mice*.

123. Stanley to W. T. Astbury, 6 Sep 1938, Stanley papers, carton 6, folder Astbury, William Thomas; Astbury to Stanley, 28 Sep 1938, Stanley papers, carton 1, folder Sep 1938; Stanley to Astbury, 23 Nov 1938, Stanley papers, carton 1, folder Nov 1938.

124. Stanley to Neurath, 7 Dec 1937, Stanley papers, carton 1, folder Dec 1937.

physical-chemical approaches in-house. Lauffer set up an apparatus to determine double refraction of flow to follow up on William Takahashi and R. E. Rawlins's earlier observations of the birefringence of TMV solution, which suggested an elongated shape for the virus particles.[125] By the end of 1938, the Rockefeller Institute machine shop was also fabricating a Tiselius electrophoresis apparatus for the virus work.[126] To the degree that these various instruments offered up different depictions of TMV as a molecule, the coherence of TMV as an epistemic object depended on both a preexisting hierarchy of authority among the instruments (in general, the ultracentrifuge had priority) and work to reconcile the various representations.

Presenting Laboratory Viruses to the World

As Stanley's laboratory was becoming crowded with specialized physical-chemical instruments, he was on the lecture circuit promoting his research on purified virus. After presenting invited lectures to the Colloid Symposium and to the American Society of Naturalists in 1937,[127] he spent the following year as a Sigma Xi lecturer, giving a general talk, "Recent Advances in the Study of Viruses," at sixteen universities throughout the country.[128] He began drafting the first review on viruses to appear in the *Annual Review of Biochemistry* and wrote a chapter extolling the ultracentrifugal approach to viruses for Robert Doerr's *Handbuch der Virusforschung*.[129]

Stanley was keenly aware of the scientific multivalence of viruses; as he saw it, the "virus protein work" had a significant impact upon several "branches of science," including medicine, genetics, and physics, in addition to plant pathology and biochemistry.[130] He publicized the biomedical relevance of the virus work as widely as possible, speaking at the

125. Takahashi and Rawlins, "Method for Determining Shape," "Rod-shaped Particles," "Stream Double Refraction Exhibited by Juice," and "Relation of Stream Double Refraction to Tobacco Mosaic Virus." Whereas Lauffer was setting up a double-refraction-of-flow apparatus to follow up Takahashi and Rawlins's work, within a few years, Rawlins also set up an ultracentrifuge (Rawlins to Stanley, 17 Jul 1942, Stanley papers, carton 11, folder Rawlins, R. E.). Stanley had collaborated with John Edsall two years earlier on measuring the double-refraction-of-flow behavior of TMV; see note 77 above.

126. Stanley to Herbert Gasser, 1 Nov 1938, Stanley papers, carton 1, folder Nov 1938.

127. "Lectures by: W. M. Stanley," Stanley papers, carton 20, folder "List of talks given by Stanley."

128. Stanley to Harry B. Weiser, 12 May 1937, Stanley papers, carton 1, folder May 1937; Stanley, "Reproduction of Virus Proteins." The 1938 Sigma Xi lecture ("Recent Advances") was published the next year with those of the other lecturers in a volume, *Science in Progress*.

129. J. Murray Luck to Stanley, 25 Nov 1938, Stanley papers, carton 10, folder Luck, James Murray; Stanley, "Biochemistry of Viruses," and "Biochemistry and Biophysics of Viruses."

130. Stanley to Frazier, 1 Dec 1938, Stanley papers, carton 1, folder Dec 1938.

Third International Cancer Congress in 1939 as well as the International Congress of Microbiology.[131] Stanley also cultivated lay interest in virus research, beginning with a talk for the Kaufmann Department Stores in Pittsburgh in May of 1937, the same month Stanley was featured in *Scientific American*.[132] His claim for the wide significance of his results with TMV and other viruses relied on the link between his experimental systems and actual diseases in plants, animals, and humans. In Stanley's view, the ultracentrifuge gave epistemological access to actual infectious agents. A reporter for the *New York Times* stressed the opportunities for battling diseases opened by the ultracentrifuge:

> Before Dr. Stanley did his work, the viruses were recognized only by their effects on living plants and animals. Since 1935 he has been isolating them, whirling them in centrifugal machines, and concentrating them just as cream is separated from milk in any dairy separator. A tremendous forward step was taken when these concentrates (the first was that of the tobacco mosaic virus) were crystallized and found to be huge protein molecules. A way has been cleared not only to study such mysterious crippling diseases as infantile paralysis and encephalitis but perhaps to prevent and cure them.[133]

Not all virus researchers shared Stanley's conviction that viral diseases were best understood by eliminating the host in order to isolate a macromolecular agent. In particular, Bawden and Pirie opposed Stanley's reliance on physical-chemical instrumentation. In their first full scientific article on TMV, they confirmed many of Stanley's results but disagreed with his interpretation of the status of purified virus as representing the agent in the plant: "From the results of experiments comparing the optical properties, activity, and filterability of crude infective sap with those of solutions of purified virus, it is more probable that the virus is in a different condition [i.e., smaller] in the clarified infective sap, and that in the isolated product it has become aggregated."[134] Stanley, responding privately to Bawden early in 1938, made it clear that he accepted their chemical corrections concerning nucleic acid and crystals but would not concede that centrifugation produced artifacts:

131. "Lectures by W. M. Stanley," Stanley papers, carton 20, folder "List of talks given by Stanley."

132. Newman, "Smooth Slide Up to Life." On the department store talk, see Stanley to Irwin Wolf, 25 May 1937, Stanley papers, carton 1, folder May 1937. Stanley was featured in Kaufmann's "Peaks of Progress" window display as the centerpiece of a series of busts of great scientists, including Pasteur, Eberth, Walter Reed, Loeffler, Noguchi, and Koch. A picture of the window is enclosed with a letter from Irwin Wolf to Stanley, 17 Jun 1937, Stanley papers, carton 13, folder Wolf, Irwin Damasius.

133. "New Disease Clue Stirs Scientists: Dr. W. M. Stanley's Talk Dealing with Giant Molecule Has Far-Reaching Implications," *New York Times,* 7 Nov 1937, sec. 2, 7:4.

134. Bawden and Pirie, "Isolation and Some Properties," 306.

> I've never had any argument with you as to whether or not the tobacco mosaic virus protein is a nucleoprotein. I thought that the nucleic acid was not necessary for virus activity. We have gone into that matter and I appear to be wrong, for we can't get the nucleic acid off without inactivating the protein.... I'm quite willing to let the question as to whether or not the 'needle crystals' are really crystals rest until some specialist comes along.... However, I'm not willing to go along with you on your aggregation idea. We have centrifuged virus protein from infectious juice and can secure no evidence that centrifugation causes aggregation. I agree that sufficient chemical treatment does it, but mere centrifugation—no.[135]

Conventional accounts of Bawden and Pirie's critique of Stanley's work have usually focused on the contention over whether TMV is a protein or a nucleoprotein.[136] But this representation (aided by Pirie's later criticisms of Stanley)[137] is anachronistic insofar as disputes about the role of the nucleic acid moiety were not foremost in the 1930s. Stanley, having already assimilated the British researchers' observation of nucleic acid, was instead most concerned with the status of ultracentrifuge-produced virus. Was sedimented TMV an authentic molecule, and was it the actual virus?

The multiplicity of instruments being used to determine the physical-chemical nature of viruses made the reconciliation of experiments between laboratories even more challenging. Stanley endeavored to bring the results from various instruments into accord with the representations generated by the ultracentrifuge to support a molecular view of TMV. In a letter to the editor in *Nature,* Stanley and his fellows laid out their strongest evidence that sedimentation of virus did not cause its artificial aggregation, as demonstrated through "before and after" measurements of double refraction of flow and filterability. They concluded, on the basis of their results and Lauffer's "theoretical considerations," that "the size and shape of the particles of purified virus are the same as those of the virus particles in the juice."[138]

Lacking an ultracentrifuge, Bawden and Pirie could only suggest reasons for the discrepant results. But they argued strongly against Stanley's idealized molecular TMV: "Even if centrifugation does not cause aggregation, completely unaggregated preparations could be expected

135. Stanley to Bawden, 19 Jan 1938, Stanley papers, carton 1, folder Jan 1938.

136. See, for example, Olby, *Path to the Double Helix,* 157.

137. For examples of Pirie's caustic retrospective evaluations of Stanley's work, see "Recurrent Luck in Research," esp. 500–505 and "Viruses" (*Scientific Thought*), in which Pirie remarks, "Stanley's paper is widely quoted by people who do not seem to have read it. Preparations with properties more closely resembling those now ascribed to the virus were made by Bawden and Pirie in 1936" (223).

138. Loring, Lauffer, and Stanley, "Aggregation."

only from plants that have been recently infected, for aggregation occurs naturally in the sap of plants that have been long infected."[139] Stanley, for his part, raised questions as to the sensitivity and accuracy of Bawden's apparatus for measuring double refraction of flow.[140]

To complicate these disputes, many virus researchers, including Stanley himself, were questioning the accuracy of determinations of TMV's molecular weight from the analytical ultracentrifuge. Svedberg's original measurement of approximately 17 million daltons assumèd that the virus exhibited a length-to-width ratio of no more than 10 to 1 (i.e., a disymmetry constant of 1.3). Work by Irving Langmuir on films of virus proteins (supplied by Stanley) suggested that TMV might be much longer than previously thought.[141] (This work on virus films was continued by Dorothy Wrinch, who experimented with Stanley's purified TMV in efforts to bolster her cyclol theory of protein structure.)[142] Bawden and Pirie argued that the birefringence of TMV increased upon centrifugation, suggesting that the length of TMV was not a stable characteristic, but reflected the state of linear aggregation induced by laboratory instruments.[143]

Stanley, facing questions in hydrodynamic theory far beyond the range of his expertise, turned to Svedberg and John Nichols for advice. They were equivocal.[144] As Kai Pedersen confessed to Stanley, "The problem of the relationship between the shape of the molecule and its frictional constant is yet, I am sorry to say, far from being generally solved. Only in very few cases we are [sic] able to say something definitely about the

139. Bawden and Pirie, "Contribution to Aggregation" (response to Loring, Lauffer, and Stanley, "Aggregation"), 843.

140. Stanley to Bawden, 2 Nov 1938, Stanley papers, carton 1, folder Nov 1938.

141. Stanley to Irving Langmuir, 25 Mar 1937, Stanley papers, carton 1, folder Mar 1937; Stanley to Langmuir, 12 Nov 1937, Stanley papers, carton 1, folder Nov 1937.

142. Dorothy Wrinch to Stanley, 1 Jun 1937, Stanley papers, carton 13, folder Wrinch, Dorothy M.; Stanley to Wrinch, 8 Jun 1937, Stanley papers, carton 1, folder Jun 1937. For more on Wrinch's scientific contributions and career, see Abir-Am, "Synergy or Clash."

143. "The physical properties of virus preparations and the X-ray measurements on them are interpreted on the theory that in purified preparations the constituent particles are rod-shaped, and it is suggested that these rods are built up by the linear aggregation of smaller units. There is evidence that, in the plant, part at least of the virus is not aggregated, for filters which pass an infectious filtrate with untreated plant sap do not do so with purified preparations" (Bawden and Pirie, "Isolation and Some Properties," 319).

144. See Stanley to Svedberg, 26 Mar 1937, Stanley papers, carton 12, folder Svedberg, Theodor; Stanley to Williams, 19 Nov 1937, Stanley papers, carton 1, folder Nov 1937; and Stanley to Svedberg, 24 Nov 1937, Stanley papers, carton 12, folder Svedberg, Theodor. Stanley sent two long letters asking Svedberg about what the analytical ultracentrifuge results revealed about the shape of TMV before receiving a letter of response, and then it was written by Pedersen, not Svedberg (letter to Stanley, 23 Dec 1937, Stanley papers, carton 11, folder Pa–Pe misc.).

shape of the large molecules, viz. when the molecule is spherical and unhydrated."[145]

By 1938, most specialists were convinced by the birefringence data that TMV was a rod, not a sphere, making a direct calculation of molecular weight from the sedimentation constant infeasible. Some researchers, such as Lauffer and the Cornell chemist Vernon Frampton, turned to viscosity measurements to estimate molecular weight, but this technique had its own limitations. As Lauffer noted, "The interpretation of such viscosity data is seen to be fraught with a considerable degree of uncertainty."[146] He argued for a molecular weight of 42.6 million, corresponding to particles 12.3 mμ (nm) in diameter and 430 mμ in length.

Debates over the size and aggregation state of purified TMV were complicated by an ongoing dispute between protein chemists and colloidal chemists over protein structure: Were the larger proteins single, covalently bonded entities or aggregates of smaller units? Could proteins be considered to exist in true solutions or in colloidal suspensions? Should a virus be viewed as a molecule or as an aggregate of molecules?[147]

Svedberg contended that all proteins were composed of molecular units of 35,000 daltons and offered evidence that TMV dissociated into units of 140,000 and 70,000 at high pH.[148] Wrinch's cyclol theory of protein structure could account for the Svedberg unit of 35,000 in terms of a ring structure containing 288 amino acids, an explanation further buoyed by Carl Niemann and Max Bergmann's observed regularity of 288 amino acids per protein molecule.[149] Bawden and Pirie's assertion that centrifugation promoted aggregation of TMV could be used to support either a cyclol or colloidalist viewpoint. Stanley opposed both of these interpretations and energetically defended the existence of a single-molecular TMV. As he wrote to colloidal chemist Ross A. Gortner,

> When we speak of a protein being homogeneous from the standpoint of the centrifuge, we simply mean that it is homogeneous with respect to molecular weight or, if you prefer, particle size. I prefer the term "molecular" because the proteins with which we have worked behave as molecules. . . . I quite agree that a material homogeneous with respect to molecular weight may consist of many different chemical species, but when a material is homogeneous with respect to

145. Pedersen to Stanley, 23 Dec 1937, Stanley papers, carton 11, folder Pa–Pe misc.

146. Lauffer, "Viscosity," 444.

147. See Bawden, *Plant Viruses and Virus Diseases,* 198.

148. Svedberg, "Ultra-Centrifuge and the Study"; Stanley to Wrinch, 8 Jun 1937, Stanley papers, carton 1, folder Jun 1937.

149. See Kay, *Molecular Vision of Life,* 112–16; and Fruton, *Molecules and Life,* 131–79, esp. 162 ff.

molecular weight and in addition is homogeneous with respect to
chemical tests, then I think that there is considerable justification for
regarding the entity as a molecule.[150]

Stanley's interpretation was contested on empirical as well as theoretical grounds. Early in 1939, Vernon Frampton at Cornell claimed that
urea caused TMV to dissociate into units of molecular weight 100,000,
and he argued that calculations of higher molecular weight relied on
false assumptions that the solution does not deviate from Fick's, Stoke's,
and Poiseuille's laws.[151] Each researcher drew the demarcation between
natural and artifactual virus differently. Whereas Bawden and Pirie had
argued that Stanley's ultracentrifuge-produced TMV was an artifact of
preparation, Stanley now argued that Frampton's small subunits were
just breakdown products of the real virus.[152] More substantively, Stanley
questioned the worth of Frampton's activity measurements, claiming
that his samples were old and impure.[153] In reply, Frampton repudiated
Stanley's claim to have pure virus at all:

> It is pertinent to state that any chemist who is honest with himself
> must make such an admission with regard to any protein preparation
> that he might be working with. The admission is a necessary con-
> sequence of the fact that there is no device whereby a chemist
> may ascertain the purity of a protein preparation. . . . It is true that
> a standardized method of preparation will yield a product of es-
> sentially constant physical properties. But is not the acceptance of
> any one particular method as *the* method of preparation somewhat
> arbitrary?[154]

150. Stanley to R. A. Gortner, 17 Mar 1937, Stanley papers, carton 1, folder Mar 1937.
Stanley defended his work as follows: "The general situation, therefore, is that Dr. Frampton
wishes to conclude that, because solutions of tobacco mosaic virus do not obey certain laws
in a rigid manner, these laws can not be validly applied to such solutions. It would be very
much the same as stating that, because certain gases, which as a matter of fact probably
should include all gases, fail to obey the perfect gas law, the use of the perfect gas law is
not valid. As you well know, we recognize the deviations of various gases from the perfect
gas law, but nevertheless use the law in our calculations to a very good purpose. We have
attempted to use the various available equations in a similar manner in connection with
the solutions of tobacco mosaic virus" (Stanley to W. Mansfield Clark, 9 Mar 1939, Stanley
papers, carton 1, folder Mar 1939).
151. Frampton and Saum, "Estimate of the Maximum Value."
152. Stanley to Vernon Frampton, 9 Jan 1939, Stanley papers, carton 1, folder Jan 1939;
and Loring, Lauffer, and Stanley, "Aggregation."
153. Reviewing Frampton's paper on the topic for the *Journal of Biological Chemistry,* he
wrote the editor, "Frampton's studies on urea solution refer . . . only to inactive degradation
products and not to tobacco mosaic virus" (Stanley to W. Mansfield Clark, 9 Mar 1939,
Stanley papers, carton 1, folder Mar 1939).
154. Frampton to Stanley, 1 Feb 1939, Stanley papers, carton 8, folder Frampton, Vernon
Lachenous; emphasis in original.

Frampton published another letter attacking Stanley in *Science,* citing the impossibility of determining a real molecular weight for viruses with the ultracentrifuge.[155] Bawden responded sarcastically to Stanley: "I liked Frampton's letter in *Science* showing that the molecular weight of tobacco mosaic virus was between 0 and infinity. I had often thought that it probably was."[156] But Bawden, too, questioned Stanley's belief that the ultracentrifuge could certify molecular authenticity. As he summarized the state of affairs in a 1939 monograph, "The values for the size and weight of tobacco mosaic virus particles quoted above have not been given because of their intrinsic value (there is no real reason to believe that any one of them is accurate), but to show the uncertainty that exists in spite of the large amount of work that has been done."[157]

The rising number of sophisticated machines for doing physical-chemical experiments on macromolecules did not by itself create consensus as to the nature of viruses or proteins.[158] In fact, the proliferation of ultracentrifuges in the late 1930s—Stanley speculated in 1938 that "before long they will be as common as an ordinary laboratory centrifuge"—made the problems of authority and consensus more acute.[159] The difficulties researchers faced in achieving accord among results in various laboratories, even from the "same" instrument, is illustrated well by the research on tomato bushy stunt virus. First isolated by Bawden and Pirie in 1938, bushy stunt virus attracted attention because its crystals, unlike those of TMV, were beautiful dodecahedrons—"genuine" crystals by Bernal's standards (see fig. 4.8).[160] However, when Stanley used differential centrifugation to prepare the virus, he isolated a material both more active than Bawden and Pirie's and with a different sedimentation constant. He attributed the change in activity to Bawden and Pirie's use of heat in purification, which may have inactivated the virus without changing its ability to crystallize.[161] But accounting for the difference in

155. Frampton, "Molecular Weight."
156. Bawden to Stanley, 7 Dec 1939, Stanley papers, carton 6, folder Bawden, Sir Frederick Charles.
157. Bawden, *Plant Viruses and Virus Diseases,* 197.
158. Joseph Fruton has claimed that "in retracing the stumbling steps of protein chemists during that period [1920–1940], we shall see that the confusion was compounded by a variety of factors; foremost among them was the fact that the available experimental methods were not adequate to resolve doubts or shake adherence to inherited belief" (Fruton, *Molecules and Life,* 148).
159. Stanley to W. J. Elford, 12 Oct 1938, Stanley papers, carton 1, folder Oct 1938.
160. Bawden and Pirie, "Crystalline Preparations."
161. The explanation of activity loss due to heat inactivation also served to undermine the relevance of true crystals to physiological activity. As Stanley explained, "The general situation with respect to bushy stunt virus is somewhat analogous to that which prevails in the case of tobacco mosaic virus. Since the crystals of the inactive material are indistinguishable

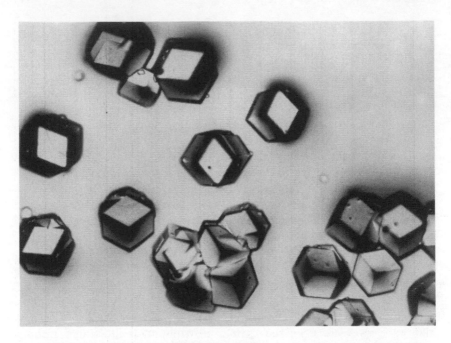

FIGURE 4.8 Crystals of the tomato bushy stunt virus ×224. Stanley prepared the virus by differential centrifugation prior to its crystallization. (From Stanley, "Purification of Tomato-Bushy-Stunt Virus." Reprinted courtesy of the Bancroft Library, University of California, Berkeley.)

sedimentation constant proved impossible. Bawden and Pirie had collaborated with McFarlane in London, who possessed one of Svedberg's analytical ultracentrifuges; he had determined a sedimentation constant of 158 for Bawden and Pirie's material. Stanley sent samples of his preparation of the virus to McFarlane in London, who obtained a sedimentation number of 132.[162] Bawden's original stock of pure bushy stunt virus was accidentally frozen and thus was lost.[163] Until Stanley obtained results so divergent from those reported, bushy stunt virus had been notable for the

---

from those of fully active material, it is obvious that crystallinity cannot be used as a criterion of purity" (Stanley to Kenneth M. Smith, 7 Aug 1940, Stanley papers, carton 12, folder Smith, Kenneth M.).

162. Stanley to A. S. McFarlane, 3 Apr 1940, Stanley papers, carton 1, folder Apr 1940. Stanley confessed to McFarlane in this letter, "I have been somewhat concerned over the constancy of the sedimentation constants of viruses in general. There has been quite a spread in certain of the tobacco mosaic virus measurements, and there is, of course, the spread from 138 to 158 in your values for the bushy stunt virus."

163. Bawden to Stanley, 13 Jul 1940, Stanley papers, carton 6, folder Bawden, Sir Frederick Charles.

constancy of its properties, particularly in comparison with TMV. The fate of bushy stunt virus reveals both the biological variability of the materials the researchers sought to fix with physical-chemical measurements and the difficulty of calibrating machines between laboratories.

The arrival of the electron microscope further complicated the task of achieving commensurate results across different instruments and laboratories. German researchers G. A. Kausche, E. Pfankuch, and H. Ruska published the first electron micrographs of TMV (the first virus to be so visualized) on the eve of World War II.[164] Ruska had been working for Siemens since 1937 to develop the first commercial electron microscope, which was introduced late in 1939.[165] The German pictures of TMV showed long, straight, rod-shaped structures, in remarkable accord with observations of its birefringence. In commenting on the recent publications for a review, Stanley emphasized the way in which the new results supported the physical portrait of the virus Lauffer had deduced from indirect means: "These photographs demonstrated not only the existence of discrete particles, but also that these particles are about 15 mμ in cross section and about 330 mμ in length, and hence must have a molecular weight of the order of 50,000,000."[166] However, the results did not so clearly resolve his disagreement with Bawden and Pirie over the "natural" length of the virus, as Stanley observed in a private letter:

> Previously, through work from this laboratory, it was established by indirect means that tobacco mosaic virus was a rod-shaped entity having a width of the order of 12 mμ and a length of the order of 400 mμ. Our findings were confirmed in general by the results obtained by the German workers with the electron microscope. Unfortunately, however, the preparation which they used was one obtained by chemical means and contained, therefore, several molecular species. You will note from the description contained in the reprint which I enclose herewith that the chemical treatment causes the production of double, triple, etc. molecules of this virus, and in their photographs they obtained definitions of entities having lengths of 150, 300, 600, and 1,200 mμ, respectively. In view of this heterogeneity, it is difficult to interpret their results.[167]

Vladimir Zworykin, Stanley's correspondent and head of RCA's electronics research, was at that time working to develop a commercial

164. Kausche, Pfankuch, and Ruska, "Die Sichtbarmachung von pflanzlichem Virus im Übermikroskop."

165. Marton, *Early History of the Electron Microscope*, 31; Rasmussen, "Instruments, Scientists, Industrialists." For more on the history of the electron microscope, see Rasmussen, *Picture Control*.

166. Stanley, "Biochemistry of Viruses," 554.

167. Stanley to Zworykin, 12 Jun 1939, Stanley papers, carton 13, folder Z misc.

electron microscope. Stanley and Stuart Mudd, a bacteriologist at the University of Pennsylvania, began providing biological samples to RCA for their development efforts.[168] Once the prototype was fully operational, Zworykin arranged for RCA to fund a special postdoctoral fellowship through the National Research Council (NRC) Committee on Biological Applications.[169] The committee, which included Stanley, Mudd, and several other prominent biologists, selected Thomas F. Anderson, a physical chemist turned biologist, to collaborate from RCA's Camden facility with biologists.[170]

Stanley's aim was to employ the electron microscope to confirm the interpretation of the size and shape of TMV he and Lauffer had advanced. Despite the priority of the German microscopists, Stanley saw an opportunity to perform a definitive experiment using the RCA machine. As he wrote Zworykin after a visit to Camden, "As I told you yesterday, the photograph of tobacco mosaic virus, which was taken so casually, is by far the best of all of the photographs of this virus which have been made."[171] He left newly prepared samples for Anderson and urged that he take "a great number of photographs of different dilutions of each of the preparations . . . so that they may be interpreted properly and unambiguously."[172] Nicolas Rasmussen has shown that Zworykin and the NRC Committee actually required researchers to get approval for publications resulting from these RCA-sponsored collaborations with Anderson. Even committee members like Stanley were not exempt from having to request permission for the presentation and interpretation of micrographic results, although given the state of disagreement about which forms of TMV were active and which were artifactual, Stanley may have found the cautious oversight justified.[173]

Stanley and Thomas Anderson published two papers in 1941 and one in 1942 on electron microscopy of purified viruses (see fig. 4.9). By obtaining TMV through the ultracentrifuge before mounting it in the electron microscope, Stanley sought to eliminate from his photographs the artifacts

168. Stanley to Zworykin, 29 Mar 1940, Stanley papers, carton 13, folder Z misc.; Rasmussen, "Making a Machine Instrumental," 314–17; and "RCA and the War Years," in *Picture Control*.

169. Rasmussen, "Making a Machine Instrumental," 328.

170. Rasmussen lists the other members of the NRC Committee on Biological Applications as Milislav Demerec from the Carnegie Institution laboratory at Cold Spring Harbor, Charles Metz from University of Pennsylvania's Department of Zoology, James Kempton from the Bureau of Plant Industry, and biophysicist Caryl Haskins from Union College (ibid.).

171. Stanley to Zworykin, 26 Sep 1940, Stanley papers, carton 13, folder Z misc.

172. Ibid.

173. Mudd to Stanley, 26 Feb 1941, Stanley papers, carton 10, folder Mudd, Stuart; Rasmussen, "Making a Machine Instrumental," and *Picture Control*, 46–69, esp. 54.

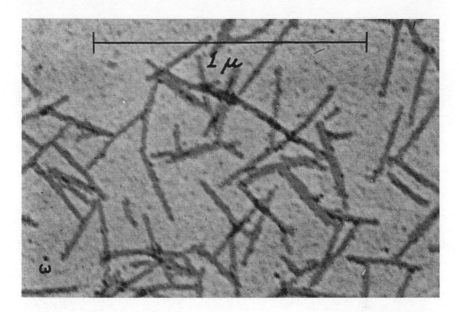

FIGURE 4.9 Electron micrograph of TMV ×55,000. The aqueous solution photographed contained 0.2 mg/cc of ultracentrifugally isolated tobacco mosaic virus. Area is near edge of mount; length of one micron is indicated. (Image and caption from Stanley and Anderson, "A Study of Purified Viruses with the Electron Microscope" [1941], plate 1. Reprinted by permission of the American Society for Biochemistry and Molecular Biology.)

he felt appeared in the German images. Also, Stanley and Anderson demonstrated that an increase in solution pH caused the physical disintegration of TMV rods into granular material, reinforcing Stanley's rebuttal of Frampton's claim that active TMV could be found in small units. However, using the electron microscope to pin down which rod-shaped structure was the "natural" virus proved difficult, and Stanley fell back on ultracentrifugal evidence:

> The aggregates do not appear to represent the natural form of the virus, for when carefully prepared samples of virus or virus in the freshly expressed, untreated, infectious juice are examined by means of the analytical ultracentrifuge, no evidence for the existence of the aggregates is obtained, whereas following treatment with salt the same samples show either a second sedimenting boundary, presumably due to a component formed by the end-to-end aggregation of two particles, or a more rapidly sedimenting diffuse boundary indicative of even more extensive aggregation.[174]

174. Stanley and Anderson, "Study of Purified Viruses," 329.

Measurement of lengths of fifty-eight particles in the micrographs showed that over half of the rods were 280 mμ long, and Stanley adopted this value, along with the 15-mμ width based on Bernal's X-ray diffraction work, as the new standard size of TMV.[175] Even as the results from the electron microscope motivated Stanley and Lauffer to adjust their estimated size of "normal" TMV, the new instrument helped to stabilize the dimensions of the virus, and indeed the notion that TMV was a discrete, elongated macromolecule.

Stanley took the collaboration with Anderson beyond simple size and shape measurements for TMV in three respects. First, they visualized the TMV-antibody precipitin reaction by imaging the virus in the presence of its antiserum. These micrographs made the specificity of virus-antibody interaction visible; they served to "demonstrate the usefulness of the electron microscope."[176] In this respect, the paper fulfilled the evangelical objectives of the NRC Committee on Biological Applications. The images were also reproduced for public appreciation by *Time* magazine, although under the rather vague if not misleading label "the first clear portrait ever made of individual molecules and the first detailed glimpse of a chemical reaction."[177]

A second research collaboration between Stanley and Anderson involved obtaining and comparing electron micrographs of TMV, bushy stunt virus, edestin, and three hemocyanins. These were all molecules whose molecular weights had been determined through the analytical ultracentrifuge, and for which other techniques had been used to assess whether they were spherical. The purpose of visualizing these proteins and viruses in the electron microscope was to enable the co-calibration of these methods—establishing the "degree of correlation between the molecular sizes and shapes estimated by indirect methods and those obtained by direct mensuration" (i.e., the electron microscope).[178] The third project Stanley took up with Anderson failed completely. In conjunction with Alfred Marshak at the Radiation Laboratory at Berkeley, Stanley and Anderson investigated the effects of X-ray and neutron radiation of TMV on its behavior in the ultracentrifuge and electron microscope. Whereas irradiated samples showed a decrease in viral activity, electron

175. Ibid., 330.

176. Anderson and Stanley, "Study by Means of the Electron Microscope," 344.

177. "Historic Pictures," *Time,* 19 May 1941, 76. Two weeks prior, *Time* had also covered Stanley's projection of the electron micrographs of TMV, taken by Anderson, at a lecture at the American Philosophical Society: "Some of the ablest U.S. scientists . . . gaped in awe, for they were seeing something never before distinctly seen by man" (*Time,* 5 May 1941, 50).

178. Stanley and Anderson, "Electron Micrographs of Protein Molecules," 25.

micrographs did not reveal any noticeable structural differences from untreated virus.[179]

Even as Stanley promoted—and benefited from—RCA's development of a commercial electron microscope, the hierarchy of instruments in his laboratory remained unchanged. The ultracentrifuge remained fundamental, because it provided the means of making as well as visualizing pure virus.[180] As he wrote a researcher interested in acquiring the latest machine:

> So far as I can see, there should be no competition between centrifuges and electron microscopes, for each is quite important in its own way, and there is no overlapping of the fields. As a matter of fact, in our particular case it is necessary to have the centrifuge in order to secure good preparations for examination by means of the electron microscope. If I were you, I should continue to urge the installation of the centrifuge irrespective of what happens in connection with the electron microscope.[181]

By 1940 the representation of TMV as a homogeneous macromolecule had been stabilized, although the issue of aggregation and the possibility of smaller subunits continued to be an important leitmotif in research on TMV's structure (and is taken up in chapter 7).

World War II and New Uses for Ultracentrifuges

In the days following Japan's attack on Pearl Harbor, Stanley offered the wartime services of his laboratory in letters to Roger Adams, his graduate adviser and a member of the National Defense Research Committee of the Office of Scientific Research and Development (OSRD), and A. N. Richards, the head of the Committee on Medical Research (CMR) of OSRD.[182] Stanley submitted a formal application the following April to Francis Blake, chair of the National Research Council's (NRC) Subcommittee on Infectious Diseases, for an OSRD grant to work on influenza virus. "I am proposing, in short, to work with influenza virus, using the general methods which have proved so successful with the plant viruses."[183]

179. Stanley to Alfred Marshak, 6 Jan 1941, Stanley papers, carton 1, folder Jan 1941; Stanley to Marshak, 11 Nov 1941, Stanley papers, carton 1, folder Nov 1941.
180. For more on the relationship between ultracentrifugation and electron microscopy, see Gaudillière, "Molecularization of Cancer Etiology."
181. Stanley to W. J. Nungester, 15 May 1940, Stanley papers, carton 1, folder May 1940.
182. Stanley to Adams, 10 Dec 1941, Stanley papers, carton 1, folder Dec 1941; Stanley to Richards, 16 Dec 1941, Stanley papers, carton 11, folder Richards, Alfred N.
183. Stanley to Francis G. Blake, 30 Apr 1942, Stanley papers, carton 1, folder Apr 1942.

The NRC recommended an initial one-year contract for Stanley of $9,050 to the OSRD in May.[184] Stanley drew on TMV to provide a model for investigating influenza virus through "biochemical and biophysical" methods, particularly the ultracentrifuge, electrophoresis apparatus, and chemical analysis.[185] Moreover, work on influenza constrained the continued research on TMV to those aspects of plant virus research which "may be of benefit in connection with the influenza problem."[186] Stanley's use of the ultracentrifuge for war work thus illustrates the way in which TMV served as a model system for investigating other objects and problems—in this case, research on vaccine development.

Stanley had already expressed his interest in extending his laboratory's purview to include animal viruses when the mobilization for war created an opportunity to do so.[187] But the priorities of military medicine meant that Stanley's research was aimed for the first time at producing a therapeutic agent rather than generating knowledge. The laboratory was not significantly retooled for war work; rather, the tools at hand were put to work on the influenza project. Stanley's workers spent the first months optimizing the conditions for growing influenza virus in chick embryos and the agglutination assay (with red blood cells) for measuring titer.[188] Once the system of virus growth was standardized, Stanley sought to use the ultracentrifuge to isolate the virus for vaccine production. The intent was to treat influenza virus like TMV—using the ultracentrifuge to purify it, initially for physical-chemical characterization. His laboratory took advantage of the PR8 strain of influenza virus, which Thomas Francis Jr.

184. Francis G. Blake to Stanley, 26 May 1942, Stanley papers, carton 6, folder Blake, Francis Gilman.

185. Stanley, "Condensed Review of Studies on Influenza Virus Carried Out During the Period June 1942 to November 1945 at the Rockefeller Institute for Medical Research, Princeton, N. J., Under Contract No. OEM-cmr 158" (Stanley papers, carton 20, folder "List of talks given by Stanley," 3).

186. Stanley to Waldo R. Flinn, 16 Sep 1942, Stanley papers, carton 1, folder Sep 1942.

187. "I should perhaps mention that for some time I have felt that it would be advisable to carry along a study from the standpoint of comparative biochemistry of one of the less vicious viruses affecting animals. Since I have had some experience with the Shope rabbit papilloma virus, I should like to secure a quantity of this virus and compare its properties with those of the plant viruses. I think that much would be gained from such a study" (Stanley to Kunkel, 18 Jan 1940, Stanley papers, carton 1, folder Jan 1940).

188. Scientific Reports, RAC RU 439, box 5, vol. 32, 1943–44 (report from late 1943), 222–24. As Ton van Helvoort has pointed out, both the growth technique and the titer technique were recent innovations when Stanley adopted them: F. MacFarlane Burnet developed the chick embryo passage technique in 1936, based on earlier work of Ernest Goodpasture and Alice Woodruff on the chorio-allantoic membrane of the embryonated chicken egg as a good culture medium. The agglutination assay was developed by George Hirst and independently by Laurella McClelland and Ronald Hare in 1941. See Helvoort, "Bacteriological Paradigm in Influenza Research," 13–14.

isolated in 1934 and F. MacFarlane Burnet subsequently cultivated in chicken eggs.[189] The ability to grow PR8 in eggs provided Stanley's group with a culture medium that could be handled in large volume.[190] Ultracentrifugation of PR8 grown up in chicken eggs produced a concentrated influenza virus solution, the basis for an inactivated virus vaccine. Chemical purity was the key advantage of Stanley's method: Whereas the virus solution obtained through other methods consisted of 80 percent nonvirus protein, the centrifuged influenza virus was largely one near-homogeneous component, presumably the virus itself.

In order to be able to scale up virus production, Stanley's laboratory experimented with using commercial laboratory centrifuges rather than their custom-built ultracentrifuge. Once again, TMV provided the model system for developing the new protocol. Late in 1942, Stanley published a method for concentrating and purifying TMV in the Sharples Super Centrifuge, a development that no doubt cheered the centrifuge company.[191] Although this was not the first attempt to use a more conventional centrifuge to purify virus,[192] the recent development of the continuous flow bowl contributed to the success of Stanley's new method, by which 10–15 grams of TMV could be prepared from sap through four consecutive sedimentations over ten hours.

These improvements proved directly applicable to influenza virus.[193] Stanley's vaccine recipe involved four stages: infecting chick embryos for two days, removing the extra-embryonic fluids, adding formalizing agents and preservatives, and purifying the inactivated virus through three centrifugations in the Sharples centrifuge.[194] Stanley was convinced of the superiority of his method:

189. Francis, "Transmission of Influenza"; Burnet, "Influenza Virus on the Developing Egg."

190. Scientific Reports, RAC RU 439, box 5, vol. 32, 1943–44 (report from late 1943), 224–25.

191. Stanley, "Concentration and Purification of Tobacco Mosaic Virus." In 1945, Stanley wrote the Sharples Corporation about his influenza vaccine, estimating that the six pharmaceutical houses would likely purchase 25–50 centrifuges to fulfill their vaccine production contracts for the military (Stanley to James T. Costigan, 9 Apr 1945, Stanley papers, carton 2, folder Apr 1945).

192. Stanley cites three earlier attempts: McKinney's attempts with TMV, "Quantitative and Purification Methods"; Schlesinger's belatedly recognized isolation of bacteriophage, "Reindarstellung eines Bakteriophagen"; and McIntosh and Selbie's work with vaccine virus and bacteriophage, "Application of the Sharples Centrifuge."

193. Stanley, "Efficiency of Different Sharples Centrifuge Bowls." For another account of Stanley's vaccine development, see Helvoort, "Early Influenza Virus Vaccines." Helvoort attributes Stanley's characterization of influenza in 1944 to discrepant reports in the literature, but as the documents cited here show, Stanley's commitment to the project was much earlier and derived from his desire to contribute to the war mobilization.

194. Scientific Reports, RAC RU 439, box 6, vol. 34, 1945–46, 260–61.

> It is difficult to over-emphasize the advantages which the centrifuge method possesses over all other methods thus far described for the concentration and purification of influenza virus for use in vaccines. . . . It is obvious that the centrifuge method represents the method of choice for the manufacture of influenza vaccine on a large scale.[195]

The Sharples Corporation made plans to run an advertisement for their Super Centrifuge based on Stanley's successful separation.[196]

Despite these rapid technological advances, identifying a molecular particle as the agent of influenza was not straightforward. Two complications arose in Stanley's own study of the virus. First, the influenza virus Stanley purified exhibited a variable sedimentation coefficient, casting uncertainty upon both its integrity as a particle and its size. "At times the virus activity appeared to sediment at a rate comparable with that of particles about 80 to 120 mμ in diameter, at other times at a rate comparable with that of particles about 10 mμ in diameter, and at still other times the bulk of the activity appeared to sediment at a rate comparable with that of the larger particles and the residual activity at a rate comparable with that of the smaller particles."[197] The appearance of the smaller particle varied depending on the strain of influenza; it was more abundant in isolated F12 strain than in PR8 material.

Stanley and his associates settled on an identification of influenza virus with an approximately 100-mμ particle identified in the electron microscope that corresponded with a sedimenting species in the analytical ultracentrifuge. (However, this identification was uncertain enough that Lauffer was also a coauthor on a paper by a group at Penn arguing that the 10-mμ particle was the actual influenza virus.) Unfortunately, Stanley's 100-mμ particle seemed to react in some immunological assays as a chicken protein would, suggesting that it was derived from the host rather than from an extrinsic agent:

> Detailed quantitative immunochemical experiments were made, and the results of these have forced the conclusion that the 100 mμ virus particles, regardless of their source, possess a common antigenic structure which is characteristic of influenza virus, but possess in addition an antigenic structure that is characteristic of the host from which the virus is obtained.[198]

195. Ibid., 261–62.
196. James T. Costigan to Stanley, 3 May 1945, Stanley papers, carton 17, folder Sharples Corporation.
197. Scientific Report to the Director, RAC RU 439, box 5, vol. 32, 1943–44, 225.
198. Stanley, "Condensed Review of Studies on Influenza Virus," 3 (full citation in note 185 above).

To the degree that there seemed to exist a contradiction between bio-physical and immunochemical criteria for the distinctiveness of the virus, Stanley privileged the physical-chemical indications of its purity.

This preference was typical of Stanley—from the outset, his contributions to virology had been predicated on the chemist's conviction that purer is better. Establishing that TMV was a molecule of precise molecular weight and shape underwrote, for Stanley, a view in which homogeneous, purifiable entities are fundamental to life itself. Stanley carried this chemical vision into his war project on influenza virus—and was rebuked by epidemiologists and clinical researchers for his naïveté. Thomas Francis offered this blunt critique in response to Stanley's advocacy of ultracentrifugation for vaccine production:

> [It appears that] your beliefs that the extraneous proteins are harmful are scarcely more than assumptions which have no basis in your evidence or experience; they are opposed by the experience in vaccination against yellow fever, typhus, and equine encephalomyelitis, where whole [chick] embryo rather than allantoic concentrates are used; they are opposed by the information obtained in various laboratories as to the antigenicity of the different portions of the egg.... While the Commission is interested in any procedures which may offer advantages, it would in no way [be] committed to adopt what you call the "CMR" vaccine unless it is shown clearly to have advantages in immunizing capacity, stability, in reducing reaction, and in practical production.[199]

Notwithstanding these criticisms, Stanley's centrifuge method for influenza vaccine was eventually accepted by the Army as an alternative to the vaccines produced by the red cell adsorption and elution method used during the first half of the war. Manufacture of the centrifuge-type vaccine began in mid-1945 for Army distribution and early in 1946 for civilian use.[200] The American Chemical Society awarded Stanley its Nichols Medal for his development of a new vaccine against influenza. On the basis of this recognition, the International Nickel Company took out an advertisement (see fig. 4.10) in *Chemical and Engineering News* touting Stanley's new vaccine under the heading "A Virus and a Centrifuge."[201]

199. Thomas Francis Jr. to Stanley, 21 Mar 1944, Stanley papers, carton 8, folder Francis, Thomas. Also see Bayne-Jones to Stanley, 14 Jan 1946, Stanley papers, carton 6, folder Bayne-Jones, Stanhope. On the role of extraneous antigens in the immunizing power of vaccines, see Rasmussen, "Freund's Adjuvant."

200. Scientific Reports, RAC RU 439, box 6, vol. 34, 1945–46, 260.

201. Advertisement, International Nickel Company, Inc., "Let us now Praise Famous Men," *Chemical and Engineering News* 25, no. 4 (13 Oct 1947): 2990.

FIGURE 4.10 "Let us now Praise Famous Men." (From *Chemical and Engineering News* 25 [1947]: 2990.)

In 1947, Stanley's vaccine also received a ringing endorsement aimed at physicians in the *Journal of the American Medical Association*. The three other vaccines on the market contained egg proteins that caused allergic reactions in some recipients. "With respect to potential allergic reactions, the centrifuge-type vaccine has definite advantages over the other three types. . . . It seems obvious therefore that the most highly purified vaccine, the centrifuge-type, should be used whenever possible."[202] However, that same month, an epidemiological study published by Francis, Jonas Salk, and J. J. Quilligan Jr. at the University of Michigan showed that influenza vaccination failed to protect college students vaccinated the previous winter against influenza.[203] Further studies of the winter of 1947 influenza outbreaks pointed to strain variability as the problem affecting the efficacy of recent vaccines.[204] Thus the hopes Stanley held that his contribution to military medicine might revolutionize postwar public health were disappointed.

Irrespective of the merits of Stanley's influenza vaccine, the use of the ultracentrifuge to isolate viruses became central to virology.[205] Not that the method always worked. In 1944, Stanley employed his virus purification scheme in an attempt to develop a vaccine for Japanese B encephalitis, but centrifugation failed even to concentrate the virus activity into a soluble, purified product.[206] On the other hand, vaccinia virus and papilloma virus proved eminently suitable to purification and study with the ultracentrifuge.[207] As Jean-Paul Gaudillière has noted, by 1944 most research groups investigating pure influenza virus as part of the war effort were using preparative ultracentrifugation.[208] Few alternatives existed for work aimed at purifying viruses that afflict humans, which became an urgent concern during the war. As Joseph Beard concluded in 1948,

202. "Influenza Vaccines," 1177.

203. Francis, Salk, and Quilligan, "Experience with Vaccination against Influenza."

204. Sigel et al., "Influenza A in a Vaccinated Population."

205. Review articles from the late 1940s attest to the importance of ultracentrifugation in virus research. See Pirie, "Viruses" (*Annual Review of Biochemistry*), in which he states, "Ultracentrifugation plays such an important part in the separation of all but the most stable viruses that examinations of the technique are of great importance" (588); Beard, "Review: Purified Animal Viruses"; Stanley and Lauffer, "Chemical and Physical Procedures."

206. Scientific Reports, RAC RU 439, box 6, vol. 33, 1944–45, 230–32. The fact that Stanley was recruited to work on the encephalitis virus does show the confidence that his centrifuge method (and his expertise with viruses) had inspired. See Francis G. Blake to E. Cowles Andrus, 29 Sep 1944, copy in Stanley papers, carton 6, folder Blake, Francis Gilman; Stanley to Gasser, 5 Oct 1944, carton 1, folder Oct 1944.

207. Pickels and Smadel, "Ultracentrifugation Studies on the Elementary Bodies of Vaccine Virus"; Smadel, Pickels, and Shedlovsky, "Ultracentrifugation Studies on the Elementary Bodies of Vaccine Virus. Part 2"; Smadel, Rivers, and Pickels, "Estimation of the Purity"; and Neurath et al., "Molecular Size, Shape, and Homogeneity."

208. Gaudillière, "Molecularization of Cancer Etiology," 146.

"Purification of the animal viruses has been effected thus far almost entirely by simple ultracentrifugal procedures."[209]

Postwar Commercialization: The Spinco Models E and L

The scientific mobilization for war expanded the market for the new physical-chemical instruments brought into existence through the Rockefeller Foundation's sponsorship.[210] For example, the vaccine programs and other virus research associated with the war effort created a demand for ultracentrifuges among military and industrial users as well as university and institute researchers. The Navy sought to acquire ultracentrifuges for its own research efforts and consulted Edward Pickels in their construction.[211] Military medical projects supported by the OSRD also involved ultracentrifuges. The ultracentrifuge at Harvard Medical School's Department of Physical Chemistry was being used around the clock to characterize the blood-derived medical products manufactured by Edwin Cohn and his coworkers in the large OSRD-sponsored Plasma Fractionation Project.[212]

The onset of demobilization increased rather than decreased the number of requests Pickels received for custom-built ultracentrifuges.[213] However, his full-time job as a virus researcher at the Rockefeller Foundation's International Health Division Laboratory constrained his ability to develop and build new ultracentrifuges. Writing to a former colleague about other job prospects, Pickels reflected:

> I feel that I can make my greatest contribution, best employ my experience and natural abilities, and best uphold my meagre professional standing only if my primary interests continue to be the development of specialized laboratory equipment and techniques. . . .

209. Beard, "Review: Purified Animal Viruses," 52.

210. Gaudillière, "Molecularization of Cancer Etiology."

211. Elzen, *Scientists and Rotors,* 368. Prior to the war, Kahler at the Navy had purchased a centrifuge from Beams, and in 1938 Bauer recommended to Kahler that he send "one or two of the better type of instrument makers from the gun shops of the Washington Navy Yards" to learn how to make driving mechanisms with Pickels (J. H. Bauer to E. G. Pickels, 26 Jul 1938, RAC RF 5, 4, box 22, folder 260). One of Stanley's technicians, a young chemical engineer and physical chemist named Howard Schachman, enlisted in the Navy and helped procure Pickels's assistance in designing ultracentrifuges for the military's research program; see Howard Schachman to Stanley, 22 May 1945, Stanley papers, carton 12, folder Schachman, Howard Kapnek; Hakansson to Stanley, 29 Apr 1946, Stanley papers, carton 9, folder Hakansson, Erik Gösta; Pickels to Stanley, 15 Apr 1946, Stanley papers, carton 13, folder Williams, Robley Cook. Schachman was working at the Naval Medical Research Institute in Bethesda (interview with Howard Schachman, Jun 1992).

212. See Creager, "Producing Molecular Therapeutics from Human Blood."

213. "Machines That Sort and Weigh Molecules," Excerpt from Trustees Bulletin, RAC RF 1.1, 252, box 1, folder 13, 27.

> Commercial development of specialized laboratory equipment might represent a more promising future for me, and I have been giving serious thought to this possibility.[214]

At about this time, Stanley's technician, Howard Schachman, introduced Pickels to Maurice Hanafin of the Glass Engineering Laboratories. Hanafin's commercial glassware venture had built a preparative ultracentrifuge for the Hooper Foundation at the University of California Medical School in San Francisco.[215] Pickels did some consulting for the Hooper Foundation when Hanafin's machine malfunctioned, and Pickels and Hanafin decided to join forces to develop and sell a commercial analytical ultracentrifuge. In 1946 they founded the Specialized Instrument Corporation, or Spinco, based in the San Francisco Bay Area.[216] Pickels hired a draftsman and a machinist and set up shop in a garage; Hanafin began the marketing efforts.[217]

Pickels designed an electrically driven centrifuge based on his previous air-driven models. It could be used with either preparative or analytical rotors; Pickels and Hanafin hoped this versatility would attract purchasers (see fig. 4.11).[218] The machine had a gearbox with thirty rpm settings and precise speed control. Temperature was also well-regulated, an important consideration for biological materials. The optical system allowed for constant viewing by the operator as well as automatic recording of sedimentation patterns on photographic plates. This ultracentrifuge was called the Model E. The first production model was installed in Stanley's

214. Pickels to Thomas Francis Jr., 24 Oct 1945, RAC RF 5, 4, box 22, folder 260, 2. Pickels was considering a job offer at the University of Michigan, where Francis had gone. In Pickels's letter to R. M. Taylor to get permission to visit Francis's lab, he wrote: "The I.H.D. [International Health Division] does not have as a primary interest the development of specialized scientific equipment per se, that is, the design and construction of major equipment for which there may be no immediate need in the I.H.D. laboratories. As an example, there is a definite need in the scientific world for a greatly simplified, less expensive, and better standardized ultracentrifuge. The development of such equipment I consider to be a specialty of mine. Actually, my professional reputation, such as it is, is largely based on such work. Most of my correspondence and most outside requests for help deal with it" (Memo, 23 Jan 1946, RAC RU 5, 4, box 22, folder 260, 2).

215. See Pickels's resignation letter to RF I.H.D., RAC RU 5, 4, box 22, folder 260; Elzen, *Scientists and Rotors,* 368–69. Schachman has referred to himself as serving as the "midwife" of the company by introducing Pickels and Hanafin to one another (interview with author, Jun 1992).

216. Elzen, *Scientists and Rotors,* 369.

217. See Spinco's announcement of the availability of the preparative "Model B" and analytical "Model D" (soon superseded by the Model E), dated 8-1-46, enclosed with letter from Pickels to Stanley, 20 Nov 1947, Stanley papers, carton 11, folder Pickels, Edward Greydon.

218. Elzen, *Scientists and Rotors,* 370. Elzen gives more detail on Pickels's design choices for the Model E.

FIGURE 4.11 Photographs of the Model E ultracentrifuge and its rotors from a technical magazine feature. *Above:* Edward Pickels placing a rotor into the ultracentrifuge chamber. Various rotors and cells are laid out beside him. *Facing page:* The preparative rotor (left) and analytical rotor (right) for the Model E ultracentrifuge. The flat metal piece in the lower center is the lid for the preparative rotor, which is hermetically sealed when in operation with a rubber gasket. The analytical rotor carries one test cell and a dummy for counterbalance; these cells are below and to the left of the rotor. The test cell, laid out in the lower right, is composed of a centerpiece with a rectangular slot for the fluid under investigation and two end caps which hold quartz windows. In operation, a high-intensity light beam is projected through the sample, and observations or photographs are made of the movements of boundaries formed by the sedimenting particles. (Images and captions from "Centrifuge for Speeds to 70,000 RPM," *Product Engineering* 20 [1949]: 92. Reprinted by permission of Miller Freeman Books.)

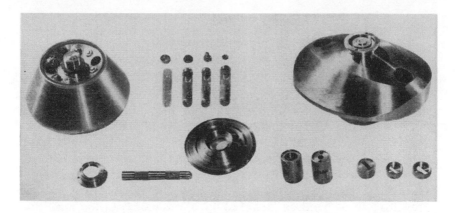

FIGURE 4.11 (continued)

postwar Virus Laboratory at the University of California, Berkeley (see chapter 5).[219]

Within a year, Pickels and Hanafin sold seven Model E machines, at $13,200 apiece. The receipts were not sufficient to support the company, and at the end of 1948 Pickels laid off half the staff. Hanafin, meanwhile, noticed that more researchers were interested in preparative than analytical machines. In response, Pickels designed an ultracentrifuge for preparative purposes only, called the Model B (and later, the Model L). This machine spun a rotor to a maximum of 40,000 rpm. It was designed to be as "simple to operate as a washing machine: open up the lid, put in the sample, and run it by turning the switch."[220] The controls were straightforward, with interlocks and safety devices to prevent operating errors.[221] A photograph advertising the machine showed it alongside a woman in a white lab coat, using prevailing gender assumptions to suggest that even untrained workers and technicians could operate the machine. The price of the Model B was only one-quarter of the price of the analytical Model E, and sales orders came in briskly. Gradually, orders for the analytical machine also increased, but Pickels reflects that the preparative machine saved the company.[222] By the mid-1950s the company began manufacturing other instruments, such as a commercial electrophoresis apparatus.

219. Interviews with Howard Schachman, Jun 1992, Berkeley, and with Edward G. Pickels, 25 Feb 1993, Cupertino, California.

220. Elzen, *Scientists and Rotors,* 372. Spinco actually had a preparative Model B, at least on paper, in 1947, but unlike the later, user-friendly Model B, this one was identical to the analytical version except that it lacked the optical system. See Pickels to Stanley, 20 Nov 1947, Stanley papers, carton 11, folder Pickels, Edward Greydon.

221. Elzen, *Scientists and Rotors,* 372–73.

222. Edward G. Pickels to author, 26 Feb 1993; Interview with Pickels, 25 Feb 1993.

The Model E and its competitors, the MSE Mk II and Centriscan analytical ultracentrifuges, spurred the dissemination of the ultracentrifuge as a machine for visualizing biological macromolecules. As Schachman reflected, "The availability of an extremely reliable commercial instrument coupled with the establishment of grant programmes in the United States led in the following decade to an almost explosive proliferation in the number of research workers using the ultracentrifuge as their principal tool."[223] At the end of World War II, around fifteen analytical ultracentrifuges were in operation in the world; by 1951, seventy Spinco Model E ultracentrifuges had been sold. In 1955 Beckman Instruments bought Spinco; they continued to produce the Model E (see fig. 4.12).[224] Over the next fifteen years the number of Model E ultracentrifuges in use grew from three hundred (1958) to one thousand.[225] At the same time, the ubiquity of preparative ultracentrifuges reinforced the utility of the analytical machines: virus researchers, cell biologists, and biochemists could make samples using the preparative machine that were, by virtue of their isolation method, suitable for characterization in an analytical ultracentrifuge.

The standardization of the ultracentrifuge did not eliminate the necessity of tinkering with the devices and experimental interpretations: researchers continued to modify not only their ideas about macromolecules, but the optical system and cell design of their chief tool. Researchers also extended the range of biological materials analyzed. In one ingenious adaptation, Matthew Meselsohn and Frank Stahl used cesium chloride at high concentration to separate nucleic acids, enabling an elegant resolution of the DNA replication problem in the ultracentrifuge.[226] Increasing interest in characterizing DNA and RNA led to the revival of ultraviolet absorption optics methods for viewing sedimenting species (first used by Svedberg in the 1930s).

Research on protein structure also spurred continued innovation. Schachman, who ran the Model E in Stanley's Virus Laboratory in Berkeley in the 1950s, developed new optical methods and new cells (such as the layering cell and synthetic boundary cell) for protein studies in the ultracentrifuge. As a consultant to Spinco, he saw some of his innovations brought into commercial production. Even as the ultracentrifuge was standardized and commercialized, so that researchers all over the

223. Schachman, "Is There a Future for the Ultracentrifuge?" 7.

224. Schachman, *Ultracentrifugation in Biochemistry*, 1, and "Development of the Ultracentrifuge."

225. Elzen, *Scientists and Rotors*, 373.

226. Meselsohn and Stahl, "Replication of DNA." For a fine-grained account of this experiment, including an excellent discussion of the DNA replication problem in the 1950s, see Holmes, *Meselsohn, Stahl, and the Replication of DNA*.

FIGURE 4.12 Advertisement for Beckman Spinco ultracentrifuges. (From *Journal of Bacteriology* 75 [1958]. Reprinted courtesy of Beckman Coulter, Inc.)

world were running the "same" instrument, variability and innovation in practice continued, feeding back into new commercial possibilities. As Boelie Elzen puts it,

> Ever since Svedberg had published his first results on studies with the ultracentrifuge, there had been a considerable interest from the scientific world. But it was not until Pickels started manufacturing the Spinco machines that a substantial spread of ultracentrifuges to other laboratories began. This did not imply that the ultracentrifuge had been fully developed by then. As more scientists started using them for different applications, new problems with respect to the apparatus were identified. The development never really stopped.[227]

At the same time, formalizing the connection between research and engineering through the commercial manufacture of ultracentrifuges affected the distribution of scientific credit. The development of new cells and optical systems for the Model E was an academic pursuit in Schachman's laboratory, which was producing Ph.D. dissertations such as "Studies with a Synthetic Boundary Ultracentrifuge Cell and Properties of an Abnormal Protein Found in Mosaic-Diseased Tobacco Plants" (1955), and "Applications of the Rayleigh Interferometer to Ultracentrifugation" (1960).[228] Pickels's continuing work in the field was relegated to technical development, and he grew to resent the lack of credit given him for his creation of a machine used by thousands of biochemists.[229] One of the consequences of the building of a commercial infrastructure to replace previous collaborations with academic physicists and chemists was the effacing of those contributions by scientists and engineers on the commercial side of the research divide.

Since the late 1930s, Pickels's machines had been favored due to their amenability for virus research, and postwar virologists continued to be avid users of ultracentrifuges. Virus biochemist Dean Fraser offered this perspective on the significance of the commercial ultracentrifuge:

227. Elzen, *Scientists and Rotors*, 380.

228. List of titles of Ph.D. dissertations courtesy of H. K. Schachman laboratory.

229. An application submitted by Lawrence Cranberg for Edward Pickels, nominating him for the National Medal of Science, includes the following observation about the 7,600 publications in the scientific literature up to 1971 which relied on ultracentrifugal analysis: "Although Pickels is beyond question the creative scientific mind that made those publications possible, my investigation reveals that in almost every instance the indispensable instrument created by Pickels is credited to 'Spinco.' Ironically, that is the name that he coined, and it has served to efface the name of its creator!" My point is not to argue for Pickels's unappreciated contributions to biology but to draw attention to the way in which credit is parceled between the commercial and the academic contributors to biochemical research efforts (document courtesy of Edward G. Pickels).

In 1949 a development occurred ... which completely revolution-
ized centrifugation and hence viral concentration and purification.
This was the production of the Spinco (now Beckman) [Model L] ultracentrifuge. . . . In my opinion this machine has done more than any
other single instrument to advance the study of viruses, yet Pickels
has received almost no attention for these remarkable achievements.
Previously, ultracentrifuges had to be run in the subbasement behind
three-foot-thick walls of reinforced concrete. They were individually
built; the scientists who operated them were considered slightly in-
sane. Centrifuges required constant attention, the technical skill of a
Kettering, a willingness to spend most of your time flat on your back
in a puddle of oil doing plumbing and mechanical repairs. Explosions
that wrecked the machine were to be expected at fairly regular in-
tervals. The Spinco preparative centrifuge, for contrast, is about the
size of a family washing machine, and anyone can learn to run it in
ten minutes.[230]

Conclusions

The incorporation of physical-chemical instruments into Stanley's experimental system with "crystalline" TMV caused a radical reordering of
the material and its representations. Stanley's early claims about the
chemical composition and behavior of TMV were superseded by a revised
view of the virus as a molecule more complex, both in terms of chemical composition and structure, than most proteins. The physical-chemical
characterization of TMV as a rod-shaped macromolecule refuted early
speculations that the virus multiplied through enzymatic autocatalysis,
but shed no new light on TMV's mechanism for self-reproduction.

Stanley's willingness to abandon his published views in the face of
new evidence, on this and other issues, left him open to some ridicule.
As Thomas Rivers remarked in his 1939 Lane Lectures, "From Stanley's
numerous writings, it is rather difficult to determine what he actually
thinks of the nature of viruses. Apparently he does not believe them to be
absolutely inanimate molecules, nor does he think of them as completely
alive; according to him they are transitional forms between the living and
nonliving, some of them being relatively simple, others more complex."[231]
At the same time, Rivers granted that Stanley's considerable rhetorical

230. Fraser, *Viruses and Molecular Biology,* 38–39.
231. Rivers, *Lane Medical Lectures,* 98. Along similar lines, Bawden and Pirie castigated
Stanley for his many conflicting reports about the size and shape of tobacco ringspot virus.
They composed a letter to *Science* calling attention to the blatant contradictions between
statements about the virus in seven of Stanley's papers, but withdrew it upon learning that
Stanley had just published a study of tobacco ringspot in the *Journal of Biological Chemistry*
(Bawden to Stanley, 25 Jul 1939, Stanley papers, carton 6, folder Bawden, Sir Frederick
Charles; Stanley to Loring, 11 Oct 1939, Stanley papers, carton 1, folder Oct 1939).

skills facilitated his easy shift in positions: "In still another respect [beyond being a chemist engaged in medical research], Dr. Stanley is like Pasteur; he is a master of exposition and knows well how to defend his work."[232]

As Stanley dropped his initial simplistic claims about the chemical nature of viruses, his continued work on the structure of TMV led the way toward the wide use of biophysical instrumentation in the purification and characterization of viruses.[233] The ultracentrifuge, which had previously been designed to visualize sedimenting materials, was adapted as the main means for preparing viruses as macromolecules. Medical researchers, biologists, plant and animal pathologists, and military scientists took up this approach to their own virus subjects, and TMV stood as a key exemplar and reference system for this style of research.

The use of TMV as a virus standard was pragmatic as well as conceptual. In 1944, Ralph Wyckoff requested the favor of some purified TMV from Stanley from the University of Michigan: "I have recently come out here to try and develop some teaching and research in the biophysics of viruses and other macromolecules. As a standard for some of the equipment we are setting up, it would be nice to use the tobacco mosaic protein."[234] The relative invariability of TMV in laboratory preparations was critical to its use as a standard. Not all viruses behaved as good macromolecules—yellow fever virus was hard to concentrate, whereas influenza virus lacked molecular homogeneity.[235] The difficulty of distinguishing viruses from the constituents of ordinary cells proved particularly vexing, threatening to undermine the authenticity of some newly isolated viruses (see chapter 6). Even so, the ultracentrifuge provided the best means for manipulating viruses as molecular entities, exemplified by the work at the Rockefeller Institute on TMV.

In explaining how these laboratory materials and tools interacted, one must be able to speak of TMV and the ultracentrifuge as real but not reified objects: Both TMV and the ultracentrifuge were materially reconfigured over the course of their interaction, as the function of the ultracentrifuge as virus-producer and the existence of TMV as macromolecule were mutually stabilized. This depiction of the dynamic nature of the virus-instrument system agrees with the general picture of biological experimentation that Hans-Jörg Rheinberger has offered through his historical analysis of research on protein synthesis. An experimental system, in his view, must be constantly shifting in order to be useful for

232. Rivers, "Infinitely Small in Biology," 144.

233. See Creager and Gaudillière, "Experimental Arrangements"; Gaudillière, "Molecularization of Cancer Etiology."

234. Ralph Wyckoff to Stanley, 6 Mar 1944, Stanley papers, carton 13, folder Wyckoff, Ralph Walter Graystone.

235. Pickels and Bauer, "Ultracentrifugation Studies"; Friedewald and Pickels, "Centrifugation and Ultrafiltration Studies"; Lauffer and Stanley, "Biophysical Properties."

research, so as to make visible novel effects and inscriptions (i.e., new results).[236] In this respect, it is the dynamic *instability* of an experimental system that makes it productive in research; once a system is standardized, it has been reduced to serving as a mere technique. This vision of experimental epistemology is particularly powerful for understanding the generation of scientific novelty at the most local level. But this emphasis on instability poses a problem for the historian wanting to explain more general patterns of change. How can one account for the reliable transfer of instruments and experimental objects to new situations while acknowledging the highly contingent and variable nature of working with a particular experimental system?

In seeking to explain the formation of consensus and the successful circulation of tools and experimental findings, historians and sociologists have tended to stress the role of standardization in twentieth-century biology. Robert Kohler argues that by genetically standardizing *Drosophila melanogaster*, the laboratory fruit fly, Thomas Hunt Morgan and his students created an instrument that could circulate widely. Moreover, the specialized purpose built into the system, in this case its utility for chromosomal mapping, enabled (and constrained) the growth of genetic research in *Drosophila* laboratories far from Morgan's.[237] Karen Rader has demonstrated how the genetic standardization of the mouse as a laboratory object facilitated its wide adoption in biomedical research.[238] Joan Fujimura speaks of "standardized experimental systems" in referring to "inbred animals [and] tissue culture materials and methods" that are used by many different laboratories. She argues that the succession of different theories of cancer can be seen in terms of the dissemination of "theory-practice" packages, often rooted in standardized experimental systems.[239]

Can one speak of TMV as a standardized experimental system in Fujimura's sense? It may have provided a kind of macromolecular standard, as Wyckoff's request for TMV indicates, but the experimental system itself kept on changing. By 1938, isolating a virus in Stanley's laboratory entailed different methods, instruments, and materials from those he had used in 1935. The representation of TMV as a unitary macromolecule, which Stanley's laboratory worked so hard to secure in the late 1930s, was again revised after the war as new biochemical methods and questions focused attention on the constituent pieces of TMV (see chapter 7). Even the ultracentrifuge, having been domesticated as a

236. Rheinberger, "Experiment, Difference, and Writing," esp. 305–6, and *Toward a History of Epistemic Things*.
237. Kohler, *Lords of the Fly*.
238. Rader, *Making Mice*.
239. Fujimura, "Standardizing Practices," 3; see also her *Crafting Science*.

purification apparatus in virology by 1937, was not standardized into a commercial instrument until the late 1940s.

Yet I have argued throughout this chapter that even though TMV continued to be altered and knowledge about it revised, it served as a benchmark in the broader scientific community, such that experiments performed with TMV were successfully extended to research on other viruses. The process of reconfiguring research objects based on an exemplar did involve the adoption of certain tools and the repetition of particular techniques, but the term *standardization* may not be sufficiently flexible to account for the work and adaptation involved in extending these approaches to new situations. I prefer to draw a distinction between the experimental system, as the highly localized arrangement of materials and instruments, and the model system, to refer to the "virus" in its more public guise as a point of reference.

There are two senses in which TMV served as a key model system for work on other viruses. First, TMV was intensively studied as a substitute for other pathogens and as a representative of the category of viruses. When Stanley sought to devise a protocol for purifying influenza virus in a high-speed Sharples centrifuge, he first developed the procedure using TMV as the starting material. He then extended this method to influenza virus and used it as the basis for preparing a new vaccine. The second sense in which TMV was a model system is somewhat more abstract. When virologists sought to prepare animal viruses in the ultracentrifuge, they followed the material example given by TMV, but this example was often conveyed via the scientific literature, in "materials and methods," representations of the virus, and the laboratory inscriptions that showed TMV to be a macromolecule. In this sense the model system refers to the idealized object of study, based upon the actual concrete example but not restricted to its material instantiation. The model system of TMV circulated widely, and was adapted for various purposes by virus researchers distant from Stanley's laboratory.

The next chapter further elaborates the diverse functions of TMV as a model system, following it beyond the realm of research laboratories altogether. The relevance of virus research to the health of citizens, an issue which the practical emergencies of World War II brought into clear relief, enabled TMV to become a representative virus, a promissory note showing how virologists could understand (and ultimately conquer) viral enemies such as poliomyelitis. In turn, the expansion of biomedical research in the postwar period intensified the use of laboratory model systems as research tools in fighting disease on behalf of the public.

5

The War against Polio

During the past ten years there has converged on a group of small, infectious, disease-producing entities, known as viruses, an array of scientific talent almost as diverse in nature as the Allied forces that were brought to bear on the Axis powers.

W. M. Stanley, *Currents in Biochemical Research*, 1946

The mobilization of scientists such as Stanley during World War II brought a generation of researchers into a new relationship with the U.S. government.[1] Within a decade, public funding became the principal source of support for basic and clinical biomedical researchers. However, this new pattern of patronage did not follow inevitably or immediately from the war mobilization. Voluntary health agencies played a critical—and largely unexamined—role in adapting the example of the scientific mobilization from World War II to peacetime. Lay activists, galvanized by the devastating polio epidemics, raised funds for therapy, education, and basic virus research. They supported not only research on human pathogens, but on model systems such as TMV as well, in the hope of uncovering knowledge and methods applicable to poliovirus. The analogies drawn between TMV and poliovirus in Stanley's laboratory illustrate the way a model systems approach prospered in the kind of research-based fight against diseases that increasingly characterized postwar biomedicine.

The science-promoting activities of lay health organizations offer a new perspective for understanding the postwar emergence of public commitment to funding research. Conventional interpretations of the legacy of World War II for science focus almost exclusively on scientists and politicians as the brokers of postwar policy.[2] In his influential 1945 manifesto, *Science—The Endless Frontier*, Vannevar Bush advocated the initiation of federal government funding of basic research in universities and institutes to ensure public welfare, national security, and advances in medical care.

1. Dupree, "Great Instauration."
2. For an account that stresses the role of business, see Kleinman, "Layers of Interests."

Although historians of science have tended to view the consequences of Bush's vision in terms of the natural (and especially physical) sciences, the continuities with the wartime mission were most evident in medical research: Bush urged that life science be supported to fight the "war against disease" in peace.[3] The Committee on Medical Research of the Office for Scientific Research and Development (OSRD) took credit for having funded four of the most dramatic medical advances during the war: antibiotics, DDT, improved antimalarial drugs, and plasma fractionation products.[4] Public demand for more such scientific solutions to medical needs was high, a sentiment not lost on politicians.[5] As historian of medicine Richard Shryock observed in 1947,

> The war has stimulated public interest in research. Newspaper editors demand funds for studies of the more dreaded diseases, pointing out that medical scientists now "labor against inexcusable odds." . . . Public enthusiasm for continuing a war-scale program in medical research has even been carried to a point somewhat disturbing to medical leaders. There is some fear that legislators, fired by the achievements of the Committee on Medical Research, will assume that solutions for all present problems can be promptly provided by similar methods.[6]

Despite this popular motivation to continue some kind of mutually beneficial partnership between government and scientists after the war, translating this vision into coherent national policy proved politically impossible in the late 1940s. As Daniel Kevles has shown in his classic article, Bush's report was an essentially conservative response to Senator Kilgore's proposed legislation for a public research foundation: "Kilgore wanted a foundation responsive to lay control and prepared to support research for the advancement of the general welfare; Bush and his colleagues wanted an agency run by scientists mainly for the purpose of advancing science."[7] Debates over the nature and range of a government foundation for the support of research delayed the establishment of a National Science Foundation (NSF) until 1950. Many scientists objected

3. Bush, *Science*. On the significance of the mobilization for the physical sciences after the war, see Kevles, *Physicists,* and "Scientists." Reingold argues that the Public Health Service (NIH) did not figure much in Bush's forecast of the relevant federal agencies: "That was a serious blind spot" ("Bush's New Deal," 314).

4. Bush, *Science,* 49. See also Richards, "Impact of the War."

5. See Strickland, *Politics, Science, and Dread Disease.*

6. Shryock, *American Medical Research,* 318–19.

7. Kevles, "National Science Foundation," 16. See also Reingold, "Bush's New Deal"; Greenberg, *Politics of Pure Science;* Wang, "Liberals"; and the documentary selections compiled by Donald C. Swain in Part 3, "Postwar Planning for Science, 1945–50," of Penick et al., *Politics of American Science.*

to legislation that gave politicians control over the selection of research topics and projects, because they feared this would interfere with the freedom of scientific inquiry.

For experimental biologists, these political deliberations were fraught with other problems. As Stephen Strickland and Harry Marks have demonstrated, considerations of federal funding for biomedical research were part of larger debates about government involvement in health care and medical education.[8] The Public Health Service Act of 1944 enlarged the authority of the agency to "pay for research to be performed by universities, hospitals, laboratories, and other public or private institutions" beyond the area of cancer investigations;[9] the Division of Grants Research was soon expanded to assimilate unfinished OSRD contracts. However, critics of the Public Health Service (many in private medical schools) vigorously opposed interference by the federal government in medical research and education.[10] Officials at the National Institutes of Health (NIH), the Public Health Service's research arm, had to defend the legitimacy of government funding to medical scientists at the same time as they had to protect "their new authority to conduct a broad extramural program" from being assimilated into a new national research agency.[11]

While politicians, President Truman, federal agencies, newspaper editors, and scientists debated peacetime science policy, voluntary health organizations launched a research-based "war against disease" in the postwar period. In particular, two health agencies advanced their causes by funding research on a large scale by the conclusion of World War II. The American Cancer Society (ACS), founded in 1913, began funding medical research at the end of World War II after lay activists took control of the organization.[12] In the 1940s, the dreaded polio epidemics provided the other obvious and immediate context for the extension of military language and research organization to the domain of biomedical research. The National Foundation for Infantile Paralysis, or NFIP, publicized the value of disease-related research at university laboratories in its yearly fund-raiser, the March of Dimes.[13] The development of the Salk and

8. Strickland, *Politics, Science, and Dread Disease;* Marks, "Leviathan and the Clinic."

9. U.S. Congress, P.L. 410, 78th Congress, 2d session, 1944, as quoted in Strickland, *Politics, Science, and Dread Disease,* 19.

10. Marks, "Leviathan and the Clinic."

11. Fox, "Politics of the NIH Extramural Program," 460. On the history of the NIH, see Harden, *Inventing the NIH.* On the significance of private funding between the war and 1948, see Mills, "Distribution of American Research Funds."

12. See Patterson, *Dread Disease,* 170–71. On the postwar involvement of voluntary health organizations in the medical research scene (and some of the resulting conflicts), see Marks, "Cortisone, 1949."

13. Benison, "History of Polio Research."

Sabin polio vaccines of 1955 and 1958, both fully funded by the NFIP, richly vindicated the organization's strategic emphasis on laboratory research.

In the immediate postwar years, the NFIP and the ACS patronized biomedical research on roughly the same scale as the extramural program of the NIH, managing massive national public collection campaigns to raise millions of dollars for research. While these organizations were private, they were markedly different from other philanthropies (most notably the Rockefeller Foundation) in their support of research. The tens of thousands of volunteers in the ranks of the health agencies and the millions of donors held concrete expectations for medical breakthroughs from laboratory research; they supported a directed and coordinated research enterprise. The meanings and constraints of postwar "basic research" were negotiated in the context of these lay crusades against specific diseases, not only on the floors of Congress.[14] The president of the NFIP claimed that "the national voluntary health agencies have caused the American people to become 'research minded,'" particularly in support of "basic research in the life sciences."[15] At the end of the 1940s, the expansion of funding for basic biomedical research through new NIH organ- and disease-named institutes mirrored the voluntary health agencies' successful disease-based campaigns.[16]

An overview of the polio-related activities in Wendell Stanley's laboratory illustrates how the NFIP fostered public appreciation for breakthroughs in fundamental research—as well as expectations for vaccine development—by drawing metaphorically on the recent experience of victory in war. Stanley's program also shows how the reliance on model systems worked in practice: He designed his research objectives for poliomyelitis virus after the successful example of his work with TMV, culminating in the crystallization of poliovirus in 1955. (Stanley's NFIP

14. On the importance of elucidating the connections between science patronage in the postwar period and mass culture, see Reingold, "Science and Government."

15. Basil O'Connor, "A Special Message," National Foundation Annual Report 1960, NFA, 3.

16. The funding of biomedical research through the NIH's disease- or organ-based institutes is termed the "categorical approach." The National Cancer Institute was founded in 1937; to it were added a National Heart and Lung Institute (1948), a National Institute of Dental Research (1948), a National Institute of Allergy and Infectious Diseases (1948; this included the former intramural microbiologists), a National Institute of Arthritis, Metabolism, and Digestive Diseases (1948), a National Institute of Mental Health (1949), and a National Institute of Neurological and Communicable Disorders and Stroke (1950). Five more categorical institutes were added between 1962 and 1974, including the non-disease-specific National Institute of General Medical Sciences. See Mider, "Federal Impact," 854. On the attempt to have disease-based funding of medical research in the NSF, see Appel, *Shaping Biology*, 34.

grant was also the major source of funding for his laboratory's ongoing research on TMV, to be taken up in the next two chapters.) Bacteriophage was another prominent laboratory model for viruses, and the NFIP funded phage researchers at several universities. Once again, the model system provided a concrete example for the visualization and characterization of poliovirus. At Caltech, Renato Dulbecco developed a bacteriophage-inspired plaque assay for poliomyelitis so that quantitative virological techniques could be used to study polio.[17] Thus, the links between "basic" virology and laboratory-based solutions to disease were not simply discursive or political but were also material, at the level of making and reshaping local experimental systems. Because of the dual nature of viruses—the fact that these pathogens were also regarded as elementary units of life—the NFIP's investment in basic virus research contributed significantly toward the development of molecular biology as well as to the conquest of polio.[18]

The March of Dimes and Virus Research

On 3 January 1938, President Franklin Delano Roosevelt inaugurated an organization that would have a great impact on virus research. He and a group of committed lawyers, philanthropists, and businessmen brought into being the National Foundation for Infantile Paralysis (NFIP) "to lead, direct, and unify the fight against every aspect of the killing and crippling infection of poliomyelitis."[19] This foundation provided Roosevelt with a new venue for continuing his dozen years of activity in polio philanthropy. In 1926, Roosevelt had purchased a small resort in Warm Springs, Georgia (where he had experienced some therapeutic relief as an adult victim of polio), and developed it into a foundation for the treatment of polio patients.[20] This was not a savvy business investment. Basil O'Connor, Roosevelt's law partner and personal attorney, helped keep the operation afloat by conducting fund-raising in the business community.[21] As Jane Smith observed, "If it had not been for the financial burden Roosevelt assumed when he bought the Warm Springs property, Basil O'Connor would never have started the great fund-raising machine of the National Foundation for Infantile Paralysis."[22]

17. Kevles, "Renato Dulbecco."
18. On the NFIP's role in supporting molecular biology, see ibid., esp. the footnotes on 420–21.
19. From the Certificate of Incorporation of the National Foundation for Infantile Paralysis, Inc., Pursuant to Membership Corporation Files, as quoted in Benison, "History of Polio Research," 320.
20. See Walker, *Roosevelt,* for a colorful and uncritical view of Warm Springs.
21. See Roosevelt, "Tribute to Basil O'Connor."
22. Smith, *Patenting the Sun,* 59.

When Roosevelt was elected to the presidency in 1932, the mechanism for supporting the Warm Springs Foundation shifted to yearly President's Birthday Balls. The first was held on 30 January 1934, on the president's fifty-second birthday, and it raised more than one million dollars. However, balls in the succeeding years brought in less money, and in 1937 O'Connor decided that their polio philanthropy needed a new focus, one that did not appear to endorse the White House's occupant.[23] The resulting organization, the NFIP, would rely on local communities in its appeals for funds and would support laboratory research as well as medical treatment for polio.

The Presidential Birthday Ball Commission (PBBC) had given modest support to polio research, but their precedent proved to be a cautionary tale for the NFIP. Paul de Kruif, author of *The Microbe Hunters* (1926), served as secretary to the PBBC's scientific advisory board and directed their funding decisions.[24] His understanding of science derived from his work as a bacteriologist at the Rockefeller Institute for Medical Research, before being dismissed by Simon Flexner for having penned an anonymous attack on the Institute in a magazine.[25] None of the Birthday Ball funds were given to Institute virologists; Saul Benison has argued that the decision to exclude Institute researchers reflected de Kruif's differences with Flexner.[26]

Instead, de Kruif targeted money at development of a polio vaccine. The largest grant went to support vaccine testing by William H. Park and Maurice Brodie of the New York City Health Department. Spurred by the competition with John Kolmer of Temple University, who was developing a different vaccine against polio, Brodie tested his vaccine on a small group of children, and then extended the immunization to three thousand children in 1935. Kolmer, too, administered his experimental vaccine widely that summer. As it turned out, the poliomyelitis virus in the

23. Ibid., 73.

24. De Kruif had also worked closely with novelist Sinclair Lewis on the development of the McGurk Institute and its characters for *Arrowsmith,* published the same year as *Microbe Hunters*.

25. Benison, "History of Polio Research," 321; Corner, *History of the Rockefeller Institute,* 160. In his autobiography, de Kruif says that he resigned of his own volition, even as he referred to resignation as "our American euphemism for firing" (*Sweeping Wind,* 55–56).

26. Benison ("History of Polio Research") observes, "De Kruif viewed medical research as a kind of organized empiricism where heroic workers in the mold of Jenner, Pasteur, Koch, and Martin Arrowsmith by brilliant insight found cures to baffling diseases. It was not really necessary to know anything about poliovirus to solve the problem of polio. After all, what did Jenner know about smallpox and vaccine virus when he developed vaccination against that disease?" (320–21). Thus de Kruif's philosophy of research was (by this time) distinctly antithetical to Flexner's (and thus the Institute's). For a different—and more romantic—perspective, see de Kruif, *Sweeping Wind.*

Brodie-Park and Kolmer vaccines was only incompletely inactivated. The Brodie-Park vaccine was ineffective and induced severe allergic reactions in some recipients; the Kolmer vaccine was even more dangerous, causing many cases of polio and several deaths.[27] Although the PBBC had not supported Kolmer, the public was not discriminating in its blame.

When the NFIP superseded the PBBC in 1938, several medical committees were constituted to implement the NFIP's research agenda and granting policies. A General Advisory Committee, composed of prominent medical researchers, the president of the American Medical Association, the surgeon general of the U.S. Public Health Service, and NFIP officials, was the apex of the committee system.[28] The Committee on Education funded a variety of groups, including the Georgia Warm Springs Foundation, the American Physiotherapy Association, the National League of Nursing Education, and numerous publications.[29] The Committee on Epidemics and Public Health gave emergency epidemic aid as well as supporting some epidemiological studies.

The other two medical committees were concerned solely with research. These served as peer-review boards for the grants, since NFIP policy specified that the awarding of any grant required a vote of the relevant committee.[30] The Committee on Research for the Prevention and Treatment of After-Effects made grants toward medical advancements in the clinical management of polio.[31] The Committee on Scientific Research provided grants to medical investigators, and it was outspending all of the other committees by the second year.[32] This committee supported virus research almost exclusively; indeed, it was renamed the Committee on

27. Smith, *Patenting the Sun,* 72; Paul, *History of Poliomyelitis,* 254–56.

28. See Cornwell B. Rogers and George H. Jones, chapter 2, "Medical Administration, 1938–1943," in vol. 1 of *History of the National Foundation for Infantile Paralysis Through 1953,* 67 (Unpublished Historical Monographs, NFA).

29. These examples are selected from the Annual Report of the NFIP, 1944, NFA.

30. This policy was not always adhered to; see Rogers and Jones, "Medical Administration," cited in note 28 above.

31. As Rogers and Jones observe, "The *General Statement of Plans* of June 21, 1938, indicated that 'in the "after-effect" field' grants must 'redound to the medical benefit of all those afflicted with the disease.' In other words, the National Foundation should undertake, in technological and methodological ways, to raise the general level of care" ("Medical Administration," 70–71; see note 28 above). To the degree that the awarding of grants followed these initial recommendations, the research funded by this committee was much more constrained to applied and clinical projects than was the research funded by the Committee on Scientific Research. Until 1940, all thirteen members of the Committee on Research for the Prevention and Treatment of After-Effects were orthopedic surgeons.

32. In 1944, the grants awarded by the Committee on Research for the Prevention and Treatment of After-Effects surpassed the amount spent by the Committee on Virus Research for the first time since 1939, due to the award of three large, five-year grants to medical schools; see Annual Reports of the NFIP, NFA.

Virus Research in 1940.[33] Thomas Rivers served as chair of this committee, de Kruif acted as secretary, and three other researchers sat on the committee, along with three NFIP administrators, including O'Connor. These scientific advisers recommended funding research on all viruses, rather than simply on poliomyelitis, in an effort to develop comparative approaches to controlling polio.[34] More significantly, their indirect association with the disastrous vaccine trials of 1935 motivated the new foundation to eschew further clinical trial programs and favor fundamental laboratory research.[35]

The attention and funds that the NFIP devoted to basic medical research represented a shift in the activities of lay health philanthropies. Early in the century, the emergence of voluntary health organizations reflected progressive America's advocacy of efficiency and education in battling social problems, including disease. Reflecting the growing status of professionalized medicine, such organizations often facilitated the collaboration of physicians with community leaders to prevent disease through education and early intervention. The National Tuberculosis Association (NTA; today's American Lung Association), founded in 1904, exemplified this kind of progressive volunteer organization, with its aim of public education and public health. As James Patterson has noted, the association's effectiveness "relied in part on arousing fear of the disease . . . [by providing saloonkeepers with] exhibits of tuberculous lungs in formaldehyde and big painted skeletons showing the damage done by the disease."[36] The Easter Seal campaigns of the NTA raised substantial sums of money in the 1920s. For example, in 1923, the Seal sale brought in more than $4.25 million.[37] The vast majority of these funds went to local chapters to support their efforts in hygiene and prevention, although the organization disbursed limited funding through its Committee on Research (established in 1915). Similarly, the American Society for the Control of Cancer (ASCC), founded in 1913, emphasized public education and early surgical treatment of cancer, and the members of the society were largely physicians.[38]

The NTA was the most visible and wealthiest of the many organizations founded to fight specific diseases. In 1917 the Rockefeller Foundation's International Health Board counted fifty-seven such groups in a report

33. That same year two additional advisory committees were formed: a Committee on Publications and a Committee on Nutrition. The second was short-lived (Rogers and Jones, "Medical Administration," 67, 81 [see citation in note 28 above]).

34. Rogers and Jones, "Medical Administration," 69–70 (see note 28 above).

35. Smith, *Patenting the Sun,* 132.

36. Patterson, *Dread Disease,* 53.

37. Shryock, *National Tuberculosis Association,* 187.

38. Patterson, *Dread Disease,* 72.

critiquing the fragmented grass-roots approach to improving public health. (The Rockefeller Foundation's president, George Vincent, proposed merging these organizations into one in an article entitled "Teamplay in Public Health."[39] Not surprisingly, the Rockefeller Foundation's own board, with its vast funding and popular hookworm campaign in the South, provided his preferred example of effective philanthropy.)[40] Apart from the Rockefeller Foundation, these health philanthropies aimed not to fund medical science but to disseminate it through education campaigns and clinics.

In the 1930s, however, voluntary health organizations began to develop a new relationship with laboratory researchers. Mouse geneticist Clarence Cook Little became head of the ASCC and shifted the focus of their educational efforts from the public to doctors themselves, promoting the value of medical research aimed at cancer to physicians.[41] The leadership of the ASCC was instrumental in using magazine articles to build popular support for the passage of the 1937 Congressional bill that established the National Cancer Institute, providing the first federal extramural grants to medical scientists. Many doctors remained skeptical of the government's new role as a patron of medical investigation; the editor of the *Journal of the American Medical Association* warned of "the danger of putting the government in the dominant position in relation to medical research."[42] In reality, the New Deal barely extended to medical research. Roosevelt did little for the Public Health Service's research efforts; its funding in 1938 was barely over 10 percent of the research allocations for the Agriculture Department.[43] However, if Roosevelt's activities as president did not particularly benefit medical researchers, his symbolic leadership of the NFIP contributed to the public funding of disease research, albeit through private channels.

39. Shryock, *National Tuberculosis Association,* 184, citing Vincent, "Teamplay." According to Jane Smith, "In the first quarter of the century more than seventy-five other groups were established to raise money for specific diseases. The National Tuberculosis Association had been founded in 1904, the American Cancer Society in 1913, the National Society for Crippled Children (now the Easter Seal Society) in 1919, and the American Heart Association in 1924. The Red Cross and the Salvation Army had become quasi-national causes during World War I, when they held rallies side-by-side with the sellers of government bonds and pledged themselves to care for the American soldiers overseas" (Smith, *Patenting the Sun,* 69). John Paul credits the National Tuberculosis Association as the first disease-oriented agency to support laboratory research as part of its mission (*History of Poliomyelitis,* 301).

40. See Kohler, *Partners in Science,* 46–48.

41. Shaughnessy, *Story of the American Cancer Society,* esp. ch. 8; Patterson, *Dread Disease.* The ASCC also created a Women's Field Army in their "War against Cancer" in 1936, which distributed pamphlets and enlisted new supporters, but not until 1946 did the ASCC directly enroll researchers in its crusade against the disease.

42. "American Foundation Proposals."

43. Patterson, *Dread Disease,* 120.

Conquering Polio through Research

From the outset, the NFIP looked beyond short-term answers to polio. At the first meeting of the NFIP's Committee on Scientific Research, held on 6 July 1938, Rivers led a discussion on the "nature of the principal unsolved or not completely solved problems which should be attacked in a long-range research program." The group settled on eleven research questions, one of which reflected the widespread regard for Stanley's work on TMV: "Nature of the virus. Bearing in mind the crystallization of mosaic viruses, and the new physical methods of concentrating and separating viruses, effort should be concentrated on developing this inquiry on multiple fronts."[44] TMV stood as the implicit exemplar for this line of research, and the evident means of concentrating viruses was the ultracentrifuge. It bears mentioning that Rivers was not an unequivocal fan of Stanley; in fact, following one of his lectures in which he criticized Stanley, the latter complained to Rivers that he felt "you were having a round with me out back in the woodshed."[45] But even for Rivers, Stanley's physicochemical characterization of a plant virus showed by example how poliovirus might be investigated. In this respect, from the beginning of the NFIP's activities, TMV served as a key model system or laboratory referent for the development of experimental research on polio.

On 9 November 1938, the committee met again and discussed how to advance its research objectives. Rivers recommended that they fund a postdoctoral fellowship for a well-trained worker to spend three years working in The Svedberg's laboratory on the chemical and physical nature of the virus (or viruses) causing infantile paralysis.[46] However, this plan failed to materialize (the virus worker Rivers had in mind was unavailable to take the fellowship), and Rivers suggested instead that the NFIP itself establish a central laboratory for poliomyelitis research like the Rockefeller Foundation International Health Division's Virus

44. Minutes, Meeting of the Committee on Scientific Research and the Committee on Public Health (Epidemics) and Committee on Education, 6 Jul 1938, 2. Bound Minutes, NFA. The eleven lines of inquiry were prioritized the next spring: "(1) Pathology of poliomyelitis in human beings; (2) Portal of entry and exit of virus; (3) *Purification and concentration of the virus* [emphasis added]; (4) What is to be called poliomyelitis? (5) Mode of transmission of virus from man to man? (6) Transmission of virus along nerves? (questions four, five, and six received identical average ratings); (7) Further attempts to establish poliomyelitis in small laboratory animals; (8) Settlement of question of chemical blockade; (9) Chemotherapy of poliomyelitis; (10) Relation of constitution to susceptibility; (11) Production of good vaccine" (Minutes, Meeting of the Committee on Scientific Research, 18 Apr 1939, Bound Committee Minutes, NFA, 3).

45. Stanley to Rivers, 15 May 1939, Stanley papers, carton 11, folder Rivers, Thomas M.

46. Minutes, Meeting of the Committee on Scientific Research, 9 Nov 1938, Bound Committee Minutes, NFA, 5. Rivers recommended Albert Sabin of the Rockefeller Institute as a "suitable virus worker."

Laboratory (where Bauer and Pickels were working).[47] This laboratory was to use physicochemical techniques, such as analytical ultracentrifugation and electrophoresis, to study poliomyelitis virus, but several factors precluded its realization.[48]

The committee did begin to fund biochemical research on polio through extramural grants. For example, Stanley's former student Hubert Loring, who was in the Department of Bacteriology and Experimental Pathology at Stanford, received funds to attempt to isolate poliomyelitis virus in the ultracentrifuge.[49] In 1943, the NFIP's medical director reported optimistically on chemical approaches to poliomyelitis, "since techniques on recovering the virus by means of the ultracentrifuge have been improved. Also, it would seem that the virus has been purified to a greater extent than ever before, and this should be of immense value in future experimental research work. Several workers are now using the electron microscope in conjunction with their projects, and the results so far obtained are very encouraging."[50] By this time, however, the NFIP had to compete with the government for scientific talent, as virus research was also a high priority in the war effort.

During the demobilization period, the NFIP was poised to attract many medical researchers who had been working on war projects to initiate investigations pertinent to polio. An upsurge in NFIP research grants included an increase in support for biophysical virus research. In 1946, the NFIP granted the University of Southern California and Yale University School of Medicine sufficient funds to acquire an ultracentrifuge and an electron microscope respectively. [51] The Committee on Virus Research's overall appropriations grew from $304,000 in 1945 to more than

47. Minutes, Meeting of the Committee on Scientific Research, 18 Apr 1938, Bound Committee Minutes, NFA, 9.

48. The lack of a suitable investigator to head such a laboratory, particularly during the war, made Rivers's plan difficult to realize, but in addition, Basil O'Connor, president of the NFIP, "always refused to implicate the Foundation directly in a research enterprise." See John Storck, chapter 6, "NFIP Organization: The Board of Trustees and Its Direct Instruments, 1938–53," in vol. 1 of *History of the National Foundation for Infantile Paralysis Through 1953,* 206 (Unpublished Historical Monographs, NFA). For Rivers's proposal, see Minutes, Meeting of the Committee on Scientific Research, 9 Nov 1938 and 18 Apr 1938, Bound Committee Minutes, NFA.

49. See Annual Report of the National Foundation for Infantile Paralysis, 1939 and 1940, NFA; Minutes of the Joint Meeting of the Committees on Virus Research and Epidemics and Public Health, 12 May 1945, Bound Committee Minutes, NFA.

50. NFIP Report of the Medical Director, 20 Dec 1943, Bound Committee Minutes, NFA.

51. Both laboratories were extending epidemiological studies of poliomyelitis to include use of the physical and chemical tools (Minutes, Meeting of the Committee on Virus Research, 12 Mar 1946, Bound Committee Minutes, NFA). Max Lauffer, who had just left Stanley's laboratory to direct a laboratory of biophysics at the University of Pittsburgh, also received NFIP funding to isolate and study poliomyelitis.

$1.26 million in 1946.[52] With the increased resources, the NFIP expanded its research funding further into basic research, often centered on viruses but encompassing current biochemical topics such as vitamin analogues, nucleic acid synthesis, and carbohydrate metabolism.[53] The enlarged grants program reflected the remarkable success of the March of Dimes campaigns, which brought in $6.5 million in 1943, $12 million in 1944, and $19 million each in 1945 and 1946, record amounts despite the economic constraints of the war. As John Storck has commented, "In fund raising the results that were achieved from January, 1942, through January, 1945, were nothing less than spectacular."[54]

Fund-raising for cancer research was similarly successful. In 1944, a group of businessmen and advertising executives, led by philanthropist Mary Lasker, took control of the ASCC from conservative doctors, renamed the organization the American Cancer Society (ACS), and began fund-raising in earnest.[55] Their annual campaigns raised more than $4 million in 1945 and $10 million in 1946 for the fight against cancer. Convinced of the necessity for "fundamental scientific research directed toward a solution of the cancer problem," the ACS established a grants program administered by a new Committee on Growth of the National Research Council (NRC). The NRC Committee on Growth was given $3.5 million to disburse in 1947–48, through its newly formed Divisions of Chemistry, Physics, Biology, and Clinical Investigations.[56] Six panels were organized within the Division of Biology. These included both a Panel on Viruses, to fund research following up new evidence of virus-induced tumors in animal models, and a Panel on Milk Factor, to further investigation of the viruslike agent responsible for mammary tumors in mice.[57] Thus the ACS as well as the NFIP targeted virus research for support.

Philanthropic support of scientific research was not novel; rather it was the dominant pattern in the interwar period.[58] However, the targeting of basic biomedical research as part of a layperson-led fight against a specific

52. See Schedule 1 of the Annual Reports of the NFIP, 1945 and 1946, NFA.

53. Joseph H. Mori, chapter 5, "Medical Research Program, 1946–1949," in vol. 3 of *History of the National Foundation for Infantile Paralysis Through 1953*, 653 (Unpublished Historical Monographs, NFA).

54. John Storck, chapter 5, "Consolidating the Early Achievements, 1942–44," in vol. 1 of *History of the National Foundation for Infantile Paralysis Through 1953*, 170 (Unpublished Historical Monographs, NFA).

55. Patterson, *Dread Disease*, 171–73; Shaughnessy, *Story of the American Cancer Society*, ch. 7.

56. American Cancer Society, "Foreword."

57. See relevant correspondence in Stanley papers, carton 17, folder National Research Council Committee on Growth. On the milk agent, see Gaudillière, "Circulating Mice and Viruses."

58. See Kohler, *Partners in Science;* Kevles, "Foundations, Universities, and Trends"; and, for a broader view, Karl and Katz, "American Private Philanthropic Foundation."

disease *was* a marked departure from the activities of large philanthropies such as the Rockefeller Foundation.[59] As Basil O'Connor scribbled in the margins of a 1944 report recommending a long study: "*No—No*—lets have a *new* philosophy of *doing* things in medicine. Let c [*sic*] how *quickly* we can do it (intelligently) and not how long we can *study* it. Remember we get money to spend—and from the people—(not like Rockefeller)—to spend intelligently—of course."[60]

The near-record number of polio cases in the summer of 1944 (the worst since 1916) underlined the need for goal-oriented action.[61] The mission of the health agencies relied on massive citizen participation and—even more strikingly—they self-consciously emulated the mobilization of science for war. The ACS report summarizing the Committee on Growth's activities in 1945–46 conveyed, in the agency's view, "a technique of co-ordinated research developed under the forced draft of war and now used in peacetime to answer a widespread demand for the control of cancer."[62] Yet the fact that the NFIP and ACS were not governmental agencies meant that critics of federal support of research did not view these voluntary, "directed" research programs as tainted by socialism.

As voluntary health organizations expanded funding for virus research, the institutional support for this field was changing in other ways. In 1947, John D. Rockefeller Jr. announced that in order to attain a more efficient organization, the Rockefeller Institute for Medical Research (RIMR) would shut down the Princeton laboratories and consolidate their research efforts in New York.[63] By this time, the Department of Animal and Plant Pathology housed many of the best-known virus workers in the country, a point driven home by the awarding of the Nobel prize in chemistry in 1946 to John Northrop and Wendell Stanley (who shared it

59. Benison claims that even among the voluntary health agencies, the NFIP was the "first to open the field of philanthropy to the so-called 'common man.' Before the organization of the Foundation, philanthropy was generally an attribute and activity of the wealthy" (Benison, "History of Polio Research," 323).

60. As excerpted in John Storck, chapter 7, "Administrative Problems and Changes, 1944–1953," in vol. 1 of *History of the National Foundation for Infantile Paralysis Through 1953*, 210 (Unpublished Historical Monographs, NFA).

61. Annual Report of the NFIP, 1945, NFA.

62. American Cancer Society, "Foreword."

63. A report by the director, Herbert Gasser, asserts that the essential features of an Institute for Medical Research include "a well-qualified staff for clinical investigation with opportunity for daily contact with patients" and "a group studying the fundamental problems of biology," grounded in chemistry and biophysics. "The foregoing argument is tantamount to saying that under the present enormous pressure for funds, the fields of animal and plant pathology should be discontinued in order to save medical research" ("Summary" of Director's Report [for the Board of Scientific Directors], 1947, RAC RU 302.4, box 1, folder 12, 17–19).

with James B. Sumner of Cornell). All three of the other full members at the Princeton laboratories, Carl TenBroeck (the director), Louis Kunkel, and Richard Shope, were prominent virologists; four of the five were members of the National Academy of Sciences and the American Philosophical Society. The administrative decision was met with incredulity and dismay in the international scientific community.[64]

At issue for the Rockefeller Institute's Board of Directors was whether current research on plant and animal diseases still required a rural situation. Stanley's program of physicochemical characterization of viruses provided a case in point that the farm setting was dispensable for first-rate research. As noted in one of Gasser's memorandums, "The nature of his interest is such that his experiments could be carried on even better in New York, where we hope to develop all the techniques which are called into play in biological research."[65] Stanley was the likely successor to Kunkel as head of the Department of Plant Pathology, but the future of the entire group at the RIMR was uncertain. As Gasser confided in his report for the board, "This fall [Stanley] came to me to report an approach to him by the President of a large western university and intimated that the conditions were so favorable that his inclination was toward accepting an offer."[66]

California's Virus Laboratory

At the University of California, Berkeley, the administration convened a faculty committee in 1946 to recommend a successor to the long-time head of the Division of Biochemistry, an appointment viewed as an opportunity for a high-profile recruitment.[67] The university president set his sights on redressing larger inadequacies in the life sciences at Berkeley, as noted by Warren Weaver: "R. G. Sproul thinks that certain aspects of science, most particularly those which center around Ernest and John Lawrence, have developed exceedingly well; but that this great University is as yet weak in biological sciences. He has chosen biochemistry, rather than any older classical area, as the place where he wishes to make a great advance."[68] The committee's choice to head biochemistry was Wendell Stanley, and they justified their decision on the basis of his research on viruses: "Dr. Stanley is by all accounts one of the three or four outstanding men in his field. He has made addresses on this campus which reveal remarkable

64. Butler, "Rockefeller Institute."

65. Undated memorandum, RAC RU 302.4, box 1, folder 12, 3.

66. Gasser, Director's Report, 12 (see citation in note 63 above).

67. Letter from Robert G. Sproul to C. B. Hutchison, 10 Jan 1946, Records of the President, series 4, box 33, folder Special Problem: Wendell Stanley, 1946–54. For more on this committee and the background of biochemistry at Berkeley, see Creager, "Stanley's Dream."

68. Excerpt from Warren Weaver diary, 14 Nov 1947, RAC RF 1.2, 205D, box 7, folder 49.

insight and originality. His work on viruses marks him as one of America's most brilliant biochemists. In the opinion of this committee, the honorary degree we gave him at the last commencement was wholly merited."[69]

Archival materials do not shed light on whether it was simply fortuitous that Stanley received an honorary degree at Berkeley while a committee was meeting on that campus to recommend a new chair for biochemistry.[70] If so, the coincidence was doubly beneficial for Stanley: a seven-hour grounding of his plane in Cheyenne en route to San Francisco, late on the night of 19 June 1946, allowed ample opportunity to talk with fellow passenger and president of the University of California, Robert Gordon Sproul.[71]

Stanley's discussions with Sproul about chairing a department of biochemistry continued over the next year, during which both Stanley's negotiating power and his incentive to move increased substantially, for reasons just mentioned: that fall he was awarded the Nobel prize, and the following summer Rockefeller announced the closing of the RIMR's Princeton laboratories.[72] Stanley and Sproul made plans to establish a new Department of Biochemistry in Berkeley's College of Letters and Science.[73] Whereas virtually all of the other prominent departments of biochemistry in the country were affiliated with either a medical or

69. Letter from committee to recommend a new chairman for the Biochemistry Division to Robert G. Sproul, 30 Jul 1946, Records of the President, series 4, box 33, folder Special Problem: Wendell Stanley, 1946–54. The ranking of Stanley above the other biochemists considered (A. Baird Hastings, Vincent du Vigneaud, and Carl Cori) provides one indication of the growing status of research on macromolecules, especially viruses, within biochemistry. See Kohler, *From Medical Chemistry to Biochemistry,* 324–35, "Epilogue."

70. Stanley had received an honorary Doctor of Laws at the commencement on the Berkeley campus of the University of California, 22 Jun 1946 (Sproul to Stanley, 18 Mar 1946, Stanley papers, carton 12, folder Sproul, Robert Gordon).

71. Stanley to Maussner, 10 Jun 1946, Stanley papers, carton 2, folder Jun 1946; Stanley to Sproul, 25 Jun 1946, Stanley papers, carton 12, folder Sproul, Robert Gordon. Stanley must have told the story of this unplanned visit with Sproul to everyone who asked why he came to Berkeley; this encounter achieved a kind of legendary status, being included in Stanley's obituary in the *New York Times* in 1971. See Fraenkel-Conrat, "Impact of Wendell M. Stanley," 254; and Williams, *Virus Hunters,* 458.

72. As of May 1947, the negotiations with Berkeley were still unfinished, as Stanley and Sproul were waiting on the state legislature to pass funding for the new laboratory building (Stanley to Gasser, 22 May 1947, carton 2, folder May 1947). The closing of the Princeton laboratories was made public in June (Stanley to Adams, 28 Jun 1947, Stanley papers, carton 2, folder Jun 1947). Stanley accepted Sproul's job offer on 24 Dec 1947.

73. Stanley to Sproul, 2 Oct 1946, Stanley papers, carton 12, folder Sproul, Robert Gordon. Biochemists from the preexisting "Division of Biochemistry" were moved with other preclinical departments of the medical school to San Francisco. Stanley was designated chair of this group of biochemists as well as of the Berkeley department, although he was never more than a titular head (see letter from Zev Hassid to H. A. Barker, 17 Nov 1952, Barker papers, box 6, folder 34).

FIGURE 5.1 Photograph of Biochemistry and Virus Laboratory. Undated, circa 1955. One of the scientists in Stanley's department (Schachman, "Still Looking for the Ivory Tower") refers to this building as "the House That TMV Built." (Reprinted courtesy of the Bancroft Library, University of California, Berkeley.)

agricultural school, Berkeley's was distinctive for being "free-standing," a trend-setting institutional arrangement for this discipline.[74]

Stanley also established the Virus Laboratory, devoted to research on viral pathogens of concern to biologists, clinicians, public health officials, and the agricultural industry in California. Both this research group and the new department were to be housed in a state-of-the-art building (completed in 1952; see fig. 5.1) with 45,000 square feet of laboratory space and specialized facilities for "X-ray analysis, stable and radioactive isotope analysis, [and] microbiological assays."[75] Stanley used a special

74. As Stanley put it a few years later, "The combination of a department of biochemistry, dedicated not to agriculture nor to medicine but to biochemistry for the sake of biochemistry as a separate discipline, with the research organization of the Virus Laboratory is, of course, in itself somewhat on the unique side" (Stanley to Chauncey Leake, 23 Nov 1955, Stanley papers, carton 18, folder U.C.S.F.). Berkeley's prior Division of Biochemistry was in the School of Medicine. See Creager, "Stanley's Dream."

75. "Information Concerning the Biochemistry and Virus Laboratory," enclosure with the letter from Stanley to Weaver, 1 Feb 1952, RAC RF 1.2, 205D, box 7, folder 50. The perceived need for the two groups to be housed together was based on the equipment both

Rockefeller Foundation grant to furnish it with the latest commercial equipment for visualizing biological materials as macromolecules. As Warren Weaver commented after touring the facilities: "The building is splendidly designed in the interior, and their instrumentation seems excellent. They have, for example, two of the standard large models of the RCA electron microscope, two analytical ultracentrifuges, and they have just obtained one of the two existing first models of the electrophoresis instruments designed and built by Pickels."[76] Stanley's program represented the vanguard of physicochemical technologies in biology, many of which the Rockefeller Foundation could take pride in having helped develop.[77]

Stanley's move to a preeminent public university in 1948 mirrored a broader transition in the United States to the research university as the normative site for biomedical research. Even the Rockefeller Institute, the representative American institution for medical research during the first half of the twentieth century, elected in 1953 to change its name to the Rockefeller University.[78] Roger Geiger argues that the rise of the "federal research economy" transformed the mission and size of universities in the postwar period. By the 1950s, university budgets were swelling due to government-sponsored research grants, which typically supported science that was at once basic and relevant to public needs.[79] Stanley's vision for biochemistry as an autonomous life science suited this postwar academic regime; it was relevant to public concerns, particularly the control of agricultural and human diseases, while exemplifying the postwar ethos of basic university research.[80]

Stanley and Sproul looked to the public to support the virus research enterprise, but not by turning to the federal government. They used widespread concerns about viral diseases in California to appeal directly for state support of the new research laboratory. This source became critical once it was evident that the Rockefeller Foundation would no longer favor capital grants to universities.[81] Their strategy to recruit state

---

would use: "This would permit both groups to make efficient use of specialized equipment and facilities such as the electron microscope, electrophoresis apparatus, infra red apparatus, microchemical room, cold and hot rooms, rooms for radio-active isotope work and animal quarters" (Stanley to Sproul, 4 Mar 1947, Stanley papers, carton 12, folder Sproul, Robert Gordon).

76. Warren Weaver diary, 10 Nov 1952, RAC RF 1.2, 205D, box 7, folder 50.

77. Kohler, *Partners in Science,* 358–91 ("Instruments of Science").

78. Corner, *History of the Rockefeller Institute,* 43.

79. Geiger, *Research and Relevant Knowledge.*

80. Stanley refers to his original intention to unify biochemistry from its dispersed campus functions into an institutionally autonomous discipline in his letter to Dean A. R. Davis, 22 Jul 1952, Stanley papers, carton 3, folder Jul 1952. By this time, he had encountered resistance from some of the biochemists he sought to unify.

81. The move away from capital grants to universities was a general policy shift for the Rockefeller Foundation. At the same time, their officers were put off by Stanley's

support was twofold. First and foremost, Stanley argued that controlling the deadly capabilities of viruses would require greater knowledge about them. The recent development of sulfa drugs and penicillin, spurred by the war, represented a victory over most infectious bacteria, yet viruses remained a threat to both the health of citizens and the strength of the agricultural industry of California.[82] The Virus Lab was to "find the cause of such diseases as measles, mumps, chicken pox, influenza, and polio," as well as offering "aid to agriculture by studying viruses which attack citrus trees, field crops, and farm animals."[83] Stanley even held out the hope of developing a vaccine for human cancers, which he maintained were caused by viruses.[84] The second rationale for the Virus Laboratory

---

presumption of support even as they approved the soundness of his plan for a virus lab: "G. [Gasser] is certainly not favorable to any building gift to California for W. M. Stanley, and W W [Warren Weaver] infers that he would not favor any large gift there for that purpose. G. emphatically dislikes the fact that California takes on Stanley with the assumption that S. has all sorts of important connections which will assure large support for him, and he equally dislikes the fact that Stanley probably himself approves of this procedure" (Warren Weaver, diary note on conversation with Dr. H. S. Gasser, 9 Jan 1948, RAC RF 1.2, 205D, box 7, folder 49). After refusing Stanley a capital grant, the Rockefeller Foundation did in fact support his research quite substantially. Despite the fact that Weaver was always annoyed by Stanley's salesmanship, he acknowledged that Stanley was "planning a scientific development at Berkeley which will be absolutely first-rate in its concept, in its personnel, and in its setting" (Interview notes, Warren Weaver diary, 15 Jun 1948, RAC RF 1.2, 205D, box 7, folder 49). Another Foundation officer echoed these sentiments: "On objective grounds it is difficult to deny that $1,000,000 devoted to a virus laboratory may produce results of as great intrinsic importance as those to be derived from the cyclotron and telescope—and of far greater importance to human welfare" (Memo on W. M. Stanley proposal by Robert S. Morrison, 30 Jan 1948, RAC RF 1.2, 205D, box 7, folder 49).

82. In the copy of the publicity release for Stanley's acceptance of the university position, the following sentence is added by hand to the description of the relevance of virus research to disease control: "The fact that penicillin and the sulfa drugs are not effective in the treatment of most virus diseases emphasizes the importance of the new approach" (University of California Office of Public Information, Records of the President, series 4, box 33, folder Special Problem: Wendell Stanley, 1946–54). On developments for the production of penicillin, see Neushul, "Science, Government, and the Mass Production."

83. Henry Palm, article from 21 Jul 1948, unidentified newspaper, from the Records of the President, series 2, box 728, folder 1948: 420-Biochem.

84. "We plan an extensive program of research on viruses affecting people, animals, plants, and bacteria. We are primarily interested in the elucidation of the mode of reproduction and mutation of viruses, but do expect to work with viruses of social and economic importance. The State of California loses about 100 million dollars annually from virus diseases of animals and plants and I am confident we will be able to do something about this. We plan work on several plant and animal viruses. Work is now under way on viruses causing tumors or cancers in animals. I expect to push this aspect because I am convinced that the viruses provide the best possible experimental approach to the cancer problem" (Letter from Stanley to Dr. Benjamin W. Carey of Lederle Laboratories, 4 Oct 1950, Records of the President, series 2, box 781, folder 1950: 420-Biochem, 2–3). Stanley became one of the key organizers of cancer research funding, serving as a director-at-large of the American Cancer Society and on the advisory panel for the National Cancer Institute.

was to supply a needed site for education and training: "Then, too, there is serious need for a center where personnel can be trained for work on viruses. I receive dozens of letters asking for individuals with training in virus research. There is no place in the world where persons can receive training in the broad aspects of virus research."[85]

In 1948, the Legislature of the State of California found Sproul and Stanley's plan compelling enough to appropriate $1,215,000 for the construction of a new laboratory building, providing an additional $500,000 for research support. Virus research merited public support because it pertained to human, animal, and plant diseases, yet control of the program remained in the hands of the university biochemists and virologists, rather than being administered through an agricultural or medical institution.[86] Through appropriations to initiatives such as Stanley's "Biochemistry and Virus Laboratory," the state of California also sought to enhance the prestige and research utility of the university. Although this trend was already evident in the support offered E. O. Lawrence and other physical scientists in the 1930s, the state's unanticipated and sustained economic growth following the war enabled tremendous increases in state appropriations for education and research, including large outlays for campus building.[87] Between 1945 and 1950, the state of California committed more than $120 million to capital construction for its statewide university system, of which $25 million was spent at Berkeley.[88] The enlarged commitment to university research was reflected in a proliferation of "organized research units," such as Stanley's Virus Laboratory, on the Berkeley campus.[89] The state provided about half of the operating costs of Stanley's laboratory in 1951; the rest of his budget was covered through a combination of grants from private philanthropies, the Public Health Service, and pharmaceutical companies.[90]

85. Wendell Stanley to Warren Weaver, 8 Jan 1948, RAC RF 1.2, 205D, box 7, folder 49.
86. As I have argued elsewhere ("Stanley's Dream"), Stanley's autonomous department marked a transition for biochemistry away from its establishment in the United States as a service discipline to medicine and agriculture in the first part of the twentieth century (Kohler, *From Medical Chemistry to Biochemistry*).
87. As Roger Geiger points out, Sproul wrote the heads of each department in 1943 asking them to plan for drastic budget cuts in the event of a postwar economic crisis in California. Instead, the postwar years became a period of sustained economic growth, enabling generous state support of the university (*Research and Relevant Knowledge,* 73–82, esp. 74).
88. California had "set aside sizable reserves of wartime tax revenues for capital construction in the postwar period" (Geiger, *Research and Relevant Knowledge,* 74).
89. As Geiger puts it, "Of crucial importance in the postwar era was the willingness of the university [of California] to underwrite organized research units. Such commitments helped to make Berkeley into the country's most fertile breeding ground of organized research units, and the initiator of the acronym, ORU" (ibid., 75).
90. See Stanley to Lester D. Summerfield, 2 Nov 1951, Stanley papers, carton 15, folder Max C. Fleischman Foundation. Stanley's private financial supporters included the NFIP,

Stanley's promises to find vaccines to protect the people of California, their animals, and their foods, drew on the same powerful public expectations that the voluntary health agencies tapped in the aftermath of World War II. Just as the physicists had been perceived as an essential national resource during World War II, Stanley posed as the preeminent life scientist, armed with physicochemical expertise to defend Americans, particularly Californians, from loss of life and crops to viruses. As the Korean War brewed, Stanley made a special plea that his laboratory should receive priority funding due to its importance in case of war emergency: "We have the most powerful virus group in the World and we should have proper [virus] facilities for work in case of a serious national emergency. I cannot help but feel that considerations such as those mentioned above, which of course do not bear on the usual University building, provide ample justification for special treatment for the Virus Laboratory building."[91] Stanley promoted his Virus Laboratory in the scientific and technological landscape of the developing Cold War economy, in which California would be central.[92] The commitment to combat the insidious invasion of viruses and their "devastating effects" (domestic and foreign) was a mission perfectly suited to the task of mobilizing public support for science, and it translated the mobilization of research from war to postwar biomedicine in precisely the same terms as the NFIP and ACS.

Stanley and the Poliovirus

Stanley's virus research crusade drew on the publicity of the war against polio associated with the March of Dimes, but he also benefited more concretely from the NFIP. The summer he moved to Berkeley, Stanley applied for a $502,200 five-year grant from the polio organization. The press eagerly picked up on Stanley's plans to study this dread disease. According to the *Oakland Post-Enquirer*'s story on Stanley's arrival at Berkeley, "The prime problem of the 43-year-old scientist will be an attempt to find a vaccine against polio."[93] With NFIP funding, members of

the ACS, the Rockefeller Foundation, and the pharmaceutical companies Lederle and Cutter.

91. Letter from Stanley to Sproul, 23 Aug 1950, Records of the President, series 2, box 781, folder 1950: 420-Biochem. Stanley is ambiguous (probably intentionally) about whether a "national emergency" requiring expert virologists involved the threat of combating bacterial warfare or the more conventional needs by the armed forces for vaccines.

92. See Rolle, *California: A History,* esp. the chapter entitled "California after World War II."

93. "Famed Scientist Arrives—Plans Polio Research in New U. C. Laboratory," *Oakland Post-Enquirer,* 20 Jul 1948, clipping in Records of the President, series 2, box 728, folder 1948: 420-Biochem. This aim was equally compelling to pharmaceutical companies. Lederle Laboratories of American Cyanamid began supporting Stanley's Virus Laboratory with $35,000 a year for five years, and Cutter Laboratories gave the Virus Laboratory $5,000

his new laboratory began to investigate poliovirus, modeled on the ongoing biochemical and biophysical work on TMV. The Virus Laboratory's
project on polio gave the rhetoric of fighting disease a material referent,
and reaffirmed the value of TMV as a model system for virus research.

In his application, Stanley asserted that the central scientific problem
to be taken up in his new Virus Laboratory was to understand the biological mechanism for virus reproduction. He argued that such knowledge
would lead to new treatments for viral diseases, particularly through
chemotherapy.[94] Chemical studies on mutant strains of TMV, revealing
that the virulence of different strains could be correlated with specific
amino acid changes in the viral protein, had already "yielded unexpectedly rich and meaningful information regarding the nature of virus
mutation." Not only were these studies pertinent to therapeutic strategies
for arresting virus reproduction, but, Stanley contended, they illustrated
"the process by means of which viruses change or adapt themselves to
new conditions," evading measures to control viral epidemics.[95]

On the basis of his group's strong record in virus research, including
its wartime vaccine development, Stanley outlined three projects for the
NFIP to fund in his laboratory, all of which involved treating TMV as
a model for investigating polio. As Stanley put it in his grant application, "If a plant virus can be used to special advantage as a model to
develop or to check a technique for later applications on poliomyelitis
virus, such a course will be followed." The first project was to purify the
three types of poliovirus in order to characterize them chemically, physically, and biologically (i.e., to discipline polio as a biochemical virus). The
second project involved "studies on the nature of the mutation process
in viruses" as assessed through amino acid composition of different virus
strains. The aim was to compare amino acid variance in polio with similar
changes in strains of TMV and bacteriophage. As Stanley argued to the
NFIP, "There is a real possibility that structural changes in the virus can
be related to changes in virulence, and this information might be useful in connection with the control of extremely virulent viruses."[96] The
third project was to apply the electron microscope to depict (literally)
the virus-host relationship. This project became the mainstay of Robley

for 1953–54 (Stanley to Benjamin W. Carey, 18 Dec 1950, Stanley papers, carton 3, folder
Dec 1950; F. F. Johnson to Stanley, 22 Oct 1953, Stanley papers, carton 14, folder Cutter
Laboratories).

94. Copy of NFIP grant sent to President Robert Sproul, 17 Jun 1948, Records of the
President, series 2, box 728, folder 1948: 420-Biochem.

95. Stanley to H. M. Weaver, 17 Jun 1948, Stanley papers, carton 13, folder We misc.

96. Application for Grant to NFIP, 1 Jul 1948, enclosed with letter from Stanley to
H. M. Weaver, 17 Jun 1948, Records of the President, series 2, box 728, folder 1948:
420-Biochem.

Williams's program of electron microscopy once he arrived at the Virus Laboratory in 1950 as a professor of biophysics. (Chapter 6 recounts the ongoing work related to the second project, concerned with the chemical differences between virus strains; chapter 7 includes more information on Williams's electron microscopy research.) The NFIP was the Virus Laboratory's most important source of funding for research on TMV.[97] My focus here is on the first of the three projects the NFIP supported in Stanley's laboratory—and the only one not conducted on TMV (even though it served as the point of reference): the biochemical purification of poliovirus.

After some negotiation with the NFIP, Stanley was awarded a five-year grant.[98] Early progress on the purification was hindered, Stanley explained, by the lack of suitable research facilities.[99] Initial efforts focused on using the ultracentrifuge to fractionate the central nervous system tissue of cotton rats infected with the Lansing polio strain, in an effort to determine which cellular fractions contained the virus.[100] Carlton Schwerdt arrived at the Virus Laboratory in 1950 as an associate biochemist to accelerate work on the purification project.[101] Differential

97. See table of funding in letter from Stanley to Benjamin W. Carey, 4 Oct 1950, Stanley papers, carton 3, folder Oct 1950.

98. The $502,000 grant was trimmed to $308,240 (1948 Annual Report, National Foundation Archives), but additional yearly expediting grants to Stanley brought the total amount of support from 1948 to 1953 to $556,049. On NFIP support to the Virus Laboratory for that five-year period, see the press release for a one-year grant of $144,473, attached to a letter from Dorothy Ducas, NFIP Public Relations Director, to Daniel Wilkes, Director of Public Information at Berkeley, 24 Jun 1954, Stanley papers, carton 16, folder NFIP 1955. On the building, see letter to H. M. Weaver, 4 Jan 1952, with "Progress Report for NFIP grant for period July 1–December 31, 1951," Stanley papers, carton 3, folder Jan 1952.

99. Special facilities for the study of poliomyelitis were planned for the new Biochemistry–Virus Laboratory Building, but construction delays interfered with the occupation of that building until the fall of 1952. In the meantime, beginning in the fall of 1950, temporary facilities were set up in the Forestry Building (see letter to H. M. Weaver, 4 Jan 1952, with "Progress Report for NFIP grant for period July 1–December 31, 1951," Stanley papers, carton 3, folder Jan 1952).

100. "Quantities of Lansing virus infected cotton rat CNS tissues are being harvested and studies of the effects of different solvents and different hydrogen ion concentrations on the extraction of virus are in progress. The material in the extract is being fractionated into nuclei, mitochondria, submicroscopic particles, and supernatant liquid by centrifugation. The virus activity and biochemical and biophysical properties of these fractions are being determined" ("Semi-Annual Report of Progress to the National Foundation for Infantile Paralysis, Inc., for the Period Ending December 31, 1950 from the Virus Laboratory of the University of California, Berkeley, California, January 15, 1951," enclosed with letter to H. M. Weaver, 15 Jan 1951, Stanley papers, carton 3, folder Jan 1951). Stanley reported on subsequent studies of the intracellular distribution of Lansing virus in "Semi-Annual Report of Progress . . . for the Period Ending June 30, 1951, July 5, 1951," carton 16, folder NFIP Progress Reports.

101. See "The Virus Laboratory," prepared for the Graduate Division's announcement in Biological Sciences, Stanley papers, carton 21, folder UC Loyalty Oath controversy, 1950–51.

centrifugation, which worked so successfully to isolate many plant and animal viruses, proved to be inadequate as a means to recover poliovirus.[102] Instead Schwerdt began to use chemical agents such as salt and butanol to enrich the infective fraction from crude cell homogenate.[103] The use of butanol extraction in conjunction with ultracentrifugation led to a 10,000 to 100,000-fold purification "without an appreciable loss of total infectivity."[104] Even more significantly, when Schwerdt compared electron micrographs of the centrifugal fractions of Lansing-infected cotton rat CNS tissue to measurements of the infectivity of each fraction, he found that virus activity was consistently correlated with the presence of a spherical particle, absent in uninfected tissue.

Late in 1953, Stanley presented these results at a meeting of the National Academy of Sciences. The university press release interpreted the significance of the finding in terms of the fight against polio:

> The human polio virus has been isolated and identified directly for the first time, in research at the University of California supported by the National Foundation for Infantile Paralysis.
> Dr. Wendell Stanley, ... Dr. Howard L. Bachrach, ... and Dr. Carlton E. Schwerdt ... have obtained the first photographs definitely identifying the human polio virus, and for the first time know its size and shape with certainty. The virus is a spherical particle, about ... one millionth of an inch in diameter. ... The new knowledge is expected to speed up the fight against polio, primarily by providing more definite facts about the virus.[105]

Their electron micrographs of the spherical polio particles, 280 Å in diameter, were widely publicized (see fig. 5.2), and similar pictures were released that month by researchers at Parke Davis and Co., a pharmaceutical house.[106] Stanley was at pains to point out that only his group

102. See Stanley to Henry W. Kumm, 1 Oct 1953, Stanley papers, carton 16, folder NFIP 1952–53.

103. Stanley, "Progress Report—July 1 to December 31, 1951," enclosed with letter to H. M. Weaver, 4 Jan 1952, Stanley papers, carton 3, folder Jan 1952; "Progress Report—January 1 to June 30, 1952," enclosed with letter to H. M. Weaver, 2 Jul 1952, carton 16, folder NFIP Progress Reports; Schwerdt and Pardee, "Intracellular Distribution"; Bachrach and Schwerdt, "Purification Studies."

104. "Progress Report—July 1 to December 31, 1952," enclosed with letter from Stanley to H. M. Weaver, 6 Jan 1953, carton 16, folder NFIP Progress Reports.

105. Enclosure with letter to Dan Wilkes, 21 Oct 1953, Stanley papers, carton 3, folder Oct 1953.

106. *Time* magazine, 16 Nov 1953; *Chemical and Engineering News* 31, no. 46 (16 Nov 1953); see also Press Release from Science Service, 9 Nov 1953, enclosed with letter from Watson Davis to Stanley, 16 Nov 1953, Stanley papers, carton 19, folder Science Service. The researcher who obtained the micrograph at Parke, Davis and Company was A. R. Taylor; Stanley corresponded with him about their results on 29 Oct 1953 (Stanley papers, carton 3, folder Oct 1953). Taylor's results were presented at the

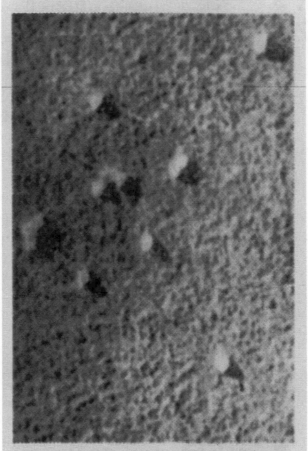

Associated Press Wirephoto

POLIO VIRUS: Pure human virus, each about a millionth of an inch in diameter, isolated and photographed for first time after being separated from test tube culture at the University of California.

FIGURE 5.2 Electron micrograph of spherical particles of Type I poliomyelitis virus. (From the *New York Times*, 13 Nov 1953, 16. Clipping reproduced courtesy of the Bancroft Library, University of California, Berkeley.)

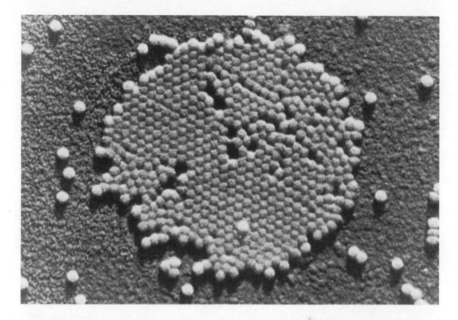

FIGURE 5.3 Electron micrograph of Type II (MEF-1) poliomyelitis virus
(×74,000). The uniformity of particle size (270 Å in diameter) is demonstrated
by the existence of regions of close packing. (Image reprinted courtesy of the
Bancroft Library, University of California, Berkeley. Caption adapted from
Schwerdt et al., "Morphology of Type II Poliomyelitis Virus.")

actually demonstrated that the infectivity was associated with the 280-Å-
diameter spherical particle, and not attributable to the array of smaller
particles also present in the micrographs.[107]

Stanley's group then turned to the other poliomyelitis strains. They
soon identified a similar 280-Å particle in cells infected with the Type II
polio, the MEF-1 strain, grown in tissue cultures of monkey kidney (see
fig. 5.3).[108] Stanley took the concordance of these results as further confirmation that his group had identified the authentic poliovirus particle.
At this time, the laboratory was shifting all of its polio work to tissue
culture (see fig. 5.4). Growing polio in cotton rat CNS tissue could not

annual meeting of the Electron Microscope Society of America a week earlier than the
National Academy of Science meeting, where the results from Stanley's laboratory were
presented.

107. Stanley to Robert F. Gould, 19 Nov 1953, Stanley papers, carton 19, folder *Chemical
and Engineering News*.

108. Only a "fraction of a milligram of the purified virus" was isolated, however (press
release attached to letter from Stanley to Daniel Wilkes, Public Information Office, UC
Berkeley, 21 Oct 1953, Stanley papers, carton 3, folder Oct 1953).

FIGURE 5.4 Stanley holding a bottle of tissue culture fluid in the Virus Laboratory. Undated, circa 1955. (Reprinted courtesy of the Bancroft Library, University of California, Berkeley.)

provide enough starting material for preparation of a sufficient amount of poliovirus for complete biochemical characterization. Stanley's laboratory began growing HeLa cells as well as monkey kidney cells in tissue culture with hopes (but little assurance) of producing a sufficient quantity of poliovirus for these studies.[109]

Crystals and Vaccines

Research on polio in Stanley's laboratory culminated in October of 1955 when Schwerdt and Frederick Schaffer obtained crystals of Type II (MEF-1) poliomyelitis virus, the first time an animal virus had been crystallized (see fig. 5.5).[110] As it turned out, the success of their crystallization relied, at a basic material level, on the contemporaneous development of a vaccine by Jonas Salk.[111] At this point, the NFIP was

109. Stanley to H. M. Weaver, 8 Jul 1953, Stanley papers, carton 16, folder NFIP Progress Reports.
110. Schaffer and Schwerdt, "Crystallization of Purified MEF-1 Poliomyelitis Virus."
111. Salk's vaccine method took advantage of another advance the NFIP had supported: John Enders's method for growing polio in cell culture. For more information on Salk

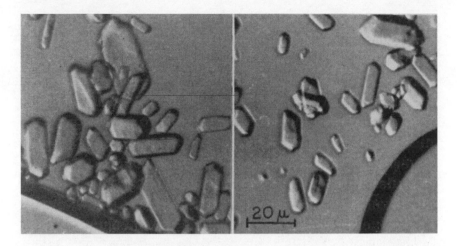

FIGURE 5.5 Crystals of Type II (MEF-1) poliomyelitis virus particles photomicrographed in visible light. (From Schaffer and Schwerdt, "Crystallization of Purified MEF-1 Poliomyelitis Virus Particles" [1955]: 1021. Reprinted courtesy of L. Van Rie Barnes Schwerdt.)

fully funding not only the development but also the testing of the Salk vaccine, which combined inactivated preparations of all three strains of poliomyelitis virus. The NFIP arranged for Connaught Laboratories in Toronto, which was producing the formaldehyde-treated polio vaccine for clinical trials, to ship to Stanley's laboratory their surplus virus fluids, approximately four hundred liters of each of the three strains. The NFIP also offered Stanley's researchers lots of rejected vaccine fluid (in which the poliovirus had already been chemically modified with formaldehyde).

The first shipment of polio Type II (MEF-1) infectious tissue culture fluid arrived in Berkeley late in the spring of 1954 (see fig. 5.6), and shipments of fluid containing Mahoney Type I poliovirus and Saukett Type III virus arrived the following year.[112] The NFIP made clear that

---

and his development of the inactivated virus vaccine against polio, see Smith, *Patenting the Sun.*

112. These amounts are based on Stanley's letter to Henry W. Kumm, 19 Apr 1955, Stanley papers, carton 16, folder NFIP 1955; the amounts received were closer to 350 liters of each strain. Stanley justified the request for the fluids to Hart E. Van Riper, Medical Director of the NFIP (25 Feb 1954): "As you probably know, we are extremely interested in the possibility of having any excess tissue culture fluid from the Toronto operation sent to us as starting material for biochemical and biophysical studies on the virus. We have developed a satisfactory purification procedure which yields virus of excellent purity, but are handicapped because of the lack of adequate amounts of tissue culture fluid. Therefore, when Toronto produces enough fluid for the present trials, we hope they will continue production for a time at least and, if necessary, on a reduced scale and let us have this material for our biochemical work" (Stanley papers, carton 16, folder NFIP 1955). Other

Department of
HEALTH, EDUCATION, AND WELFARE
PUBLIC HEALTH SERVICE
Washington 25, D. C.

In replying, address the
PUBLIC HEALTH SERVICE

Refer to: May 11, 1954

PERMIT TO IMPORT QUARANTINABLE MATERIAL

This permit should be sent to the shipper to be affixed to the outside of the container; if this is not possible, it should be presented to the Customs officer at the port of entry.

To Collector of Customs:

In accordance with section 71.156 of Foreign Quarantine regulations permission is hereby granted Dr. Wendell M. Stanley, University of California, Biochemistry and Virus Laboratory, Berkeley, California, to import into any port under the control of the United States, tissue culture fluids containing the virus of poliomyelitis from the Connaught Medical Research Laboratories, Toronto, Canada.

This permit covers all importations of the above mentioned character which may be made during the period beginning May 11, 1954, and ending June 30, 1955.

It is directed as a condition of this permit that a record of importation be maintained at the University of California, Biochemistry and Virus Laboratory. Such record shall show specifically the kind of import, type of infection, and the use to be made thereof. This record is to be available to Service officers responsible for the enforcement of section 71.156 of the Foreign Quarantine regulations.

/s/ Ralph Gregg
Ralph Gregg, Medical Director
Assistant Chief
Division of Foreign Quarantine

cc: Cal. State Dept. of Health
USQS: Boston; New York

A CERTIFIED COPY

FIGURE 5.6 Copy of permit for shipping quarantinable fluid from Toronto to Berkeley. (From Stanley papers, 78/18c, carton 16, folder NFIP 1955. Reprinted courtesy of the Bancroft Library, University of California, Berkeley.)

these fluids were an exceedingly precious commodity; as Henry Kumm reminded Stanley, "At the present cost of production in Toronto, that virus is worth more than a quarter of a million dollars."[113] Stanley viewed

correspondence related to the shipment of infectious tissue culture fluid from Connaught Laboratories to Berkeley is in the Stanley papers, carton 16, folders NFIP 1952–53 and NFIP 1955. Following the initial shipments of the Types I and III fluids, Stanley requested and obtained 150 extra liters of each of those types in June 1955.

113. Henry W. Kumm to Stanley, 27 May 1954, Stanley papers, carton 16, folder NFIP 1955. Stanley asserted from the outset that the results of the investigation would be well

the crystallization of MEF-1 poliovirus by Schwerdt and Schaffer in 1955 as clear vindication of the investment, and he proudly drew attention to the fact that he had first crystallized a plant virus exactly twenty years earlier.[114] Schwerdt and Schaffer went on to purify and crystallize viruses from both of the other polio strains, and the achievement, while primarily a technical feat, garnered substantial attention.[115] However, the public acclaim for this achievement was somewhat attenuated by the enormous public response to the Salk vaccine and the drama following its release.

Salk's formaldehyde-killed polio vaccine received unequivocal endorsement by epidemiologist Thomas Francis Jr. at the conclusion of his comprehensive assessment of the NFIP's tests on schoolchildren during the spring and fall of 1954. Francis's report was released to the press on 12 April 1955, unleashing an immediate demand for the vaccine.[116] Six pharmaceutical companies had prepared the vaccine through contract with the NFIP, so that it would be available if the Francis report was positive.

The Salk vaccine began to be distributed right away, and within two weeks, by 27 April, several cases of paralytic polio caused by the Cutter Laboratory vaccine had been reported. Parents throughout the country who had or were going to have their children inoculated were panicked about whether the vaccine would prevent or cause polio. The political fallout was substantial; the secretary of health, education, and welfare, the first woman in a Republican cabinet, Oveta Culp Hobby, resigned from her post by July. (Her department was heavily criticized for not making plans prior to the release of the Francis report for how to allocate the limited supplies of polio vaccine.) The U.S. Congress held hearings in the late spring and summer on whether the vaccine program should be continued, and Stanley testified as a member of the National Academy of Science's Panel on Polio Vaccine on 22 June 1955.[117]

Stanley joined virologists Albert Sabin and John Enders in expressing deep reservations about the vaccine. (Equally prominent supporters of the vaccine on the panel included Rivers, Francis, Frank Horsfall, and James Shannon, then assistant director of the NIH. By contrast, during the civil trial against Cutter Labs over the unsafe vaccine, Stanley testified

---

worth the high cost (Stanley to Kumm, 17 Mar 1954, Stanley papers, carton 16, folder NFIP 1955).

114. Stanley to Thomas M. Rivers, 15 Sep 1953, Rivers papers, APS BR52, folder Virus Laboratory.

115. See, for example, Peter Hayes, "Scientists Crystalize [sic] Pure Polio Virus," Washington Post and Times Herald, 4 Nov 1955; "Polio Virus Is Now Crystallized," New York Times, 6 Nov 1955; Williams, Virus Hunters, 106.

116. See Smith, Patenting the Sun; Benison, "History of Polio Research."

117. See Smith, Patenting the Sun, chapter entitled "Political Science."

on behalf of the pharmaceutical producer.)[118] Testifying to Congress, Stanley used his reasoning as a chemist and his experience with TMV to cast doubt on the complete inactivation of virus in Salk's "killed-virus" vaccine:

> I suspect you, as laymen, may be wondering about this shadow-line zone between safety and danger. Eventually [sic] what is taking place here is a chemical reaction, a combination between formaldehyde and poliomyelitis virus in a medium in which the virus makes up one part per thousand of the protein [the other 999 parts being cellular debris]. Formaldehyde reacts with all of the protein, and not only the virus. . . .
>
> This reaction is a chemical reaction. It has what we call an equilibrium constant. . . . The chemist will tell you that in such a reaction it is theoretically impossible to end up with a situation in which you have no active virus. This is a theoretical consideration. . . .
>
> But from a chemical standpoint, it is even worse than that. We found years ago, working with a plant virus, that it is possible to treat it with formaldehyde so that it is inactivated by all tests which can be applied. You can put it away and leave it on your shelf for months and test it from time to time, and it is inactive. By the simple chemical treatment known as dialysis, it is possible to get activity back again and to reactivate the material.[119]

He went on to suggest that the mass production of vaccine should wait on a better inactivating agent, such as ultraviolet radiation, which might inactivate the RNA rather than the protein of poliovirus. In fact, members of Stanley's lab were working to develop just such a method.[120] Stanley's invocation of TMV as a model for understanding polio inactivation indicated once again the importance of such laboratory models for polio research.

However, Stanley's strongest complaint regarding the Salk vaccine had to do with the peculiar haste with which the vaccine had been developed:

> This is the first time in history, so far as I know, when a scientific program has gone ahead pretty much on the basis of not completely unpublished work, but work which is not readily available to scientists generally. Scientists over the years have followed a procedure of experimentation, checking and double-checking within their own laboratory, and publication, so that the scientific world is then

118. See Stanley's trial testimony, 30 Dec 1957, Stanley papers, carton 21, folder Polio Vaccine Testimony.

119. Poliomyelitis Vaccine; Hearings before the U.S. Congress, House Committee on Interstate and Foreign Commerce, 84th session, part 2, 22 and 23 Jun 1955, 171.

120. Stanley's NFIP "Progress Report, July 1, 1953, to December 31, 1953," enclosed with letter to Henry Kumm, NFIP, 11 Jan 1954, Stanley papers, carton 4, folder Jan 1954.

able to judge the results, and [perform] checking, and rechecking in laboratories throughout the world, and then decisions [have] been made upon that.[121]

Stanley's frustration with the impatience exhibited by Salk, the NFIP, and the American public for a vaccine was shared by many virologists. But the same impatience which he bemoaned was also directed by the National Foundation's Committee on Virus Research toward their own long-term investment in basic investigations, such as the research in Stanley's lab. In the words of Rivers, who chaired the committee:

> When Stanley went out to California in 1948, he needed a great deal of money to build up his laboratories, and I backed him with the National Foundation. I honestly didn't know what he would deliver; the only thing I was certain of was that he would deliver something. By 1954 the National Foundation had spent well over a million dollars in support of his laboratory, but to that time nothing much had come out of it. A number of people on the Virus Research Committee began to ask me why I was so keen on Stanley, and did I think it was wise for the Foundation to continue giving him support? They ragged me good and proper, but I held the fort because it was obvious to me that these had been years of preparation for Stanley....
>
> In 1954 Stanley and his associates began a program of research designed to improve the purification process of all three types of poliovirus, and to see if the biochemical and biophysical properties of such pure preparations could be determined. Much of the surplus poliovirus fluids that the Connaught Laboratories had prepared for the production of Salk vaccine was sent by the Foundation to Stanley's laboratory so that he and his associates could carry on this new work. I don't remember now how many hundreds of liters of virus fluid the Foundation eventually shipped to Stanley, but I am willing to bet that at one point that Stanley had more poliovirus in his possession than any other laboratory in the world.... I don't mind admitting that some people on the Virus Research Committee were just about ready to ask Dr. Stanley, "What the hell are you doing with all the poliovirus we have been sending you?" Luckily, before they asked that impertinent question, Carlton Schwerdt presented irrefutable evidence that he and Frederick Schaffer had crystallized a purified Type II (MEF-1) poliovirus. Although several plant viruses had been crystallized before, this was the first time that an animal virus had been crystallized....
>
> Suffice it to say that, with the crystallization of poliovirus and the nucleic acid work of Dr. Fraenkel-Conrat, the Foundation got back every cent and more that it had invested in Dr. Stanley.[122]

121. Congressional Hearings, 178 (see citation in note 119 above).
122. Benison, *Tom Rivers*, 589–91. I discuss Fraenkel-Conrat's contributions, which Rivers mentions, in chapter 7.

So whether evaluating the trials of the Salk vaccine or the investments in basic research, the question (for the NFIP, at least) was one of production—rapid production of a laboratory discovery or of a wonder drug. For research underwritten by the American public (in this case through subscription, not through taxation) and whose support was sustained by intense media coverage, timely production of either a therapeutic agent or a newly discovered fact was essential.

That such expectations for results were becoming part of funding systems for basic research troubled many scientists. One editorial writer in *Science* magazine expressed his opposition to the growing requirements for project reports for "design" research as opposed to "free research."[123] Nonetheless, a mode of accountability of research, to the scientific manager or peer-review body of the granting agency, and more generally to the public, was being set in place by the 1950s. The degree of scientific freedom entailed by a "basic" research contract remained nebulous. One NSF official stated, "The concept 'basic research' may comprise the systematic endeavor, without preconception, to increase our knowledge and understanding of nature. It is the kind of research that some of our colleagues characterize as 'pure science.' If it is indeed pure, it derives that quality from uncompromising objectivity, unconcern over specific aims, and absence of intent to exploit results."[124] By contrast, the research director at Shell Oil Company differentiated pure from basic research: "For my requirements," he wrote,

> I suggest three categories. We have pure research, which I define as the inquiry after knowledge for its own sake, without consideration or hope of practical gain. We also have applied research, the investigation carried out in response to immediate, direct, and obvious needs. Basic research is in between.
>
> By basic research, then, I mean the scientific inquiry carried on, not under pressure of immediate needs or in hope of quick profit, but with reasonable hope of some eventual payout.

He recognized that these usages were context-dependent: in the case of Shell-sponsored research grants to university chemists and chemical engineers, he said, "These men are working on problems of their own selection; to us it is basic, to them it is pure."[125]

123. Richter, "Free Research versus Design Research." On the pre–World War II history of scientists' derogatory comments about "design" research and "project" science, see Dennis, "Accounting for Research"; and Hollinger, "Free Enterprise and Free Inquiry." Marks, "Leviathan and the Clinic" discusses these debates as they played out for biomedical researchers situated in private medical schools. On the role of Michael Polanyi and the British (and anti-Marxist) Society for Freedom in Science in promoting the ideals of autonomous scientific research, see Oreskes and Rainger, "Science and Security," 327–28.
124. Klopsteg, "Role of Government," 781.
125. Spaght, "Basic Research," 785.

The NFIP's own investments in biomedical research enabled wide-ranging contributions to biological knowledge about viruses, proteins, and nucleic acids. Indeed, their understanding of the domain of basic polio research was remarkably broad, even contradictory, as assessed by their in-house historians of the late 1950s. For example, during the same year, 1949, the NFIP launched both its most expensive project to date, Salk's virus typing program, which had highly specific objectives and applications, and also its large-scale grant to Stanley's laboratory "in support of the physical, chemical, and biological studies of viruses with the *ultimate* aim of delineating the basic characteristics of polio virus." As one of their historians noted, the second "was a long-term undertaking, broadly exploratory in nature, and of a kind that precluded any possibility of 'a yes or no answer.' We are forced to conclude that in this, as in other matters, the men who guided the Foundation's research program were supremely elastic in their practice."[126] In fact, it depended on the context whether the director of research boasted of or belittled these sorts of investments.[127]

Similarly, the NIH grants system was being presented to the Congress as fostering directed research toward specific diseases and organs, even as it was simultaneously viewed as a mechanism for funding open-ended, basic research. As Kenneth Endicott and Ernest Allen summarized the development of NIH funding in 1953,

> Although the Public Health Service awarded a few grants for cancer research every year from 1937 on, the broad program began in 1946 with the transfer of 50 projects from the Office of Scientific Research and Development when that agency was dissolved. The new program had as its objective the improvement of the nation's health through the acquisition of new knowledge in all the sciences related to health. In the sense that new knowledge is sought for the purpose of improving health this program is one of applied research, but many of the grantees consider their projects basic. Those who established the program believed that maximum progress can be achieved only if the scientists enjoy freedom to experiment without direction or interference, and they drew up policies and procedures accordingly.... Congress imposes a degree of control and direction when it appropriates funds earmarked for research on a designated disease or a specific organ.... In 1953 about 80 per cent is earmarked. In actual practice, however, it has been possible to provide reasonably equal opportunity for scientists, regardless of their specialty in the health field, since the earmarked areas are broad and overlap

126. Joseph H. Mori, chapter 5, "Medical Research Program, 1946–1949," in vol. 3 of *History of the National Foundation for Infantile Paralysis Through 1953,* 657 (Unpublished Historical Monographs, NFA).
    127. Ibid.

to a considerable degree, especially with regard to the basic medical
sciences.[128]

Along similar lines, G. Burroughs Mider recalls that when "the NSF asked
the NIH to identify its basic research projects, . . . the NIH replied, 'none';
all of its awards were essential to its mission."[129]

Not surprisingly, the achievement of an effective vaccine in 1955
changed the rhetoric of urgency surrounding the foundation's generous
research funding. The days of the NFIP's directing millions of dollars
toward basic virus research were numbered; the development of an
attenuated live-polio oral vaccine (achieved by Sabin in 1958) became
a primary objective, and the foundation ceased funding virus research
altogether in 1964, having shifted its attention to birth defects. (It now
funds basic research in developmental biology.) During this time, the
public's partnership with basic biomedical researchers was broadened
through the remarkable expansion of research funding by the Public
Health Service's NIH, an increasingly important source of funding for
biochemistry and molecular biology. The NIH extramural budget expanded from $180,000 in 1945 to $4 million in 1947, $28 million in 1950,
$40 million in 1955, $244 million in 1960, $589 million in 1965, exceed-
ing $1 billion in 1973.[130] (Only in the immediate postwar years was the
NFIP keeping pace; its grants budget was around $2 million in 1947.)
The Atomic Energy Commission and the NSF, with their own rapidly
growing budgets, provided additional sources of support for biological
research.[131]

In the 1950s, as scientists assessed the bountiful development of postwar federal funding, they continued to worry that public expectations
of improved health, having justified phenomenal growth in biomedical
research funds, would put unrealistic expectations on basic research.
Stanley participated in the work of an NSF Special Committee on
Medical Research in 1955,[132] and their report criticized the mode of
funding coordinated research against diseases:

> This so-called "categorical approach" to the problems of certain
> diseases has been justified in some quarters (a) because it is believed
> that the support of Congress and the people is more easily obtained

128. Endicott and Allen, "Growth of Medical Research," 341.
129. Mider, "Federal Impact," 848.
130. "NIH Obligations and Amounts Obligated for Grants and Direct Operations," *NIH 1999 Almanac,* http://www.nih.gov/about/almanac/index.html.
131. On AEC funding, see Beatty, "Opportunities for Genetics"; on NSF funding of biology, see Appel, *Shaping Biology.*
132. On the political origins of this committee, see Marks, "Leviathan and the Clinic"; and Strickland, *Story of the NIH Grants Programs,* ch. 7.

for research in diseases for which no cure is known and which all individuals in consequence fear, since they identify them in terms of themselves or their families; (b) the widely held belief, no doubt fostered by certain wartime successes, that the solution of pressing national needs is best met by drafting all available talent into the pursuit of the desired objective.

The Committee believes that it must draw attention to the dangers inherent in accepting these reasons as justification for the "categorical approach" in medical research. In the first place, while the citizens of this country have been extraordinarily generous in their support of such programs, both through government and private agencies, they have been led to believe, consciously or unconsciously, that the donation of sufficient sums of money is all that is needed to eradicate diseases which have plagued mankind for centuries. Such a belief is contrary to experience.... Mere numbers of investigators and countless dollars will not in themselves ensure success in a search where, in point of fact, the seeker does not know what to look for. *This is the essential difference between the production of a new weapon of war and a new weapon for the eradication of disease* [emphasis added].[133]

However much these elite scientists objected to the framing of scientific research in terms of a war against disease, they were unable to dislodge this fundamental paradigm for the support of postwar biomedicine. And the voluntary health agencies, while their proportionate share of research funding declined (due mostly to the growth of NIH funds) in the 1950s, continued to play an important political role in ensuring steady increases in research funding. Mary Lasker and her associates lobbied Congressmen to allocate ever more funds to the NIH and NCI at the same time as they sought more generous donations from the public.[134] With their public relations apparatus and solicitations in places such as movie theaters, the voluntary health agencies consistently advocated an image of laboratory research as the key to conquering disease. Indeed, one poster by the ACS depicting money as "the cure for cancer" (see fig. 5.7) articulated the very assumption that both benefited and worried scientists.

Changing Political Meanings of the War against Disease

In Berkeley, local circumstances made Stanley's program on polio difficult to sustain after 1955. Since his arrival there, a rift had developed

133. "NSF Special Committee on Medical Research Report," Dec 1955, 28–30, Stanley papers, carton 17, folder National Science Foundation, Special Committee on Medical Research.

134. American Foundation, *Medical Research*, 1: 687 ("American Medical Research: In Principle and Practice"). For a critique, see Drew, "Health Syndicate."

FIGURE 5.7 "The Cure for Cancer" fund-raising advertisement. (Reprinted from *Cancer News* 24, no. 1 [1970]: 1 by permission of the American Cancer Society, Inc.)

between the faculty members of the Department of Biochemistry and the researchers of the Virus Laboratory. Both groups shared the same building, but the virus researchers were on soft money and did not generally have teaching responsibilities—only a few were on faculty. There was constant bickering over laboratory space, and the battles were intensified by the assimilation of a group of distinguished plant biochemists, which aggravated space constraints.[135] To the teaching faculty, Stanley's program of research on polio represented a use of space that did not contribute to the educational functions of the department. (Biologists at Caltech expanding their work into animal virology and polio studies faced similar space conflicts.)[136]

In 1958 the conflict between biochemists and virologists came to a head. The biochemists broke off formally from the Virus Laboratory, and both Stanley and the new chair of biochemistry, Esmond Snell, agreed that "interdepartmental relations" would be best served by moving the few faculty members among the virus group from the Department of Biochemistry to a new Department of Virology.[137] Stanley, voicing his frustrations in 1959 to a dean, said he would have to let six postdoctoral men go—including Schaffer, who had crystallized the polio virus— due to conflicts with the biochemists over space.[138] The Department of Virology mutated into a Department of Molecular Biology in the early 1960s, partly in response to university administrators' unease in accounting for a Department of Virology in a College of Letters and Science. By this time Stanley's Virus Laboratory had left polio behind for the race to solve the genetic code, a quest for which TMV seemed, for a time, the best model system (see chapter 7).

During the same period, Stanley focused efforts on another medical problem that demonstrated the medical relevance of his Virus Lab's basic

135. See Creager, "Stanley's Dream."

136. "A second urgent need is additional space to take care of the virus work now being carried on in Kerckhoff. There are two reasons for this. First, because of the present crowded condition in Kerckhoff, it is difficult to expand the animal virus program in the way justified by the gratifying progress made during the past year. Second, a segregated virus laboratory should be available, but cannot be provided in the present building in any economically sound way" (Biology 1952 at the California Institute of Technology: A Report for the Year 1951–1952 on the Research and other Activities of the Division of Biology, California Institute of Technology Archives, 11).

137. Memo by Acting Chancellor James D. Hart, "Subject: Establishment of a Department of Virology," University of California Committee on Budget and Interdepartmental Relations, Berkeley Campus, 6 Feb 1958, Records of the Chancellor, box 39: 27, folder 1959: 400-Vir.

138. Memo of conversation between E. W. Strong and Wendell Stanley, 3 Feb 1959, Records of the Chancellor, box 39: 27, folder 1959: 400-Vir.

research. Stanley was frequently giving lectures on polio and poliovirus research throughout the early 1950s, but a marked shift occurs in his lecture titles around 1956. After giving a host of talks with titles such as "Where We Are Today in Polio and Virus Research," most of the lectures he gave after 1955 were on viruses and cancer. For example, in 1957, at the Penrose Memorial Lecture at the American Philosophical Society in Philadelphia, he gave a speech entitled, "On the Nature of Viruses, Cancer, Genes and Life—A Declaration of Dependence."[139] Researchers were not the only audience for his new enthusiasm: in 1956 (and again in 1958) he testified before the Senate on the importance of funding basic research in biochemistry toward breakthroughs in cancer research. (He had also spent the past four years on the National Advisory Cancer Council and was joining the board of directors of the American Cancer Society.) Basil O'Connor, president of the NFIP, also perceived cancer as the next enemy for assault by biomedical research. In an international poliomyelitis meeting in 1955 he declared, "Studies of viruses may give us in time victory over many diseases in addition to poliomyelitis.... It is within the limits of possibility, therefore, that, out of researches on viruses—perhaps the virus of poliomyelitis—a new understanding of cancer will emerge."[140]

Indeed, by the 1960s the cancer problem provided a new nexus of biological and medical funding for virologists, molecular biologists, and biochemists.[141] For cancer even more than for polio, the construction of relevant basic research involved a careful choice of research materials and problems which manifested both the "nature of life" and the mechanisms (most often, molecular mechanisms) of disease. Cancer, unlike polio, is an umbrella term for many diseases, and all tumor viruses being intensively studied were from nonhuman sources. Thus the relevance of tumor virus work to actual human cancer was mediated through animal models and reasoned scientific extrapolation. A tight policy-making community linked the ACS, the National Cancer Institute, pharmaceutical industries, and the U.S. Congress, resulting in a vast research effort toward sorting out a viral etiology for cancer, but unlike in the case of polio, an effective vaccine was not forthcoming.[142]

While the metaphors of military mobilization were as central to the organization and funding of cancer research as they had been to the fight

139. List of Stanley's lectures, Stanley papers, carton 20, folder "List of talks given by Stanley."
140. O'Connor, "Research and Physical Energy," xxiii.
141. Gaudillière, "Molecularization of Cancer Etiology."
142. Ibid.; and Gaudillière, "Circulating Mice and Viruses."

against polio, analysis of the two campaigns also reveals interesting differences. The critical years for the establishment of the March of Dimes
were during World War II; the organization was founded in 1938 and
managed to raise increasing amounts of money throughout the war. (In
like manner, with the lay activist takeover of the leadership of the ACS,
public campaigns raised millions of dollars, increasing each year through
the 1940s.) There is a sense in which the rhetoric of the NFIP *was* the
rhetoric of war; the obvious enemy was polio, which crippled American
children at the same time as American soldiers were being maimed and
killed in Europe and the Pacific. Perhaps even more significant than any
symbolic resemblance between the campaigns against the Axis forces and
against polio was the fact that both efforts shared the same spokesperson,
Franklin Roosevelt. In a public letter in 1944, Basil O'Connor, President
of the NFIP, wrote Roosevelt that only "unremitting research will provide
the key which will unlock the door to victory over infantile paralysis."
Roosevelt's response, written in the closing year of World War II and
only four months before his death, called for the deployment of all-out
research:

> We face formidable enemies at home and abroad.... Victory is
> achieved only at great cost—but victory is imperative on all fronts.
> Not until we have removed the shadow of the Crippler from the fu-
> ture of every child can we furl the flags of battle and still the trumpets
> of attack. The fight against infantile paralysis is a fight to the finish,
> and the terms are unconditional surrender.[143]

This exchange of letters provided the basis for a National Foundation
news release, 8 December 1944, building toward the collection of funds
in January 1945.

The fight against polio thus embodied the transfer of both rhetoric
and research organization from war mobilization to basic postwar science, a model made all the more powerful because the biomedical war
against polio was "won" by scientists in 1955. When Stanley moved to
Berkeley in 1948, his use of military metaphors in newspaper publicity
alluded to the ongoing campaigns against polio as well as to World War II.
One newspaper gave the following account in July 1948: "History's most
concentrated attack on virus diseases—one of man's deadliest enemies—
will get under way within two months on the Berkeley campus of the
University of California. This was announced yesterday by Dr. Wendell
M. Stanley, 43, world's top authority on the virus, who arrived at the

143. Quoted in Paul, *History of Poliomyelitis*, 319.

university to take command."[144] That same year, the NFIP entitled their summary of research sponsored through the March of Dimes, "Weapons of Defense":

> The challenge of epidemics and their aftermath claimed much of our attention in 1947. These were dramatic front-line activities designed to minimize the weight of the polio attack. Meanwhile, behind the scenes, the year-round job of forging new defensive weapons continued without interruption as twenty-four branches of science prosecuted their relentless search for answers to the mystery of polio virus under National Foundation sponsorship.[145]

During the same years that the establishment of a National Science Foundation was being debated (and stalled) by the federal government, lay activists and journalists elaborated the continuities between wartime and postwar science.

Here the comparison with the subsequent campaign against cancer is revealing. The March of Dimes' fight against polio drew strength from the perceptions that scientists had won World War II and that basic research was critical to the victory over polio in the mid-1950s. This was Roosevelt's final legacy against "the Crippler" at home. The ACS fostered similar expectations for novel, research-based treatments or cancer vaccines and contributed similarly to the new regime of public support for biomedical research. However, the hopes for a cure for cancer were prolonged, and ultimately disappointed, despite Nixon's War on Cancer, launched in 1971. Vigorous criticisms of the cancer establishment were part of broader critiques of the authority of science and medicine in the 1960s and 1970s. As one writer assessed the changed situation, "By comparison with the fight against polio, the war on cancer is a medical Vietnam."[146] This sentiment expresses the profound cultural shift since World War II in popular attitudes toward science as well as the unanticipated challenges biologists faced in understanding cancer.

Even when it did not produce cures and vaccines, cancer funding, like polio funding, undergirded much of what we now regard as the premier molecular biology of its generation. After taking the directorship at Cold Spring Harbor in the late 1960s, when its financial situation was bleak, James Watson directed most of the research toward the cancer problem, so that it became largely supported by federal NCI grants.[147] The

144. Henry Palm, article from 21 Jul 1948, unidentified newspaper, Records of the President, series 2, box 728, folder 1948: 420-Biochem.
145. NFIP, "Tenth Annual Report, Period Ending December 31, 1947," NFA, 36.
146. Kennedy, "Animal Research," 84; see Patterson, *Dread Disease,* 252.
147. Watson, *Houses for Science,* 175.

framework for eukaryotic gene regulation came to be largely constructed from experimental systems that were once again both relevant to medical concerns and also models of gene expression, namely, tumor viruses.

Conclusions

In 1947, George Gamow wrote Stanley of the abundant funding available through the Office of Naval Research:

> There seems to be an epidemic among the physicists, "maladia bio-
> logica" you may call it.... The Office of Naval Research (in which I am
> a consultant in [the] physics section) spends, as you may have heard,
> millions of dollars on pure research.... The second division became
> interested in "aperiodic crystals" (in the sense of Schrödinger's book
> *What Is Life?*) or, in plain language, genes and viruses. So they want to
> spend money on subsidizing research in this direction on very broad
> lines. On the lines of physics and biology collaboration of course. The
> moral: may I ... come to Princeton sometime the end of March to
> talk to you about the way to spend few hundred thousand dollars? It
> isn't joke [*sic*], it is serious![148]

As it turned out, the Office of Naval Research did not become a major source of funding for Stanley's Virus Laboratory.[149] However, scientists and historians have recurrently invoked the cross-disciplinary move captured by Gamow's diagnosis of physicists' "maladia biologica" to explain the emergence of molecular biology.[150] Several former physicists, including Max Delbrück, Francis Crick, Maurice Wilkins, Leo Szilard, and Seymour Benzer, made notable contributions to postwar biology, some (Wilkins, Szilard, Crick) after participating in military research during the war. In addition, Gunther Stent has claimed that Schrödinger's little book *What Is Life?* drew the attention of disenchanted young physicists to the unsolved mysteries of biology.[151] The conjuncture of the "physicists' war"

148. Gamow to Stanley, 24 Feb 1947, Stanley papers, carton 8, folder Gamow, George.

149. I can only find evidence of a grant from the Office of Naval Research to Howard Schachman; see Stanley to H. S. Thompson, 26 Nov 1951, Stanley papers, carton 3, folder Nov 1951. Stanley himself does not appear to have received any research funding from military agencies for the Virus Laboratory.

150. See Fleming, "Emigré Physicists"; Olby, *Path to the Double Helix;* Judson, *Eighth Day of Creation;* Keller, "Physics and the Emergence." The role of Max Delbrück is central to this historiography; see essays in Cairns, Stent, and Watson, *Phage and the Origins*.

151. Stent, "That Was the Molecular Biology That Was," 347. Critical historical reassessments of Schrödinger's book as a catalyst for molecular biology include Yoxen, "Where Does Schrödinger's *What Is Life?* Belong?"; and Olby, "Schrödinger's Problem." Keller offers a different perspective on Schrödinger's book in "Molecules, Messages, and Memory," ch. 2 of *Refiguring Life.*

with the "revolution in biology" has served as an enduring motif in the historiography of molecular biology, one that stresses the importance of those scientists who turned from secrets of death to secrets of life, in Evelyn Fox Keller's memorable phrasing.[152]

There are recognized problems with this genealogy. For one, the scholarship addressing the Rockefeller Foundation's 1930s Natural Sciences program has shown that the recruitment of physical scientists into biology predates World War II.[153] In addition, as biochemists have complained since the 1960s, the origins story about physicists fails to acknowledge the key contributions of many other individuals and disciplines (most of which were already identified with the life sciences) in the emergence of molecular biology.[154] Recently, Nicolas Rasmussen has reinterpreted the role of physics in terms of cultural and disciplinary changes ushered in by the atomic bomb. He argues that the growth in the 1940s of biophysics, as the key precursor to molecular biology, should be understood as "a biomedical silver lining in the mushroom cloud that menaced life everywhere."[155]

I prefer a different interpretation altogether of the legacy of World War II for biomedical researchers—one that takes into account the role of groups less illustrious than physicists. This chapter has emphasized the role of public expectations and lay activists in shaping the economic and political possibilities of postwar life science.[156] As it turns out, the war against polio was especially consequential for molecular biologists, who were already using viruses as key experimental tools (for reasons that will

152. Keller, "From Secrets of Life to Secrets of Death." On the "physicists' war," see Kevles, *Physicists*. The most widely read account of the "revolution in biology" is Judson, *Eighth Day of Creation*.

153. Kohler, "Management of Science"; Abir-Am, "Discourse of Physical Power"; Kay, *Molecular Vision of Life*. Chapter 4 discusses this historiography of the Rockefeller Foundation in more depth.

154. See Abir-Am, "Politics of Macromolecules."

155. Rasmussen, "Midcentury Biophysics Bubble," 245; see his *Picture Control* for a fuller elaboration of his argument. Rasmussen uses Stanley's Virus Laboratory as one of his several institutional case studies for biophysics in *Picture Control,* and I should point out that much of my interpretation of Stanley's work agrees with his—for instance, Rasmussen also highlights the NFIP's role in supporting the Virus Laboratory. He tends to emphasize the biophysics in Stanley's enterprise, and my work ("Stanley's Dream") stresses its rootedness in biochemistry, but clearly both disciplinary orientations were important. Our greatest point of divergence concerns the cultural resonances we see as important; he argues that scientists used worries associated with the atomic age to further molecular approaches to understanding life, whereas I view the resources of the campaigns against disease as essential.

156. These two kinds of explanations intersect in suggestive ways. As Stuart Feffer has demonstrated, the Atomic Energy Commission was recruiting physicists into biomedical research as part of their own "war on cancer" ("Atoms, Cancer, and Politics").

be examined in the next chapter). Delbrück's and Stanley's laboratories received NFIP funding for basic research on a long-term basis; other notable grantees included Linus Pauling and James Watson, whose labors on the structure of DNA at Cambridge were sustained by a polio foundation fellowship. The NFIP also underwrote the 1953 Cold Spring Harbor symposium on viruses, which is most often now cited as the meeting where Watson first presented publicly the double-stranded model he and Crick had devised for DNA.[157]

The boost to molecular biology was only one outcome of the growing public enthusiasm for supporting biomedical research. Voluntary health agencies played a critical role in upholding the model of mobilized science after the war (despite the apprehension of many scientists) and adapting it to include support for basic laboratory research. Other contingent factors reinforced the value of channeling public monies toward biomedical research. As Paul Starr argues, the demise of Truman's 1949 legislative proposal for universal health insurance gave the funding of medical research (and of hospital construction) a new political function; it served as the federal government's visible commitment to improving health.[158] By 1950, politicians realized that the "war against disease" was an almost invulnerable political formula, resulting in average yearly increases of almost 25 percent through that decade to the NIH's grants budget.[159] Scientists who had protested having their research priorities determined by laypeople and politicians found that they could live with the flexible terms of disease-relevant funding.

Finally, this chapter has pointed to the concrete consequences of the new system of support for biomedical research. The crystallization of poliovirus by Schwerdt and Schaffer is a case in point. The achievement drew on local biochemical expertise and the precedent of TMV, but it would never have been accomplished if the NFIP had not supplied both the funding and the actual starting material—extra vaccine fluid. At Caltech, polio research followed the exemplar of bacteriophage, with Dulbecco's development of the plaque assay for polio and the rapid growth of quantitative animal virology. Within this framework of disease-relevant basic research, model systems such as TMV played an important role. Stanley himself justified the use of plant viruses as models in polio research on the basis of scarcity of materials: "It would be wasteful to use small and

157. Demerec, "Foreword."
158. Starr, Social Transformation.
159. Strickland, Politics, Science, and Dread Disease. The average yearly increase is based on numbers from "NIH Obligations and Amounts Obligated for Grants and Direct Operations," NIH 1999 Almanac, http://www.nih.gov/about/almanac/index.html.

very valuable amounts of purified poliomyelitis virus in the development of analytical techniques proposed, ... when such development could be accomplished with plant viruses, which are available in large amounts and in highly purified form."[160] In fact, model viruses were more than substitutes for poliovirus and other pathogens; they showed how to make the enemy visible.

160. Stanley to H. M. Weaver, 17 Jun 1948, Stanley papers, carton 13, folder We misc., 3.

6

Viruses and Genes: Phage and TMV as Models

It is true, the temptation is almost overwhelming to pour all into one pot: growth, and its change from benign to malignant; bacterial transformations; the transmission of hereditary properties; the multiplication of virus and phage; the synthesis of inducible enzymes; the differentiation of tissues; the formation of antibodies—everywhere an interplay between nucleic acids and proteins, a spinning wheel in which the thread makes the spindle and the spindle the thread. But if it is hard to recognize the similarities between what is ostensibly dissimilar, it is even more difficult to make out the differences between what appears so similar; and moderation still is among the tools of our science.

Erwin Chargaff, 1955

There can be little doubt that the field of virology grew in the 1940s and 1950s due to the abundant research support available through the NFIP. In addition, as we have seen, virus research benefited from the new technologies developed for biology, especially through the sponsorship of the Rockefeller Foundation. But one more development solidified the growing status of virus research, especially within the life sciences: many biologists viewed viruses as good experimental tools for investigating the physical nature of heredity. This chapter traces this trajectory, rooted in the 1920s, in order to show how experimental developments with particular model systems shaped the image of viruses as material referents for genes. Understanding how viruses, like genes, duplicated themselves was at once a biological problem, with relevance to the general mechanisms of heredity, and a medical problem, with potential significance for stopping epidemics. Thus amid the larger, disease-oriented context that sustained basic virus research was a flourishing niche in biology that viruses came to occupy, representing as they did bare elements of life.

The rendering of viruses as experimental systems for understanding the "secrets of life" was somewhat paradoxical; its advocates never claimed that viruses were alive. Rather, the simplicity of viruses as parasites, the fact that they seemed to be *merely* autonomously multiplying molecules, invited their investigation to understand the process of reproduction, and so to illuminate the physical and chemical nature of genes. Research on the physical basis of genetics through the 1930s had failed to recover the gene as a material entity; in the wake of Stanley's isolation of TMV, viruses seemed to provide a favorable means to fulfill this ambition. At the same time, to define life as self-reproduction marginalized vital processes that had been intensively investigated during the first half of the twentieth century, such as metabolism and development. Even Schrödinger, in his 1944 book *What Is Life?*—usually noted for its notion of a genetic "code-script," pointed to metabolism and development as well as to reproduction as distinguishing features of life.[1] The interest in viruses as representatives of vital processes, on the other hand, reinforced a gene-centered view of life.

Among the many viruses under investigation, TMV was a good candidate for genetic research because variant strains of the virus were readily available. Bacteriophages, as Félix d'Herelle had named viruses that infect bacteria, provided another good experimental material, adaptable to investigating the dynamics of virus multiplication and bacterial growth.[2] This chapter compares work on these two systems: the use of bacteriophage by Max Delbrück and coworkers to study virus reproduction, and Stanley's concurrent attempts to understand replication and mutation with TMV.[3] Phage and TMV came to represent quite different exemplars for work on other viruses. During the 1950s, as we saw in the last chapter, poliovirus was characterized both as a macromolecule like TMV and as a plaque-forming genetic unit like bacteriophage. That TMV and bacteriophage provided different models was attributable not only to the divergent biological properties of the viruses, but also to the ways in which

1. Schrödinger, *What Is Life?* Schrödinger offered three ways in which living organisms seem to defy the second law of thermodynamics: (1) the fidelity of gene transmission, and of mutations, through generations (e.g., p. 49); (2) the unfolding of a complex multicellular organism from a single cell (e.g., p. 65); and (3) the ability of organisms to "avoid decay" metabolically (e.g., p. 75). Those who have given this book a prominent place in the history of molecular biology tend to stress only the first aspect, the puzzle of heredity. See Stent, "That Was the Molecular Biology That Was"; Yoxen, "Where Does Schrödinger's *What Is Life?* Belong?"; Olby, "Schrödinger's Problem"; Keller, "Molecules, Messages, and Memory: Life and the Second Law," in *Refiguring Life*. For a dissenting point of view, see Perutz, "Erwin Schrödinger's *What Is Life?* and Molecular Biology."

2. Not all bacteriologists agreed with d'Herelle that bacteriophage were viruses; on this issue, see Helvoort, "Construction of Bacteriophage."

3. Historical accounts of phage research and its relation to the emergence of molecular biology are numerous. A few of the widely cited accounts are Cairns, Stent, and Watson, *Phage and the Origins;* Mullins, "Development of a Scientific Specialty"; Olby, *Path to the Double Helix;* Judson, *Eighth Day of Creation*. More recently, Weiner, in *Time, Love, Memory,* has given a historical account of the phage group with an eye to later developments in the molecular genetics of behavior.

they were visualized and characterized—the experimental systems made with TMV and phage.

Several of the events covered in this chapter are part of the standard "pre-history" of molecular biology. Viewed from the present, these developments appear as missed opportunities—neither genetic nor biochemical approaches elicited the "riddle of the gene" from viruses in the 1940s. Furthermore, scientists failed to appreciate the significance of evidence that "transforming factors" of pneumococcus were composed of nucleic acid. Historians have conventionally attributed these failures to the widespread misconception that genes were made of proteins.[4] I will seek to complicate this explanation by offering a three-phase story of how viruses were understood as a model for genes.

In the late 1930s and through the mid-1940s, researchers attempted to elucidate the mechanism of self-reproduction by investigating viruses as more accessible representatives of nuclear genes. However, experiments with the best-studied viruses, bacteriophage and TMV, did not illuminate the physical process of genetic reproduction in the ways anticipated. In part, this was because viruses were more complex entities, biologically and biochemically, than scientists like H. J. Muller had expected.

Subsequently, in the immediate postwar years—my phase two—growing interest in the perceived similarities between viruses and genetic particles *outside* the nucleus (in the cytoplasm), such as microsomes and chloroplasts, changed the grounds for the analogy between viruses and genes. Many of these particles resembled viruses in both size and composition, being made of protein and nucleic acid. These newly emphasized commonalities reinforced a belief that the smallest viable genetic units (including those on chromosomes) were all *nucleoproteins,* but the exact relationship between putative cytoplasmic genes and nuclear genes became an area of intense debate. The postulation of a class of self-reproducing but non-nuclear particles—plasmagenes—seemed to fit with a variety of contemporary findings, including results from virus research. However, many biologists rejected the notion that there was a class of authentic genetic particles beyond the nucleus, so interest in the kinship of viruses and plasmagenes threatened the validity of the older analogy between viruses and nuclear genes.

The plausibility of an overarching framework of cytoplasmic genetic particles, which for a brief period seemed to unify processes of infection, differentiation, and non-nuclear inheritance, crumbled in the early 1950s, as the quest for the self-reproducing gene shifted to chemical

---

4. Olby, *Path to the Double Helix;* Kay, "Protein Paradigm," in *Molecular Vision of Life,* 104–20. For other accounts of the "pre-history" of molecular biology, see the references in note 1.

considerations and sequence specificity. In this third phase of genetic investigations of viruses (explored further in chapter 7), experimental systems using bacteriophage and TMV were reconfigured to follow not the reproduction of particles per se but the replication of genetic information.

This chapter emphasizes the key role of analogies between different experimental systems in shaping scientific expectations regarding genes as molecular entities. These analogies, even when vague, served an important disciplinary function: they specified the relevance of virus research to biological problems and processes. In addition, the network of experimental comparisons that grew up around the virus-gene nexus had enduring consequences for the conception of viruses. The displacement of the virus-gene-enzyme analogies of the 1920s by the virus-plasmagene-microsome analogies of the 1940s called into question the strictly autonomous nature of viruses in relation to cells, materially and genetically. By the 1950s, a notion that there were various representatives of "infectious heredity," drawing on observations of bacterial conjugation and "transforming factors," informed virus research and microbial genetics. This recasting of viruses as agents of infective heredity was particularly important in the burgeoning area of tumor virology, where the conventional differentiation between exogenous and endogenous units of transmission had always been problematic.[5] Ironically, then, although it was the autonomy of viruses that prompted researchers to use them to understand self-multiplication, their study revealed just how extensively and intimately the virus relies on its host to achieve its replication. This qualified understanding of the autonomy of viruses remained in place even after the relevance of cytoplasmic inheritance to understanding genetic self-reproduction had faded from view.

Genes—Material or Operational Entities?

The "gene" was introduced by Wilhelm Johannsen as an operational unit of heredity in 1909 just as interest in and research on "filterable viruses" was growing (see chapter 2).[6] In 1917 Leonard Troland offered his speculative theory of enzyme self-replication, drawing on colloidal explanations of enzyme action to account for the autocatalysis of genes.[7] Viruses, too, appeared to self-reproduce in an enzymatic fashion, perhaps as autocatalytic genes themselves.[8] Troland's speculations thus

5. Creager and Gaudillière, "Experimental Arrangements." For a different view, see Helvoort, "Century of Research."
6. Johannsen, *Elemente der exakten Erblichkeitslehre.* On Johannsen's contributions, see Roll-Hansen, "Genotype Theory."
7. Troland, "Biological Enigmas." For different perspectives on the significance of Troland's paper, see Olby, *Path to the Double Helix,* 147; and Kay, "Protein Paradigm," in *Molecular Vision of Life.*
8. Ravin, "Gene as Catalyst."

heightened interest in using viruses to understand heredity.[9] In particular, viruses that seemed amenable to physicochemical characterization might be used to isolate single genes.

This research endeavor was not necessarily valued by geneticists. As Thomas Hunt Morgan stated in his Nobel address (June 1934),

> What are genes? Now that we locate them in the chromosomes, are we justified in regarding them as material units; as chemical bodies of a higher order than molecules? Frankly, these are questions with which the working geneticist has not much concerned himself, except now and then to speculate as to the nature of the postulated elements. There is no consensus of opinion amongst geneticists as to what the genes are ... whether the gene is a hypothetical unit, or whether the gene is a material particle.[10]

Raphael Falk has argued that for many geneticists (not only Morgan, but also W. E. Castle and E. M. East) the gene was simply an *instrumental* entity. The success of this conception could be seen in the Morgan school of chromosomal *Drosophila* genetics, in which quantitative mapping of genes bore no necessary relation to the identification of genes as specific physical entities.[11] The geneticist who most resisted a merely operational understanding of genetics was H. J. Muller, who advocated research on the gene as a *material* object. As Muller outlined in his 1926 treatise, "The Gene as the Basis of Life,"

> What is meant in this paper by the term "gene" material is any substance which, in given surroundings—protoplasmic or otherwise—is capable of causing the reproduction of its own specific composition, but which can nevertheless change repeatedly—"mutate"—and yet retain the property of reproducing itself in its various new forms. There is clear evidence that such material is to be found in the chromatin.[12]

Muller also articulated the analogy between viruses and genes most clearly during the 1920s and 1930s, as Elof Carlson has amply demonstrated.[13] Muller's comparison of Félix d'Herelle's bacteriophage to genes, in a paper delivered to the American Society of Naturalists in Toronto in 1921, is widely cited:

> If these d'Herelle bodies were really genes, fundamentally like our chromosome genes, they would give us an utterly new angle from

9. See, for example, Muller, "Variation Due to Change."
10. Morgan, "Relation of Genetics," 5.
11. Falk, "What Is a Gene?"
12. Muller, "Gene as the Basis of Life," 897.
13. Carlson, "Unacknowledged Founding," and *Genes, Radiation, and Society;* Olby, *Path to the Double Helix.*

which to attack the gene problem. They are filterable, to some extent isolable, can be handled in test-tubes, and their properties, as shown by their effects on the bacteria, can then be studied after treatment. It would be very rash to call these bodies genes, and yet at present we must confess that there is no distinction known between the genes and them. Hence we cannot categorically deny that perhaps we may be able to grind genes in a mortar and cook them in a beaker after all.[14]

In 1936, Muller expressed confidence that Stanley's recently isolated plant virus might throw new light on the nature of the gene:

> In the past year the opportunity has also arisen of obtaining from another source material which may serve our present purpose. This possibility arises out of the discovery by Stanley and Loring, that the substance (or "organism") causing the so-called mosaic disease of tobacco, and likewise that of tomato (and doubtless of various other higher organisms), may be obtained in crystalline form, apparently as a pure protein. We judge that this material has the properties of a gene, inasmuch as it can reproduce itself, *i.e.,* it can undergo autosynthesis when present in a cell and it is probably mutable, since different "species" of it are known. We may provisionally assume, then, that it represents a certain kind of gene.[15]

The isolation of different strains of TMV showed that viruses exhibited not only the genetic property of self-duplication, but also that of the propagation of mutations. In Muller's view, these were the two distinguishing features of genes.[16]

Muller was not the only scientist to advocate the similarity of viruses to genes. The Seventh International Genetical Congress, held in Edinburgh in 1939, included a double session devoted to "Protein and Virus Studies in Relation to the Problem of the Gene." William T. Astbury, Dorothy Crowfoot, Dorothy Wrinch, Torbjörn Caspersson, G. A. Kausch, John Gowen, and H. H. McKinney contributed papers with titles such as "Behaviour of Viruses and Genes under Similar Stimuli" (Gowen) and "Virus Genes" (McKinney).[17] Researchers highlighted two shared properties: similarities between the chemical composition of chromosomes and of viruses, and self-duplication of viruses and genes. Papers presented at the 1941 Cold Spring Harbor Symposium on Quantitative Biology, on the

14. Muller, "Variation Due to Change," 48–49.
15. Muller, "Physics in the Attack," 213.
16. Muller, "Gene as the Basis of Life," 897; Falk, "What Is a Gene?" 150.
17. See Punnett, *Proceedings of the Seventh International Genetical Congress.* My summary here is indebted to Olby, *Path to the Double Helix,* 151–52.

subject "Genes and Chromosomes," also reflected this interest in viruses as genetic units. As Muller observed at the conclusion of that symposium,

> The most fundamental—in fact, the unique and distinctive— characteristic of a gene is . . . that, in its protoplasmic setting, it pro- duces a copy of itself, next to itself, and that when its own pattern becomes changed, the copy it now builds is true to its new self. . . . Theoretically, we should be able to follow this material back in evo- lution to the stage where it existed alone, without protoplasm of its own construction. Such a primitive stage is today most nearly repre- sented by the phages and other viruses.[18]

Bacteriophage and viruses were genes stripped down to their most basic property, or "naked genes," in Muller's view.[19]

Thus, for those scientists who, like Muller, wished to pursue the gene as a material entity, viruses seemed promising experimental tools. In the late 1930s and early 1940s, tobacco mosaic virus (TMV) and bacteriophage were both configured as experimental systems to register "genetic" be- havior, although what this meant differed for each virus.[20] The selection of these particular viruses for experimentation reflected their history of amenability to laboratory study (as both d'Herelle and Muller had empha- sized for bacteriophage in the early 1920s).[21] Despite their manipulability, however, using viruses to connect the gene as a conceptual entity with the genetic material took a more circuitous path than Muller had predicted. As we have already seen, within two years of Stanley's crystallization of TMV, evidence refuted his speculation that this virus self-reproduced autocatalytically, like an enzyme. Delbrück rejected the analogy between bacteriophage and enzymes by the early 1940s for other reasons. The next two sections recount attempts in the late 1930s and early 1940s to retrieve the material gene by using phage and TMV as experimental tools, paying attention to these shifting grounds for viewing viruses as genelike.

Max Delbrück and the Bacteriophage as a Quantitative Virus Replication System

To physical scientists, Wendell Stanley's crystallization of TMV served as validation (see chapter 3). Beyond the practical achievement of trans- forming viruses into chemical objects, Stanley's results suggested to physi- cists and chemists that their approaches and methods might elucidate the basic processes of life. Stanley's crystalline virus had a particular impact

---

18. Muller, "Résumé and Perspectives," 290.
19. Kay, "Stanley's Crystallization," 467; Sapp, *Evolution by Association*, 151; Watson, *Double Helix*, 17.
20. See Lederman and Tolin, "OVATOOMB."
21. Muller, "Variation Due to Change"; d'Herelle, "Sur un microbe invisible," and *Le bactériophage*.

on Max Delbrück, who in 1935 was a young theoretical physicist seeking to make his way in biology.[22] After spending 1931–32 as a Rockefeller Foundation fellow in theoretical physics with Niels Bohr in Copenhagen and Wolfgang Pauli in Zurich, Delbrück joined the Kaiser Wilhelm Institut für Chemie in Berlin-Dahlem as an assistant to Lise Meitner.[23] While in Berlin he collaborated with N. W. Timoféeff-Ressovsky and K. G. Zimmer on X-ray mutagenesis of *Drosophila,* studies drawing on Muller's discovery of radiation-induced mutations in the fly.[24] In 1935 the three published a long paper, entitled "On the Nature of Gene Mutation and Gene Structure."[25] Delbrück's contribution to the group venture was theoretical: he drew on quantum mechanics to interpret mutations in terms of shifts in atomic configuration from one stable energy state to another.[26] Despite its positive reception, Delbrück was dissatisfied with this approach of applying quantum mechanics to biology. As he wrote Bohr, this work fell short of his desire to fulfill his mentor's expectation that one could find the complementarity relation in biology.[27]

During this same period, the Rockefeller Foundation's Warren Weaver was actively involving Bohr in his interdisciplinary program for the natural sciences.[28] Through the foundation's sponsorship, Bohr hosted a small conference in Copenhagen in September of 1936 to promote considerations of the relevance of physics to biology, particularly concerning the nature and mechanism of genetic mutation.[29] Max Delbrück attended

22. For an extended account of Delbrück's early work as a physicist and his move into biology, see Kay, "Conceptual Models and Analytical Tools." See also the biography of Delbrück by Fischer and Lipson, *Thinking about Science.* Both accounts argue for the formative role of Bohr's famous lecture "Light and Life" in inspiring the young Delbrück to search for a biological example of complementarity, which might require the development of new physical theories to account for life. This argument is further elaborated in Kay, "Secret of Life." For a recent interpretation, see Roll-Hansen, "Application of Complementarity."

23. It was in Copenhagen in August 1932 that Delbrück heard Bohr deliver his famous lecture "Light and Life," which, according to his companion that day, Léon Rosenfeld, inspired Delbrück's decision to make his mark in biology; see Rosenfeld, "Niels Bohr in the Thirties."

24. Muller, "Artificial Transmutations."

25. Timoféeff-Ressovsky, Zimmer, and Delbrück, "Über die Natur der Genmutation und der Genstruktur." Olby notes that its publication in this venue meant that it "was virtually inaccessible save through reprints" (*Path to the Double Helix,* 232); thus it became known as "The Green Paper," after the color of the reprint cover.

26. See Kay, "Conceptual Models and Analytical Tools" and "Secret of Life." As Morange has noted, "On the basis of this study Delbrück and Timoféeff-Ressovsky subsequently published an article in *Nature* concluding that cosmic radiation had little effect on speciation" (*History of Molecular Biology,* 41).

27. Kay, "Secret of Life," 496. Kay cites a letter from Delbrück to Bohr with the reprint of the 1935 paper.

28. Slater, "Max Delbrück."

29. The Rockefeller Foundation funded the conference as part of a five-year plan (1935–40) for support of collaborative research between physicists, physical chemists, and

the meeting with Timoféeff-Ressovsky, and Delbrück reported that their "discussions occurred very much under the impact of the findings of W. M. Stanley reporting the crystallization of tobacco mosaic virus."[30] Delbrück subsequently wrote a private memo, entitled "Riddle of Life," which summarized his interest in exploiting viruses to study genes.[31] Delbrück took the uniformity of TMV particles as detected through electrophoresis and recrystallization as reliable evidence that viruses were molecules. He interpreted Stanley's isolation of these self-reproducing molecules as enabling direct experimental access to the atomic processes of reproduction and mutation. In particular, Delbrück argued that viruses were exogenous, not endogenous, entities; that their replication could be studied as "an autonomous accomplishment of the virus, for the general discussion of which we can ignore the host."[32]

Shortly after writing "Riddle of Life," Delbrück arrived in the United States on another Rockefeller Foundation fellowship (this time in theoretical genetics) to seek out an appropriate biological system for his approach. Among the several laboratories he visited was that of Stanley in Princeton. Theoretician Delbrück may well have found the physical machinery Stanley employed to investigate TMV daunting.[33] He was no

---

physiologists at Bohr's institute. There were to be "one or more small scientific conferences at Copenhagen to aid in the planning of the biophysical program"; the one in September 1936 was the first of these ("University of Copenhagen—Biophysics," RAC RF 1.1, 713, box 4, folder 46). The first gathering was intended to foster a general discussion of the mechanism of mutation based on the findings of participant H. J. Muller. (See letters from Bohr to W. E. Tisdale, 17 Sep 1936; H. M. Miller to Bohr, 25 Sep 1936; and Bohr to Miller, 2 Oct 1936, all in RAC RF 1.1, 713, box 4, folder 47.) Others who attended this gathering included "Delbrück of Berlin, Mohr and Bonnevie of Oslo, Dirac of Cambridge, and ... a group of about thirty Danish scientists including Bohr, Krogh, Hevesy, Sorensen, Linderstrøm-Lang, Winge, Holter, etc. The second [on metabolism and radioactive tracers] was held early in May 1938, and was attended by Joliot of Paris, Meyerhof of Heidelberg, Mr. and Mrs. Needham of Cambridge, Parnas of Lwow, Runnström of Stockholm, and others" ("University of Copenhagen—Biophysics," RAC RF 1.1, 713, box 4, folder 46).

30. Delbrück, "Physicist's Renewed Look at Biology," 1313.

31. Delbrück, "Riddle of Life" (Berlin, August 1937), published as an appendix with his Nobel prize address, "Physicist's Renewed Look at Biology." Delbrück, who found the memorandum among his papers in the 1960s, referred to the piece as a summary of the Copenhagen meeting. Olby considers that origin unlikely, since the memorandum was dated August 1937 and the meeting had been held the previous September (*Path to the Double Helix*, 236). The phage researcher Dean Fraser first translated the unpublished draft from German into English for his virology students at the University of Indiana; see Delbrück to Dean Fraser, 10 Dec 1964; Fraser to Delbrück, 23 Oct 1968; and Delbrück to Fraser, 31 Oct 1968, Delbrück papers, folder 8.2.

32. Delbrück "Riddle of Life," appendix to "Physicist's Renewed Look at Biology," 1315.

33. In this speculation I follow Lily Kay, who wrote of the visit, "One could only guess the reaction of the author of the 'Riddle of Life' to the contrast between his elegant theoretical concepts of virus research and the reality of complex laboratory methods of studying plant

more encouraged by the model system he found in the laboratory of
T. H. Morgan, where he spent several months after arriving at Caltech.
Drosophila melanogaster was much too complex and unwieldy for the
kind of simple quantitative experiments he had in mind.[34]

After several discouraging months in Pasadena, Delbrück met Emory
Ellis, a postdoctoral researcher who had been working on basic cancer
research since 1936.[35] Since some tumors had been shown to be transmissible by cell-free filtrates, and thus were virus-induced, Ellis was pursuing
knowledge about cancer via virus research.[36] He selected bacterial viruses
(bacteriophages) as a model for tumor viruses. The choice was pragmatic:
although links between bacteriophage and cancer had been postulated,
their most important feature to Ellis was that they were easier to grow
and study than mammalian tumor viruses.[37] Working from methods outlined in d'Herelle's 1926 book, *The Bacteriophage and Its Behavior,* Ellis
isolated bacteriophage from sample sewage from the Pasadena treatment
plant.[38] The host he chose was *E. coli,* a bacterial strain he obtained from
colleague (and former Morgan student) Carl Lindegren.[39] Ellis selected
phage which, upon mixing and plating with *E. coli,* produced discrete
"plaques," or holes of lysed cells, on the lawn of cells. As he noted later,

---

and animal viruses" ("Conceptual Models and Analytical Tools," 228). Delbrück discusses
his visits to several laboratories in his oral history, but does not specifically mention his visit
to Stanley's laboratory (Max Delbrück Oral History, interview by Carolyn Harding, 1978,
California Institute of Technology Archives).

34. Delbrück later recalled, "I consulted with [Calvin Bridges] for quite a bit and tried
to learn some *Drosophila* genetics, and, as I say, I didn't make much progress in reading
these forbidding-looking papers; every genotype was about a mile long, terrible, and I just
didn't get any grasp of it" (Max Delbrück Oral History, interview by Carolyn Harding,
1978, California Institute of Technology Archives, 63). See also Kay, "Conceptual Models
and Analytical Tools," 230, n. 75.

35. Ellis's fellowship was funded through a donation from the mother of Seeley W.
Mudd, an M.D. experimenting at Caltech with X-ray cancer therapies. The fullest account
of Ellis's research is given by Summers, "How Bacteriophage Came to Be Used."

36. The work in the mid-1930s on the Shope papilloma virus of rabbits, as well as earlier
work on the Rous sarcoma virus of chickens, had raised general scientific interest in the
viral etiology of cancer; see Creager and Gaudillière, "Experimental Arrangements."

37. d'Herelle had claimed that bacteriophages were associated with mammalian cancer:
d'Herelle and Peyre, "Contribution à l'étude des tumeurs expérimentales," and "Contribution à l'étude des tumeurs spontanées." See Summers, "How Bacteriophage Came to Be
Used," 259.

38. d'Herelle, *Le bactériophage et son comportement.*

39. At the time Ellis and Delbrück began publishing on bacteriophage, they referred to
the host bacterial strain they used as *Bacillus coli.* It had been officially renamed *Escherichia
coli* in 1919, but the old name continued to be used for many years. Because the renaming
played no role in the scientific developments I will be discussing—and for the sake of
clarity—I will refer to the strain by its now-familiar name, *E. coli.* For an insightful essay on
E. coli as a research tool, see Lederberg, *"Escherichia coli."*

the criteria for selection were quite simple—he sought "a phage from a plaque of a size which would be readily seen, but small enough to allow counts of 50 or more on a petri plate culture."[40] This phage could be purified bacteriologically by serial culture, and studies of the plaque numbers versus the dilution of phage showed the assay to be linear. What impressed Delbrück about Ellis's experimental system was the way he could visualize individual virus particles as plaques and the fact that the plaque method provided quantitative access to virus replication.[41]

Delbrück and Ellis collaborated to achieve the synchronous growth of bacteriophage after infection, producing a "one-step" growth curve. Their one joint publication opens with the statement, "Certain large protein molecules (viruses) possess the property of multiplying within living organisms. This process, which is at once so foreign to chemistry and so fundamental to biology, is exemplified in the multiplication of bacteriophage in the presence of susceptible bacteria."[42] Ton van Helvoort summarizes their methodology clearly:

> The method was based on the use of a virulent bacteriophage, a young and fast-growing bacterial culture, and a high concentration of bacteria. In this way, adsorption of only one particle per bacterium was obtained after which the suspension was diluted to prevent adsorption of newly produced phage to other bacteria. Growth of bacteriophage was limited to *one cycle*, so the process was prevented from becoming diffuse, and normally an experiment lasted less than two hours.[43]

Much of the joint paper focused on the linear relationship between phage concentration and plaque count, which was irrespective of temperature, bacteria concentration, and agar concentration. This correlation suggested that phage were exogenous pathogens, reproduced by the cell after infection but genetically foreign. Ellis and Delbrück defined the fraction of "infective centers" which produce plaques as the "efficiency of plating," and then they went on to show that this proportion was reproducible (varying from 0.3 to 0.5) under the conditions of their assay (see fig. 6.1).[44]

This approach to bacteriophage differed significantly from the methods of other phage researchers, as Delbrück and Ellis noted. In particular,

40. Ellis, "Bacteriophage: One-Step Growth," 57. The plaque technique was adapted from d'Herelle.
41. Delbrück Oral History, interview by Carolyn Harding, 1978, California Institute of Technology Archives, 64; Kay, "Conceptual Models and Analytical Tools," 231.
42. Ellis and Delbrück, "Growth of Bacteriophage," 365.
43. Helvoort, "Controversy," 558; emphasis in original.
44. Ellis and Delbrück, "Growth of Bacteriophage," 368.

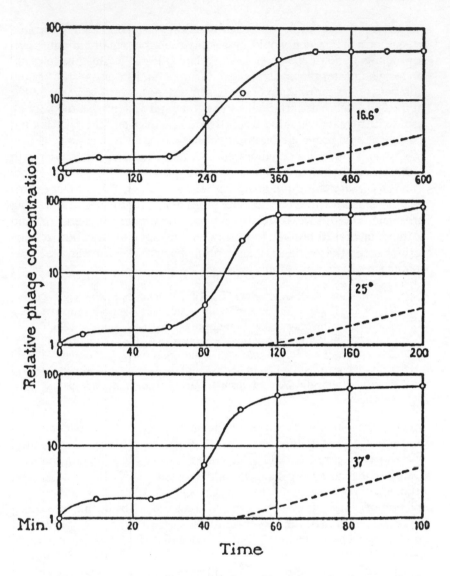

FIGURE 6.1 "One-step growth curves of phage. The curves, showing growth of phage in bacterial cultures at different temperatures, were obtained in the following manner: A suitable dilution of phage was mixed with a suspension of bacteria (containing 2×10^8 organisms per cc.). The inoculated cultures were then incubated for 10 minutes at the indicated temperature to allow more than 90% adsorption of the phage by bacterial cells. This mixture was then diluted $1:10^4$ and incubated. It was again diluted $1:10^4$ at the start of the first rise to decrease the rate of adsorption of the phage set free (by bacterial lysis) in the

Alfred Krueger and John Northrop performed studies of the kinetics of phage growth and bacterial lysis in the 1930s. They argued that bacteriophage was not an exogenous infectious agent, but rather an inactive enzyme precursor native to the bacteria. An increase in phage concentration, they argued, could stimulate this host precursor to produce more phage autocatalytically.[45] Their assay for phage growth was very different from Ellis and Delbrück's method, and they interpreted their own graphs of phage concentration versus time in terms of a chemical equilibrium between intracellular and extracellular phage. According to Krueger, adsorption of bacteriophage was a simple chemical process enabling the phage to diffuse into the bacterial cell.[46] Lysis of their host bacterial strain, *Staphylococcus aureus,* occurred when the ratio of phage to bacteria reached 125.[47] Thus, in their view, phage replication was not independent of normal bacterial multiplication, but rather characterized the growth of bacteria with metabolism gone awry. Northrop acknowledged that these different viewpoints had been part of scientific debate about bacteriophage since the earliest reports:

> The biological properties of bacteriophage have been determined by d'Herelle, who has shown that they are closely analogous to those of living cells.
> [Jules] Bordet, on the other hand, has shown that the facts may be more simply explained by assuming the autocatalytic production of phage from a normal cell constituent. It has been found recently that pepsin and trypsin, under the proper conditions, also increase, and there are, of course, many other known autocatalytic reactions, so that the property of increasing in quantity or growth cannot be considered a definite criterion for the presence of living cells. Stanley has isolated and crystallized a protein which appears to be the active agent of the tobacco mosaic, and the writer has obtained a nucleo-

45. On the d'Herelle-Bordet debate, see Helvoort, "Controversy"; and Summers, *Félix d'Herelle,* esp. ch. 5.
46. See Helvoort, "Controversy"; and Kay, "Virus, enzyme ou gène?"
47. Krueger and Northrop, "Kinetics of the Bacterium-Bacteriophage Reaction."

---

first cycle of growth. Log P/P_0 was plotted (this is the 'relative phage concentration'), P_0 being the initial concentration of infective centers and P the concentration at time t. The broken line indicates the growth curve of the bacteria under the corresponding conditions." (Images and caption reproduced from Ellis and Delbrück, "The Growth of Bacteriophage" *Journal of General Physiology* 22 [1939]: 376, by copyright permission of the Rockefeller University Press.)

protein which appears to be the bacteriophage. Some filterable
viruses, therefore, are probably enzymes which possess the property
of forming themselves under proper conditions.[48]

Northrop's interpretation of Stanley's isolation of TMV—that it was an
enzyme like pepsin or trypsin—undercut the prevailing assumption that
viruses were not derived from "normal cell constituents." This supposition
was especially important to Delbrück's selection of viruses as models for
genes. Delbrück sought to differentiate genetic from enzymatic behavior
in his debate with Northrop over the origin of bacteriophage, despite the
longstanding tendency to group together the "autocatalysis" of genes,
enzymes, and viruses.

Thus Northrop's notion that bacteriophage was an autocatalytic enzyme produced from a precursor in the bacterial cell was incompatible
with Delbrück's commitment to the bacteriophage as an autonomous and
genetic agent. But the disagreement between Delbrück and Northrop,
following in the pattern of d'Herelle and Bordet's dispute, also reflected
the different kinds of experimental systems they had constructed with
bacteriophage. Northrop followed the kinetics of phage growth, much
as he had traced the kinetic behavior of his purified digestive enzymes,
whereas Delbrück used bacteriological and statistical methods that distinguished phage growth from bacterial multiplication. In addition, unlike in
the case of TMV, whose genetic populations all over the world are highly
similar,[49] bacteriophage is far from being a unitary biological entity. There
are dozens of different phages that infect various microbes, and Ellis and
Delbrück had isolated one from Pasadena sewage that lysed *E. coli,* quite
literally a different organism from Northrop and Krueger's *Staphylococcus* parasite. Complicating the debate were the observations of lysogenic
phage, which seemed to arise spontaneously from uninfected strains of
bacteria. Delbrück rejected these observations out of hand, to the consternation of many prominent microbiologists.[50] Lysogeny did not fit with
his conception of viruses as exogenous and infectious agents.

In the fall of 1939, Delbrück reported back to the Rockefeller Foundation on the progress he was making. By this time, Ellis had been required
to give up the phage work and focus on research more directly related to
the cancer problem, to satisfy the conditions of his funding.[51] Delbrück,

48. Northrop, "Chemical Nature," 480.
49. Fraile et al., "Century of Tobamovirus Evolution"; Gibbs, "Evolution and Origins
of Tobamoviruses."
50. See Lwoff, "From Protozoa to Bacteria and Viruses," esp. 14–15.
51. In fact, the tensions between Ellis's aspiration to do basic research related to cancer
and Seeley Mudd's determination to get applicable results led to the termination of Ellis's
fellowship in 1941 (Summers, "How Bacteriophage Came to Be Used," 265).

taking charge of the phage project, stressed the virus-gene connection: "The leading idea was the belief that the growth of phage was essentially the same process as the growth of viruses and the reproduction of the gene. Phage was chosen because it seemed to offer the best promise for a deeper understanding of this process through a quantitative experimental approach."[52] Given his commitment to the problem of reproduction, Delbrück still devoted a surprising amount of time trying to reconcile his results with those obtained by Northrop and Krueger. He postulated "lysis from without" as a different mode of bacteriocide than the "lysis from within" he and Ellis had documented. The high ratio of phage to bacteria might precipitate such "lysis from without" as Northrop and Krueger had observed.[53]

Delbrück was struggling to get Northrop's staphylococcus phage to yield plaques for him when he met another phage researcher from Italy, Salvador Luria. Luria had recently immigrated to the United States from France, where he had worked with Eugène Wollman on the effects of radiation on bacteriophage.[54] Trained as a physician, Luria had a strong appreciation for physics, having conducted research in Enrico Fermi's laboratory. As Thomas Brock has pointed out, Luria had already developed a strongly quantitative approach to phage research; independently of Delbrück he worked out the Poisson distribution of phage-infected bacteria.[55] The wide interest in radiation as a tool for studying genetics supported Luria's and Delbrück's conviction that physics was fundamental to investigating heredity. In his 1940 article, "Radiation and the Hereditary Mechanism," Delbrück compared his own work on the inactivation of bacteriophage by ultraviolet light to similar studies of TMV inactivation by Alexander Hollander and B. M. Duggar, concluding that there was a "close relationship between these two types of particles."[56] In addition to the importance of new tools for inducing mutations, Delbrück and Luria took from radiation physics a statistical approach to their understanding of gene inactivation. Just as the effects of ionizing radiation had to be understood statistically, one could envision the bacterial cell as a "black box," in which the "output" from phage infection could be treated statistically,

52. "Report of Research Done During the Tenure of a Rockefeller Fellowship in 1938 and 1939," typescript dated 14 Sep 1939, Delbrück papers, folder 40.32, as quoted and cited in Summers, "How Bacteriophage Came to Be Used," 264.

53. Delbrück, "Growth of Bacteriophage"; Helvoort, "Controversy."

54. Wollman, Holweck, and Luria, "Effects of Radiation on Bacteriophage C_{16}."

55. Brock, *Emergence of Bacterial Genetics,* 123.

56. Delbrück, "Radiation and the Hereditary Mechanism," 354. The significance of radiation genetics for scientific attempts to apprehend the material gene is also apparent from the contributions to the Cold Spring Harbor Symposium on Quantitative Biology in 1941, on the topic "Genes and Chromosomes."

with the intent of distinguishing unexpected events from low-frequency but usual effects.[57]

Luria and Delbrück were already familiar with one another's publications on bacteriophage. After their initial meeting on 28 December 1940, at a meeting of the American Association for the Advancement of Science in Philadelphia, they retreated to Luria's New York laboratory to do some experiments together.[58] They joined forces again at Cold Spring Harbor the following summer (where Delbrück presented a speculative model for how genes made of polypeptides might duplicate),[59] and in 1942 Luria was awarded a Guggenheim which enabled him to spend time doing research in Delbrück's Vanderbilt laboratory.[60] Before collaborating, Delbrück and Luria had been studying different bacteriophages. They reasoned that if they could infect a suitable host with *both* of their phages, one might lyse the bacteria while the other was in the process of reproducing, revealing an "intermediate stage of virus growth" usually hidden within the cell.[61] Instead, they were surprised to find that infection with one bacteriophage (γ, later called T2) prevented the bacteria from producing another (α, later called T1), a phenomenon termed "interference." While this finding confounded their hopes of catching bacteriophage in the act of multiplying, it strengthened the argument that bacteriophages are true viruses, for interference had also been observed in plant and animal viruses.[62] If Delbrück sought to emphasize the negative analogy between bacteriophage and enzymes, he argued for a strong positive analogy between bacteriophage and viruses infecting other organisms.[63] Subsequent studies with another phage (δ, later called T7) enabled Delbrück and Luria to differentiate "mutual exclusion" (the fact that a

57. Kay, "Conceptual Models and Analytical Tools," 233.

58. Fischer and Lipson, *Thinking about Science;* Luria, *Slot Machine,* 32–37. During the war Luria was working in a laboratory at the College of Physicians and Surgeons at Columbia University.

59. Delbrück, "Theory of Autocatalytic Synthesis."

60. Delbrück accepted a faculty position in physics at Vanderbilt in 1940; see Kay, "Conceptual Models and Analytical Tools," 235–37.

61. Delbrück and Luria, "Interference between Bacterial Viruses. Part 1," 111.

62. As Delbrück wrote privately about this same time, "I will take for granted that phages and viruses are similar things, in the sense that the relation between phage and bacterium is essentially the same as that between virus and plant or animal cell. If proof of this assumption was needed, I believe that the occurrence of interference in both cases has sealed it. This assumption is helpful in limiting the possibilities for theories of the nature of phage" (Delbrück to Anderson, "A Conducted Tour by Easy Stages to the Last Paragraph, for T. F. A.," unpublished ms., [1943], Delbrück papers, folder 1.20). In plants, the phenomenon of "interference" had been termed "acquired immunity" by Louis Kunkel, and researchers searched in vain for antibodies in affected plants; see Kunkel, "Studies on Acquired Immunity"; Price, "Acquired Immunity from Plant Virus Diseases."

63. Delbrück, "Bacterial Viruses (Bacteriophages)."

single bacterium would only produce one type of virus at a time) from the "depressor effect," which referred to the observation that a bacterium infected with more than one phage produced less of the prevailing phage than it would otherwise. The interaction between different bacteriophage strains infected into one host became a tool for studying the dynamics of virus growth.

The contrast of this approach with Stanley's concurrent work on TMV is instructive. Whereas Stanley sought to understand TMV and its properties, including reproduction, in chemical terms, Delbrück and Luria focused on statistical methods to understand viral growth.[64] It was not that bacteriophage could not be studied chemically—this had been Northrop and Krueger's motivation in studying the kinetics of bacteriophage production—but that Delbrück and Luria believed that genetic and quantitative methods would reveal the underlying material mechanisms of viral self-duplication. But experimental systems can have a life of their own, as Hans-Jörg Rheinberger has pointed out, and Delbrück and Luria's bacteriophage studies, rather than cracking open the host cell to reveal the process of virus reproduction, made unexpectedly important contributions to bacterial genetics.[65] Furthermore, as we will see below, the subsequent discovery of recombination among bacteriophages took phage studies in the direction of the gene as an operational, rather than material, unit.

Luria and Delbrück had noted that after lysis, an infected bacterial culture could become turbid with cells again due to the growth of a bacterial variant resistant to the bacteriophage. Two hypotheses could explain the appearance of these variants: either (1) these were spontaneous mutations conferring resistance to bacteria that were subsequently selected for in the presence of phage, or (2) exposure to the phage was inducing specific adaptation to a resistant state. Only the first type of mutation, a true genetic change, would be passed onto subsequent generations, whereas the second, an acquired immunity, would be specific to the cells induced by the presence of phage. Luria realized (drawing on the metaphor of a slot machine) that the fluctuation in the number of resistant cultures provided the basis for an experimental test of the two hypotheses (see fig. 6.2). For true hereditary mutations, a mutation early after inoculation would result in a large number of resistant cells—a "jackpot." By contrast, acquired immunity would be expected to occur at a regular rate in bacteria that survived infection. The acquired resistance would not be shared by genetic clones. Thus one could discriminate between the two mechanisms through the degree of

64. Summers, "Concept Migration."
65. See, for example, Rheinberger, *Toward a History of Epistemic Things*, 67.

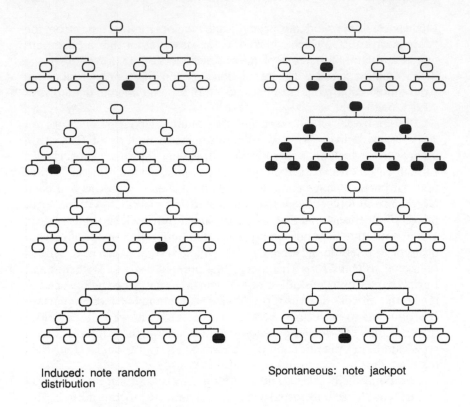

Induced: note random
distribution

Spontaneous: note jackpot

FIGURE 6.2 "Diagram explaining the theory of the fluctuation test. The early
stages in the development of several small cultures are shown. Resistant
bacteria are dark. *Left:* Anticipated results if resistant bacteria are induced by
contact with phage. In this case, the proportion of resistant cells in the
population should be more or less the same in each culture. *Right:* Anticipated
results if resistance arises randomly by mutation. Families of resistant siblings
should develop, and depending on when the mutation occurred, the sizes of
these resistant clones should vary widely." (Diagram and caption from Brock,
Emergence of Bacterial Genetics, 59.)

fluctuation in the number of altered bacteria—"large fluctuations [in the
number of resistant bacteria] are a necessary consequence of the mutation
hypothesis."[66]

As it turned out, the experiment showed a jackpot effect: fluctuations
in the number of bacteria resistant to bacteriophage among the otherwise

66. Luria and Delbrück, "Mutations of Bacteria," 494. See Luria, *Slot Machine,* 74–79,
for his account of the experiment, and see Brock, *Emergence of Bacterial Genetics,* 58–63,
for an insightful analysis.

identical cultures were significantly greater than could be accounted for by the acquired immunity hypothesis. This observation was a landmark for bacterial genetics, providing concrete evidence that inheritance in bacteria was not Lamarckian.[67] In addition, the finding cast light on antibiotics and the emergence of resistant bacterial cultures. This issue was particularly important during the war years; even at Cold Spring Harbor, defense-related research was underway on mutations to enhance penicillin production in *Penicillium chrysogenum*.[68]

However, the experiment did not advance the understanding of virus reproduction, to the evident frustration of Delbrück. As he asserted in his 1946 Harvey Lecture, "Remember that what we are out to study is the multiplication process proper; we want to get to the bottom of what goes on when more virus particles are produced upon the introduction of one virus particle into a bacterial cell. All our work has circled around this central problem."[69] By contrast, he referred to the fluctuation paper as "a side issue which is threatening to displace the main issue [of virus replication], by virtue of its explosive content of possibilities for studying bacterial genetics."[70] Indeed, the experimental system had already expanded to include the genetics of the host as well as of the virus.

In 1943 Delbrück invited another phage researcher, Alfred Hershey, to visit him at Vanderbilt. He described Hershey to Luria this way: "Drinks whiskey but not tea. Simple and to the point. Likes living in a sailboat for three months, likes independence."[71] Hershey worked in collaboration with J. J. Bronfenbrenner, a student of Bordet's, at Washington University in St. Louis. Bronfenbrenner's diffusion experiments suggested that bacteriophage particles were much smaller than Delbrück and Luria reckoned, and Hershey continued to have reservations about the question of virus size.[72] As he wrote Delbrück, "Science is very discouraging. There are a number of things I would like to do with phage, but I seem to be hung up at the moment on the first question, which is whether phage is what

67. As William Summers states, "In the views of most molecular biologists, it was the work of Salvador Luria and Max Delbrück on bacteriophage-resistant variants of *Escherichia coli* that established the validity of bacteria as organisms for genetic studies. Thus, it was not until the mid-1940s that bacterial genetics gained the respectability and scientific legitimacy long accorded to fruit flies and corn" ("From Culture as Organism," 172).

68. Brock, *Emergence of Bacterial Genetics*, 126.

69. Delbrück, "Experiments with Bacterial Viruses," 162.

70. Ibid., 180.

71. Delbrück to Luria, 3 Feb 1943, quoted in Luria, "Mutations of Bacteria," 175.

72. The size of bacteriophage was at issue not only because Northrop and Krueger understood phage as simple chemical particles (as opposed to the more biological conception of Delbrück and Luria), but also because measurements by Bronfenbrenner at Washington University had shown the phage to be much smaller than the electron micrographs revealed.

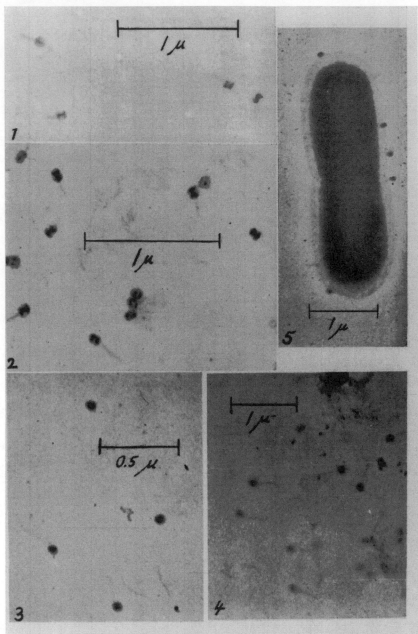

1

2

3

0.5 μ

4

1 μ

5

1 μ

(S. E. Luria, M. Delbrück and T. F. Anderson: Electron Microscope Studies of Viruses)

you and Luria think, or what K.[Kimura] and B.[Bronfenbrenner] (and most of the time I) think. I can't say I have any very brilliant ideas how to decide this. . . . You appear to regard this question as settled."[73] Delbrück confirmed Hershey's statement: "It is true that I think the issue regarding the phage size is settled, for the time being at least." He listed several reasons for believing the bacteriophage was quite large (on the order of 100 million daltons in molecular weight), but perhaps most persuasive were the pictures of phage taken that year using an electron microscope.[74]

Thomas F. Anderson was at that time working as a postdoctoral fellow funded by Radio Corporation of America (RCA) to demonstrate the value of their new commercial electron microscopes to biologists through collaborations (see chapter 4, "Presenting Laboratory Viruses").[75] The pictures he took of Delbrück and Luria's samples at Woods Hole in the summer of 1943 showed the bacteriophage to be tadpole-shaped particles, with shapes distinctive for each type (see fig. 6.3).[76] Particles of phage α, later called T1, had a round head, 45–50 mμ in diameter, attached to a tail 150 mμ long and less than 15 mμ wide. Particles of phage γ, later called T2, presented "a very peculiar aspect. To an oval head, 65 × 80 mμ, a straight tail, 120 mμ long and 20 mμ thick, is attached at one of the narrow poles."[77] The darkness of these oval heads suggested that the distribution of solid material, particularly the high-scattering phosphorus of DNA, was not uniform. But the most surprising result was that the bacteriophage seemed to remain adsorbed on the outside of bacteria, even after infection had taken place.

Delbrück and Luria were inclined to interpret this observation by analogy with fertilization—inferring that only the first virus to come into contact could enter the cell, like the first sperm to penetrate an egg. "The pictures here reproduced, if interpreted on the assumption that one virus

73. Hershey to Delbrück, 6 Oct 1943, Delbrück papers, folder 10.34.

74. Delbrück to Hershey, 9 Oct 1943, Delbrück papers, folder 10.34; Luria and Anderson, "Identification and Characterization of Bacteriophages." Similar images had been obtained first in Germany: see Ruska, "Über ein neues bei der bakteriophagen Lyse auftretendes Formelement."

75. Rasmussen, "Making a Machine Instrumental," and Picture Control, chs. 1 and 2.

76. Luria, Delbrück, and Anderson, "Electron Microscope Studies."

77. Ibid., 59.

FIGURE 6.3 Micrographs of phages γ and α. "Panel 1: Particles of phage γ ×36,000. Panel 2: Particles of phage γ ×40,000. Panel 3: Particles of phage α ×47,000. Panel 4: Particles of staphylococcus phage ×20,000. Panel 5: E. coli plus phage γ after 15 minutes contact; picture shows a bacterium with adsorbed particles of virus ×20,000." (Images and captions from Luria, Delbrück, and Anderson, "Electron Microscope Studies of Bacterial Viruses" [1943]: 69. Reprinted by permission of the American Society for Microbiology.)

particle enters the cell, would indicate that the entry of one virus particle bars the entry of other virus particles by making the bacterial cell-wall impermeable to them."[78] Such an explanation fit with the observation of mutual exclusion. But Anderson regarded the virus-as-sperm idea as highly speculative:

> The analogy with fertilization seems to me to be amusing and can be carried for some way, ... but I would not want to take it seriously until I knew a lot more of what goes on in the phage reaction, or for that matter, in the egg.... Of course the analogy breaks down completely when we consider the products of the two reactions for phage-infected bacteria are lysed with the production of more phage while the fertilized egg has never been observed to break down with the production of more sperm.[79]

Delbrück believed that Anderson's micrographs demolished the Northrop-Krueger precursor theory of phage formation. As stated in the joint paper,

> In a series of papers Krueger (1938) has proposed the idea that the bacterial cell contains a precursor of the virus particle, which, upon infection of the cell with a virus particle, is promptly converted into virus. This theory was elaborated as an analogue to the well-known relations between proteolytic enzymes and their precursors. According to this theory an uninfected bacterium of the strain here considered should contain on the average 140 precursor particles of virus α and 135 precursor particles of virus γ. The pictures show clearly that this is not the case, since bacteria lysed under the influence of virus γ show no evidence of particles resembling virus α and vice versa.[80]

Yet the analogy with sperm which Delbrück favored also contained an assumption of relatedness between host and virus, akin to Northrop and Krueger's. As Anderson put it to Delbrück, "Does [the relatedness of phage to bacterium] mean that the virus is like a sperm in relation to the bacterium—that the bacterium and the phage have a gene or more in common? If so, these genes would be virus precursors."[81] This line of analysis resurrected the idea that bacteriophage might be endogenous, after

78. Ibid., 65.
79. Letter and appended note from T. F. Anderson to Salvador Luria and Max Delbrück, quoted in Anderson, "Electron Microscopy of Phages," 67–68.
80. Luria, Delbrück, and Anderson, "Electron Microscope Studies," 65; also quoted in Anderson, "Electron Microscopy of Phages," 66.
81. Letter from Anderson to Delbrück quoted in Anderson, "Electron Microscopy of Phages," 68.

Delbrück's valiant efforts to disprove Northrop's precursor hypothesis.[82] As Anderson teased Delbrück, "If we could show that phage can give rise to the bacteria, we might have a stronger case."[83]

More broadly, Delbrück speculated: "The genes are particles of virus size which grow and divide in step with the cell and which may or may not be essential to the cell. . . . Transmissible genes and really wild growing genes are not known, but some people have suspected that the viruses are just this group of missing cases."[84] But microscopy and other structural investigations of bacteriophage were not bringing virus genes, and their mode of multiplication, into clearer view.

Luria, Delbrück, and Anderson's paper was published three years after the first electron micrographs of TMV had been published by the German researchers, and two years after the same group published the first electron micrographs of bacteriophage.[85] During these two years, Anderson had collaborated with Stanley on plant viruses (see chapter 4, "Presenting Laboratory Viruses"), and animal virologists had published electron micrographs of influenza virus, papilloma virus, equine encephalomyelitis virus, and vaccinia virus.[86] There was thus a good basis for comparing the morphologies of the viruses (despite continuing worries about the artifacts that might be generated by the harsh specimen preparation and irradiation techniques).[87] Bacteriophage seemed to have a structure even more complex than that of most animal viruses, and the apparent fact that the structure of phage did not provide a simple template for its multiplication led Luria, Delbrück, and Anderson to caution against a simple view of virus structure and reproduction:

> The pictures here reproduced give the impression of a some-
> what more complex organization than the pictures of plant viruses
> (Stanley and Anderson, 1941) had indicated. These had shown ei-
> ther straight rods or round particles [and this view was corroborated

82. See Helvoort, "Construction of Bacteriophage."

83. Letter from Anderson to Delbrück, quoted in Anderson, "Electron Microscopy of Phages," 67.

84. Delbrück, "Conducted Tour by Easy Stages to the Last Paragraph, for T. F. A.," "unpublished ms., [1943], Delbrück papers, folder 1.20; also quoted in Anderson, "Electron Microscopy of Phages," 70.

85. Kausche, Pfankuch, and Ruska, "Die Sichtbarmachung von pflanzlichem Virus im Übermikroskop"; Ruska, "Die Sichtbarmachung der bakteriophagen Lyse im Übermikroskop"; Pfankuch and Kausche, "Isolierung und übermikroskopische Abbildung eines Bakteriophagen."

86. Stanley and Anderson, "Study of Purified Viruses"; Sharp et al., "Study of the Papillomatosis Virus Protein," and "Electron Micrography"; Green, Anderson, and Smadel, "Morphological Structure."

87. On this issue, see Rasmussen, "Facts, Artifacts, and Mesosomes."

by micrographs of several animal viruses].... A word may be added regarding the tendency to speak of viruses as molecules. This tendency received its greatest momentum from Stanley's discovery in 1935 that paracrystals of tobacco mosaic virus could be obtained by simple methods, and from the great number of subsequent studies of this and of other viruses along the lines of protein chemistry.... While no harm is done by calling viruses "molecules," such a terminology should not prejudice our views regarding the biological status of the viruses, which has yet to be elucidated.[88]

In order to approach the more "biological" aspects of viruses, Delbrück tried to advance phage research in two ways. First, during the summer of 1944, Delbrück persuaded other like-minded researchers interested in phage to restrict their attention to the set of phages that he, Luria, and Hershey were studying. These were the T-odd and T-even phages, in which T stood for type. In particular, T1, T2, T3, T4, T5, T6, and T7 were distinguishable serologically and by the use of host strains resistant to each phage.[89] This "phage treaty" guided the growing literature on phages and restricted membership in what came to be called the "phage group" according to shared experimental material. Delbrück was clearly the inspirational leader of this group of researchers, and the summer "phage course" which he began at Cold Spring Harbor in 1945 helped to solidify their commitments to a particular set of materials and methods through pedagogy.

Second, for a time Delbrück sought to emulate in *E. coli* the biochemical genetics that George Beadle and Edward Tatum had achieved with the pink bread mold *Neurospora crassa*.[90] As Delbrück wrote Beadle early in 1944,

> You may have seen a little paper by Luria and myself in "Genetics," on bacterial mutants which are resistant to a phage. E. H. Anderson (van Niels student), who is working here this year, has since isolated a great number of such mutants, characterized by resistance to one or another of three phages, or to pairs of them. The idea is to study the metabolic differences between these strains.
>
> There seems to be some difference in growth factor requirements between these strains, since the stock strain grows well on a pure glucose-asparagine mineral medium, while most of the mutants grow poorer or not at all on it. We would like to try and imitate your neurospora [*sic*] work with these mutants, and I am writing to ask

88. Luria, Delbrück, and Anderson, "Electron Microscope Studies," 63–64.
89. Brock, *Emergence of Bacterial Genetics,* 128.
90. For historical accounts of Beadle and Tatum's work, see Kay, "Selling Pure Science in Wartime"; and Kohler, "Systems of Production."

you to let [us] have some reprints of your papers, particularly any that
deal with the technique of tracking down the factor requirements.[91]

Neurospora served as a model for phage researchers, illustrating my general argument about the role of successful model systems as exemplars
guiding the experimental choices of researchers on other systems. Specifically, the *Neurospora* model inspired the isolation of bacterial mutants
to illuminate the problem of virus multiplication. Delbrück also drew
the analogy between *Neurospora* genes and virus genes, using nutritional
mutants of T4 bacteriophage to try to study virus genetics biochemically.
However, this approach led Delbrück into a biochemical quagmire.[92]

Indeed, to take seriously the biochemistry of bacteriophage infection
and reproduction challenged some of the basic features of Delbrück's
original phage system, particularly his desire to ignore the host cell. As
Seymour Cohen put it,

> The early experiments begun in Delbrück's school have seemed to
> me to embody a methodology somewhat different from that which I
> adopted as the basis of chemical work. He treated a bacterium as a
> sealed container into which a virus had been inserted and from which
> virus progeny had emerged; the quantitative yield and the qualita-
> tive nature of the progeny were to permit inferences concerning the
> nature of the intracellular process. Delbrück seemed to feel that the
> major and simplest approach was that afforded by control of the in-
> put or parental virus. Indeed, I once wagered that I could prove to
> him that the period of multiplication of a virus inside its bacterial host
> (the latent period) was not independent of nutritional conditions, as
> he had suggested in his [Cold Spring Harbor phage course] class in
> 1946. In retrospect, his early view appears to contain many of the lim-
> itations of the phage biology of the period. If the bacterium is indeed
> a virtually sealed phage duplicator, essentially inert to the medium
> in which it is placed, there is really very little to be done to alter the
> experimental conditions, other than to change the input virus and
> to examine the progeny carefully. That this fundamental experiment
> is powerful in its own right cannot be gainsaid; it is in fact the key
> experiment of phage genetics, and its ingenious variations represent
> a large part of the armamentarium of molecular biology. The con-
> trary notion, that the host cells were not invariant elements but were
> responsive to environmental conditions and could be followed by nu-
> merous other parameters, including the fate of metabolites present
> in the media, was merely a part of my inheritance as a biochemist
> which I undertook to apply.[93]

91. Delbrück to Beadle, 8 Jan 1944, Delbrück papers, folder 2.10.
92. Delbrück, "Biochemical Mutants."
93. Cohen, *Virus-Induced Enzymes,* 6–8.

Thus the biochemistry of phage infection, and particularly the interaction of phage and host in the production of cellular enzymes, required a different kind of experimental system, in which environmental and cellular elements were more than background to viral multiplication.[94] Cohen was the individual who most successfully developed this line of work with the T phages; he along with other biochemists (e.g., Earl Evans, Frank Putnam, and Lloyd Kozloff) began using radioisotopes to trace changes in cellular metabolism following infection.

By contrast, what followed most directly from Delbrück and Luria's collaborative work was not biochemistry nor biochemical genetics, but a wealth of classical genetic studies of bacteriophage. In 1945, Luria used the fluctuation test methods to show that phage mutations, like those of their hosts, arose spontaneously and could be selected for genetically.[95] The following year, Hershey and Delbrück both performed experiments showing that phages could be crossed genetically, based on the observation that new phenotypes arose from mixed infection with two mutant strains. These results were interpreted initially by Delbrück as an instance of induced mutation.[96] However, Hershey soon produced unequivocal evidence for crossing-over, and in 1948 used his results with Raquel Rotman to produce the first genetic map of a virus (see fig. 6.4).[97] As Thomas Brock has astutely noted, phrases such as "mutual exclusion" that had been coined to describe phenomena in phage work were soon replaced by the classical genetic terminology of "loci" and "linkage."[98] But with this shift, the conceptualization of a phage gene also shifted from the *material* entity Delbrück had been trying to catch in the act of replicating to the *operational* entity mapped abstractly through crosses. Moreover,

94. See, for example, Cohen, "Growth Requirements."
95. Luria, "Genetics of Bacterium-Bacterial Virus Relationship"; Brock, *Emergence of Bacterial Genetics*, 131–32.
96. In Hershey's case, the phenotypes he used in his crossover experiments concerned host range and rapid lysis. As Brock notes (*Emergence of Bacterial Genetics*, 133), Delbrück was more cautious than Hershey in seeing the new phenotypes as a result of genetic crossing-over. The titles of their respective articles (published in the same volume of *Cold Spring Harbor Symposia on Quantitative Biology*) reveal the way in which the fluctuation experiment set the terms for the interpretation of the mixed phage infections: Hershey, "Spontaneous Mutations in Bacterial Viruses"; Delbrück and Bailey, "Induced Mutations in Bacterial Viruses." As Brock also points out, "The first interpretation of the phage results as crossing-over was actually made by Hermann Muller" (134), following a presentation in January 1946 by Delbrück at a mutation conference. Delbrück attempted to benefit from his foray into biochemical genetics by using the biochemical mutants to isolate new independent mutations for the mixed infection studies, but without success. See Delbrück to Hershey, 2 Sep 1946, Delbrück to Hershey, 11 Sep 1946, Delbrück papers, folder 11.1.
97. Hershey and Rotman, "Linkage Among Genes," and "Genetic Recombination."
98. Brock, *Emergence of Bacterial Genetics*, 137.

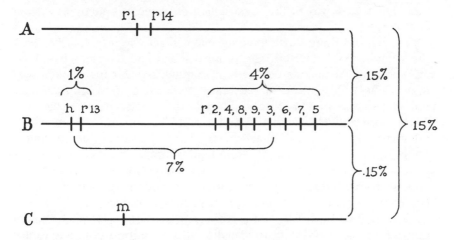

Figure 6.4 "Schematic diagram depicting linkage relations among mutants of phage T2. The percentages indicate yields of wild type in two factor crosses." (Diagram and caption from Hershey and Rotman, "Genetic Recombination between Host-Range and Plaque-Type Mutants" [1949]: 49. Reprinted by permission of the Genetics Society of America.)

the phage could no longer be considered to be a single gene; it contained a genome, if small. The implicit reference system was now *Drosophila,* the exemplar of chromosomal mapping.[99] The shift toward the instrumental gene in phage research was reinforced by the intensive work on recombination with T phage strains in the 1950s. After moving back to Caltech in 1947, Delbrück suggested to Hershey that they drop both biochemistry and genetics in order to become "gadgeteers" in the study of phage, but in fact he did not take up the structural studies of phage.[100] Thus Delbrück's early intention to use bacteriophage to study the gene as a material entity

99. See Kohler, *Lords of the Fly.*

100. From a letter from Delbrück to Hershey, 5 Dec 1947, Delbrück papers, folder 11.1: "I have a lovely and startling idea: let us not work on either genetics or biochemical mutations but instead play with the gadgets they have here on the [Caltech] campus. I have done some snooping around in Chemistry and Physics, and found that they not only have wonderful Centrifuges and Electron Microscopes and Tiseliuses etc., but also have competent people in charge of these gadgets who are anxious to do something sensible with the instruments. I think it would be nice to see whether we could not do better than the professional gadgeteers. The little run we had in Woods Hole [with electron microscopy] four years ago was quite encouraging. Now we know so much more about phage and the gadgets have become so much better and we would have plenty time [*sic*] and competent assistance; for all these reasons I think we should discover something, maybe even something about the problems we are interested in now."

qua Muller gave way to the construction of phage as a mapping system *qua* Morgan.

Throughout these developments, one can see the interplay between the suitability of bacteriophage for a particular line of research and the intentions of various researchers in shaping actual experimental systems. Robert Olby has declared that "in choosing an appropriate system in which to study replication, clearly the superiority of phage over *Neurospora* and *Drosophila* played a part."[101] This assessment of the inherent "superiority" of phage overlooks the ways in which bacteriophage was used for several different kinds of experimental systems—its growth was rendered as an enzymelike system by Northrop and Krueger, and its enzyme-inducing properties were followed by biochemists such as Cohen. The amenability of bacteriophage for quantitative studies was an achievement of Delbrück, not a given. Subsequently, Delbrück's initial purpose to use the system to visualize virus multiplication, and hence get at the material gene, was thwarted by the successes of classical genetic techniques for mapping genes. And along the way, systems such as *Neurospora* and *Drosophila* provided models for the experimental trajectory of bacteriophage.

TMV as an Alternative Model for Reproduction

During the same period, researchers sought to use TMV to understand heredity. In fact, genetic studies of tobacco mosaic disease were well established before Stanley began his chemical isolation of the virus. In a 1923 research review, B. M. Duggar and Joanne Karrer Armstrong had suggested that TMV was "a particle of chromatin or of some structure with a definite heredity, a gene perhaps."[102] Three years later, H. H. McKinney first isolated a yellow-mottled variant of TMV and demonstrated the heritability of its distinct features.[103] McKinney also gathered naturally occurring strains of TMV from the Canary Islands, West Africa, and Gibraltar and on this basis suggested that the virus could, like an organism, mutate. "A satisfactory interpretation of the yellow and green mosaic associations cannot be given at this time because so few data are available. . . . It seems entirely possible that viruses may become altered locally in the plant, thus producing mutations, to use this term in its broadest meaning."[104]

101. Olby, *Path to the Double Helix,* 227.
102. Duggar and Armstrong, "Indications."
103. McKinney, "Virus Mixtures." On McKinney's contributions, see Lederman and Tolin, "OVATOOMB," 242. In 1925, Eubanks Carsner had first reported on the attenuation of virus strains in plants, although he does not explicitly discuss it in terms of mutations; see "Attenuation of the Virus."
104. McKinney, "Mosaic Diseases," 576.

Several years later James H. Jensen, who worked alongside Stanley in Louis Kunkel's laboratory at the Rockefeller Institute, corroborated McKinney's field observations through his own experimental studies of TMV strains and documented the reversion of a yellow variant to the original mosaic strain (see chapter 3, "Bawden and Pirie").[105]

By virtue of his coworker Jensen's dozens of distinct strains of TMV, Stanley had at hand a rich store of genetic variants whose isolated virus proteins might be compared to that of generic TMV. Stanley and Hubert Loring were able to isolate the same kind of needle crystals from the yellow strain, aucuba mosaic virus, as Stanley had from TMV, and the chemical characteristics of this virus protein were similar, but not identical, to those of TMV.[106] Given both the similarities and differences between TMV and aucuba mosaic, Stanley proposed that aucuba was a genetic mutant of the regular tobacco mosaic strain:

> In view of the evidence that ordinary tobacco mosaic virus may, in the plant, mutate or in some manner become altered so that new and different strains of virus are produced, it seems logical that the new virus protein so evolved should be fairly closely related to the original tobacco mosaic virus protein. The manner in which this change from a given virus protein to a different virus protein takes place is, of course, open to much speculation.[107]

Stanley himself drew attention to the paradox of a mutation arising in a nonliving thing.[108]

As Muriel Lederman and Sue Tolin have pointed out, the study of naturally occurring mutants that could be done with a viral system was

---

105. Jensen, "Isolation of Yellow-Mosaic Viruses." In 1935 McKinney demonstrated that the yellow type of tobacco mosaic "can be propagated as a continuous pure strain" (as summarized via a Science Service release in *Nature* 137 [1936]: 540). As Lederman and Tolin have observed, the notion that plant viruses were authentically genetic was met with skepticism by some botanists ("OVATOOMB," 245).

106. Scientific Reports, RAC RU 439, box 6, vol. 25, 1936–37, 263. As mentioned in chapter 3, the protein isolated from the aucuba mosaic strain had a greater molecular weight and a more alkaline isoelectric point, and it formed larger crystals. In other respects, including chemical composition and serological properties, aucuba mosaic virus was indistinguishable from ordinary TMV.

107. Stanley, "Chemical Studies on the Virus of Tobacco Mosaic. Part 8," 339. Also, having accepted the British finding of nucleic acid, Stanley set Loring to isolate and chemically analyze the nucleic acid from TMV (Scientific Reports, RAC RU 439, box 6, vol. 26, 1937–38, 268).

108. "Mutations in Tobacco Mosaic Virus," quotation in a Science Service release in *Nature* 137 (1936): 540. See also Stanley, "Reproduction of Virus Proteins," esp. 115–16, in which he states, "Now, this ability to reproduce and to mutate only when within the living cells of certain hosts is a property that we have not hitherto ascribed to ordinary molecules."

quite distinct from the chromosomal mapping which statistical studies of crosses could achieve, as exemplified in *Drosophila*.[109] But, based on Muller's definition of the gene a decade earlier as a substance that can both mutate and duplicate, TMV was an excellent model for research on the material, rather than instrumental, gene.[110] It is perhaps not surprising that Muller advocated the study of mutations as the best strategy for achieving a biological assay for the gene.[111]

In 1936, the same year that Muller drew attention to the work of Stanley and Loring for its promise for genetics,[112] Stanley was contacted by Alfred Blakeslee, of the Carnegie Institute of Washington's Department of Genetics at Cold Spring Harbor. Blakeslee offered to collaborate with Stanley on an experiment to see if X-rays could be used to induce mutations in TMV.[113] Stanley informed Blakeslee that W. C. Price of Kunkel's laboratory was already working on the problem, and he offered his own speculation concerning the physical nature of mutations: "Since I am a chemist, I prefer to regard the phenomenon as isomerization or rearrangement within the molecule to give similar or isomeric proteins. Such isomers might be expected to have different physical and chemical properties, and hence different biological properties."[114] Price and John Gowen's study of the inactivation of TMV by X-rays suggested that "the absorption of a single unit of energy in a virus particle is sufficient to cause inactivation of the particle."[115] The survival curve they obtained for TMV resembled that observed for *Drosophila* sperm, indicating that viruses and genes were fundamentally similar:

> The virus of tobacco mosaic is composed of particles the size of which is estimated to be of the same order as that of genes. Tobacco-mosaic virus also resembles genes in other respects; both are incapable of reproduction outside of living cells, they produce similar effects, as, for instance, variegation or mottling, in plants, and they are, under natural conditions, capable of mutating to new forms which retain the ability to reproduce themselves. The virus differs

109. Lederman and Tolin, "OVATOOMB."   111. Falk, "What Is a Gene?" 151.

110. Muller, "Gene as the Basis of Life."   112. Muller, "Physics in the Attack."

113. "We have a considerable program here centered around mutations and the nature of the gene and have gotten in the habit of thinking about this problem. Dr. Demerec, as you probably know, has been particularly engaged in an intensive study of the gene by the use of *Drosophila*. We have an X-ray machine which we use in inducing mutations which we should be glad to place at your disposal" (Blakeslee to Stanley, 21 Jan 1936, Stanley papers, carton 6, folder Blakeslee, Albert Francis).

114. Stanley to Blakeslee, 22 Jan 1936, Stanley papers, carton 1, folder Jan 1936.

115. Gowen and Price, "Inactivation of Tobacco-Mosaic Virus," 536.

from genes in being able to move from cell to cell and in being capable of inoculation into the cells of healthy plants.[116]

Like Gowen and Price, Stanley put increasing emphasis on the significance of the size similarity between TMV and genes. This drew on the emphasis of his research in the late 1930s on structural studies of the virus: the analytical ultracentrifuge, double-refraction-of-flow apparatus, and electron microscope provided information on the size and shape of TMV (see chapter 4). This experimental orientation shaped the way in which Stanley approached the virus as an analogue to genes. As TMV was recast from a simple crystal to a complex macromolecule, Stanley's continuing work on virus strains took the form of amino acid analysis of TMV and its variants, in the hope of detecting chemical differences among them that would reveal the nature of mutations at the molecular level. It was a daunting task, as Stanley made clear:

> The problems of chemical structure presented by tobacco-mosaic virus appear to be about as difficult and complex as those that would be presented in an attempt to learn the inner workings and structural details of the Empire State Building by means of a casual inspection. However, when faced with such a problem, the organic chemist usually takes refuge in a study of the degradation products obtained by destroying the original structure by various means.[117]

The chemical characterization of TMV suggested that differences in strains would be detectable. A. F. Ross in Stanley's laboratory had used colorimetric assays to determine the amino acids found in native TMV: "Definite and reproducible amounts of thirteen different amino acids were found."[118]

At the 1941 Cold Spring Harbor meeting entitled "Genes and Chromosomes, Structure and Organization," Stanley and C. Arthur Knight reported on the chemical differences between strains of TMV. Their paper opens with a justification of treating the virus as a model for the gene:

> There is a striking similarity between the properties that have been found for the viruses that have been isolated in the form of high molecular weight nucleoproteins and the properties that have been ascribed to genes (Muller 1935; Demerec 1939; Stanley 1940). As may be seen from Figure 1 [fig. 6.5 of this chapter], one estimate of the

116. Ibid.
117. Stanley, "Chemical Structure and the Mutation of Viruses," 42–43.
118. Ibid.

COMPARATIVE SIZES OF VIRUSES

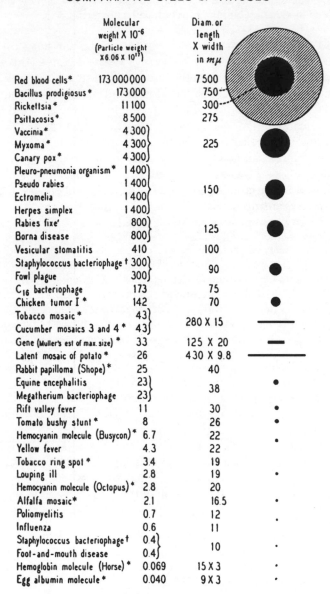

	Molecular weight X 10⁻⁶ (Particle weight X6.06 X 10¹⁷)	Diam. or length X width in mμ
Red blood cells*	173 000 000	7 500
Bacillus prodigiosus*	173 000	750
Rickettsia*	11 100	300
Psittacosis*	8 500	275
Vaccinia*	4 300 ⎫	
Myxoma*	4 300 ⎬	225
Canary pox*	4 300 ⎭	
Pleuro-pneumonia organism*	1 400 ⎫	
Pseudo rabies	1 400 ⎪	
Ectromelia	1 400 ⎬	150
Herpes simplex	1 400 ⎭	
Rabies fixe′	800 ⎫	
Borna disease	800 ⎭	125
Vesicular stomatitis	410	100
Staphylococcus bacteriophage †	300 ⎫	
Fowl plague	300 ⎭	90
C₁₆ bacteriophage	173	75
Chicken tumor I *	142	70
Tobacco mosaic *	43 ⎫	
Cucumber mosaics 3 and 4 *	43 ⎭	280 X 15
Gene (Muller's est of max.size) *	33	125 X 20
Latent mosaic of potato *	26	4 30 X 9.8
Rabbit papilloma (Shope)*	25	40
Equine encephalitis	23 ⎫	
Megatherium bacteriophage	23 ⎭	38
Rift valley fever	11	30
Tomato bushy stunt *	8	26
Hemocyanin molecule (Busycon)*	6.7	22
Yellow fever	4.3	22
Tobacco ring spot *	3.4	19
Louping ill	2.8	19
Hemocyanin molecule (Octopus)*	2.8	20
Alfalfa mosaic*	2.1	16.5
Poliomyelitis	0.7	12
Influenza	0.6	11
Staphylococcus bacteriophage †	0.4 ⎫	
Foot-and-mouth disease	0.4 ⎭	10
Hemoglobin molecule (Horse)*	0.069	15 X 3
Egg albumin molecule *	0.040	9 X 3

FIGURE 6.5 Stanley and Knight's chart presents the relative sizes of several selected viruses, including TMV and bacteriophages, as compared to those of red blood cells, *Bacillus prodigiosus,* rickettsia, pleuropneumonia organism, and protein molecules. An asterisk (*) indicates that evidence regarding shape is

approximate size of a gene would place it in the midst of the viruses, many of which are larger and many of which are smaller. Both viruses and genes may be regarded as large nucleoproteins that have the ability to perpetuate themselves by reproduction within, and only within, certain specific living cells. Both may undergo sudden changes either spontaneously or as a result of irradiation, and these changes are then faithfully reproduced in subsequent generations. Although it has not proved possible to isolate and study genes *in vitro* without loss of their viability, in the case of the viruses it is possible to remove them from their host cells, obtain them in pure form, and subject them to extensive study *in vitro* without impairment of their peculiar activity. In addition, it is possible to change the structure of a virus *in vitro* by chemical means and then determine the effect of the change by inoculating a host with the altered virus.... At the present time the points of similarity between viruses and genes appear to be sufficient to warrant an examination of some recent information on the chemical composition of variant strains of tobacco mosaic virus as a possible means of learning something of the nature of gene mutation.[119]

Believing that the various strains represented mutations "of a virus or also of a gene," they sought to correlate them with chemical differences. They first had to discern which part of the virus was altered by mutation: "It is possible that such changes might occur in the nucleic acid, or the protein components, or in both."[120] The German workers E. Pfankuch, G. A. Kausche, and H. Stubbe had reported that X-rays and gamma rays induced mutations of TMV, apparently altering the amount of nucleic acid (as detected by altered phosphorus content).[121] By contrast, Stanley and Knight localized chemical changes between naturally arising strains of TMV in the amino acid composition, inferring that "the chemical differences between strains probably lie not in the nucleic acid but rather in the protein part of the virus molecule."[122]

Stanley and Knight focused attention on the strain-dependent differences in amounts of aromatic amino acids, which were easier than other amino acids to assay using small amounts of purified virus. In addition,

119. Stanley and Knight, "Chemical Composition of Strains," 255–56.
120. Ibid., 256–57.
121. Pfankuch, Kausche, and Stubbe, "Über die Entstehung."
122. Stanley and Knight, "Chemical Composition of Strains," 257.

---

available; the dagger (†) indicates large size as determined from filtration and sedimentation of concentrated solutions and small size from diffusion of dilute solutions. (From Stanley and Knight, "The Chemical Composition of Strains of Tobacco Mosaic Virus" [1941]: 256.)

the relative amounts of these amino acids—tyrosine, tryptophan, and phenylalanine—in the various virus preparations correlated quite strikingly with their immunoreactivity to antibodies prepared against TMV. The content of other amino acids (e.g., arginine) served to differentiate strains with the same serological specificity.[123]

Results were highly suggestive, but the genetic interpretation remained uncertain. Stanley pointed out that the mechanism for the production of variants remained mysterious. "Because of the nature of the present approach, in which a sample of virus produced as a result of millions of duplications is examined, it is impossible to determine whether the changes in composition occurred in a single step or as the result of the cumulative effect of a number of successive alterations in structure during the multiplication of viruses."[124] In the discussion session following the paper, Stanley made it clear that he remained open-minded on this question:

> HUSKINS: Is it possible that we have three different categories of change, as in the genetic changes of point mutation, chromosomal aberration and ploidy?
> STANLEY: There may of course be different types of changes and I am in favor of the geneticist drawing as many analogies as possible....
> DELBRÜCK: Can the two cases [of genetic strains arising] outside the laboratory have arisen by step mutation or by a single change?
> STANLEY: One cannot be sure. However, I think that there is good evidence that they arose from tobacco mosaic virus.[125]

One strategy for addressing these questions was to induce new mutants of TMV. Developing a technique to create virus mutants could also be deployed in the fight against viral diseases, as Stanley pointed out in a 1942 lecture:

> The results already obtained indicate that eventually it may be possible, by means of definite chemical reactions carried out *in vitro*, to secure known structural changes in a virus molecule which will be perpetuated in its progeny. [The] accomplishment of this objective would have a far-reaching significance in medicine, for then one could visualize bending the will of viruses to the better service of mankind. It should be possible to eliminate the disease-producing viruses by supplanting them with innocuous viruses produced in the laboratory by chemical means from pathogenic viruses.[126]

123. See Stanley, "Chemical Structure and the Mutation of Viruses," 43–46.
124. Stanley and Knight, "Chemical Composition of Strains," 259.
125. "Discussion" following Stanley and Knight, "Chemical Composition of Strains," 261–62.
126. Stanley, "Chemical Structure and the Mutation of Viruses," 57.

The potential medical benefits provided additional incentive to advance this genetics project. However, Stanley and his coworkers were unable to create heritable TMV mutations via chemical modification. Knight treated the virus with formaldehyde, but "the remodeled virus, while still able to reproduce, gave rise to daughter molecules of the original, standard type."[127] Work on influenza virus vaccine during the war led Knight to suspend his chemical study of TMV mutations.[128]

In 1947, Knight again took up the chemical comparison of TMV strains. The recent development of microbiological techniques for quantifying the amounts of amino acids provided a more sensitive instrument than previous chemical methods (see fig. 6.6).[129] Knight identified two amino acid changes that specifically differentiated TMV from a particularly lethal variant; the variant J14D1 had more lysine and less glutamic acid than normal TMV (see fig. 6.7). The biological consequences of such small changes were dramatic:

> A comparison of the composition of TMV and of J14D1 shows in a striking manner how small the chemical differences need be between a virus which kills a given host and one which does not. If these results were translated to the field of animal viruses, it would be easy to understand the sudden conversion of a relatively mild epidemic virus to a lethal form. Such an event may possibly have occurred in the formation of the virus strain responsible for the influenza pandemic of 1918–19.[130]

Knight, in fact, sought chemical differences between strains of influenza based on the example of TMV.[131] However, his continuing attempts to "produce unequivocal heritable changes" in TMV by chemical reagents and radiation were unsuccessful.[132] Seymour Cohen, working in Stanley's laboratory during 1941–42, had isolated and characterized the nucleic acid moiety of TMV, but this line of investigation was not seen as related to the mutational or genetic studies at that time.[133]

---

127. Muller, "Résumé and Perspectives," 300. See Stanley's private summary of this work in Stanley to Doerr, 28 Oct 1941, Stanley papers, carton 1, folder Oct 1941.

128. See chapter 4, "World War II and New Uses for Ultracentrifuges," on influenza virus research in Stanley's laboratory.

129. Fraenkel-Conrat, "Protein Chemists Encounter Viruses," 310.

130. Knight, "Nature of Some of the Chemical Differences," 306.

131. Knight, "Amino Acid Composition."

132. Knight, "Nature of Some of the Chemical Differences," 306. John Gowen had been able to use X-ray radiation to produce mutants of TMV, including reversions among the aucuba strain back to wild-type ("Mutation in *Drosophila,* Bacteria, and Viruses"), but members of Stanley's group were unable to produce mutants using radiation (Stanley to Price, 16 May 1947, Stanley papers, carton 2, folder May 1947).

133. Cohen and Stanley, "Molecular Size and Shape."

Amino acid	Strain								
	TMV	M	J14D1	GA	YA	HR	CV3	CV4	M.D.†
Alanine...............	5.1	5.2	4.8	5.1	5.1	**6.4**		**6.1**	0.2
Arginine.............	9.8	9.9	10.0	**11.1**	**11.2**	9.9	9.3	9.3	0.2
Aspartic acid........	13.5	13.5	13.4	13.7	13.8	12.6		13.1	0.2
Cysteine..............	0.69	0.67	0.64	0.60	0.60	0.70	0	0	
Cystine..............	0		0		0	0		0	
Glutamic acid........	11.3	11.5	10.4	11.5	11.3	**15.5**	**6.4**	**6.5**	0.2
Glycine..............	1.9	1.7	1.9	1.9	1.8	1.3	1.2	1.5	0.1
Histidine............	0	0	0	0	0	0.72	0	0	0.01
Isoleucine‡...........	6.6	6.7	6.6	5.7	5.7	5.9	5.4	4.6	0.2
Leucine..............	9.3	9.3	9.4	9.2	9.4	9.0	9.3	9.4	0.2
Lysine...............	1.47	1.49	1.95	1.45	1.47	1.51	**2.55**	**2.43**	0.04
Methionine...........	0	0	0	0	0	**2.2**	0	0	0.1
Phenylalanine........	8.4	8.4	8.4	8.3	8.4	5.4	**9.9**	**9.8**	0.2
Proline..............	5.8	5.9	5.5	5.8	5.7	5.5		5.7	0.2
Serine...............	7.2	7.0	6.8	7.0	7.1	5.7	**9.3**	**9.4**	0.3
Threonine...........	9.9	10.1	10.0	10.4	10.1	8.2	6.9	7.0	0.1
Tryptophan...........	2.1	2.2	2.2	2.1	2.1	1.4	**0.5**	**0.5**	0.1
Tyrosine.............	3.8	3.8	3.9	3.7	3.7	**6.8**	3.8	3.7	0.1
Valine...............	9.2	9.0	8.9	8.8	9.1	6.2	8.8	8.9	0.2

FIGURE 6.6 Knight describes this table as follows: "Table of amino acid content of purified preparations of various TMV strains. TMV: common strain; M: masked strain; J14D1: lethal strain; GA: green aucuba; YA: yellow aucuba; HR: Holmes ribgrass; CV3: cucumber virus 3; CV4: cucumber virus 4. The values given in the table represent percentages of the indicated amino acids. In order to facilitate comparison, the values that are considered to differ significantly from those of TMV are in bold-faced type. †: Mean deviation of the values of single determinations from the averages given. ‡: Toward the end of the investigation, a purer isoleucine standard became available. Assays of TMV in which this standard was employed yielded an average value of 6.6% isoleucine instead of the 8.4% yielded by the earlier standard. The values for the other strains that were obtained using the earlier standard have been corrected proportionately." (Table and caption from Knight, "Nature of Some of the Chemical Differences" [1947]: 301. Reprinted by permission of the American Society for Biochemistry and Molecular Biology.)

Upon moving to Berkeley, Stanley presented virus reproduction as the central problem to be investigated by his new Virus Laboratory (see chapter 5). The results would have important medical ramifications, as Stanley wrote to the head of the NFIP: "At present it is not known whether viruses reproduce by growth and division or by means of some new and as yet undescribed process. Obviously, specific information on this point would be of great importance in any consideration of the nature of virus diseases and especially with respect to the possible utilization of

FIGURE 6.7 Photograph of tobacco leaves infected with various strains of TMV. *TMV:* common strain; *YA:* yellow aucuba; *GA:* green aucuba; *M:* masked; *J14D1:* lethal strain; *HR:* Holmes ribgrass. (Reprinted courtesy of the Bancroft Library, University of California, Berkeley.)

chemotherapy in connection with virus diseases."[134] As part of the large NFIP grant Stanley secured, Knight continued his chemical analysis of virus strains with the ultimate aim of comparing poliomyelitis virus strains (see fig. 6.8). On the basis of Knight's earlier observation that differences in amino acid composition between strains of TMV correlated with differences in virulence, Stanley made a case that similar relations should be sought in polio. "There is a real possibility that structural changes in the virus can be related to changes in virulence, and this information might

134. Stanley to H. M. Weaver, 17 Jun 1948, Stanley papers, carton 13, folder We misc.

FIGURE 6.8 C. Arthur Knight measuring out chemical reagents in the Virus Laboratory. Undated, circa 1955. (Reprinted courtesy of the Bancroft Library, University of California, Berkeley.)

be useful in connection with the control of extremely virulent viruses."[135] However, one needed pure samples of virus for this kind of precise amino acid analysis, and until the first project—aimed at isolating viruses from the three strains of poliomyelitis—was completed, pure poliovirus was unavailable.

In the absence of poliovirus as a material, Knight undertook a chemical comparison of bacteriophage strains "in extension of the work of Stanley and Knight on plant virus strains." T3 bacteriophage was selected for this project for two reasons: because "of the ease with which strains of known genetic relationship may be isolated" and also because the effort could "provide information for use in the purification of the somewhat similar poliomyelitis virus."[136] Thus TMV served as the model for this work on bacteriophage T3, which in turn was to serve as a model for purifying poliovirus, which was thought to contain DNA (and so was more chemically similar to phage). While the preparations of chemically homogeneous T-odd bacteriophages yielded beautiful electron micrographs of these viruses, few chemical differences associated with strains could be detected.[137] As purified strains of poliomyelitis virus became available, Stanley's workers finally extended the "know-how gained" with TMV to find chemical differences between polio strains, although without obtaining the clear-cut results they had sought.[138] This approach, even if not fruitful, clearly illustrates the interaction *between* experimental systems in Stanley's laboratory: TMV was an exemplar for the search for strain-specific chemical differences in influenza and T3, which

135. Application for Grant to NFIP, 1 Jul 1948, enclosed with letter from Stanley to H. M. Weaver, 17 Jun 1948, Records of the President, series 2, box 728, folder 1948: 420-Biochem.

136. Stanley, "Semi-Annual Report of Progress to the National Foundation for Infantile Paralysis, Inc., for the Period Ending June 30, 1951 from the Virus Laboratory of the University of California, Berkeley, California, July 5, 1951" (Stanley papers, carton 16, folder NFIP Progress Reports), 8, 9, 11.

137. On the electron micrographs of the T-odd bacteriophages, see Dorothy Fraser to Max Delbrück, 4 Sep 1950; Delbrück to Dean Fraser, 21 Apr 1952; and Dean Fraser to Delbrück, 11 Feb 1953, all in Delbrück papers, folder 8.2; Fraser and Williams, "Details of Frozen-Dried T3 and T7 Bacteriophages"; Williams and Fraser, "Morphology of the Seven T-Bacteriophages"; Williams, "Shapes and Sizes of Purified Viruses." On the chemical comparisons between strains, Stanley stated in 1951, "A considerable amount of work has been done on the amino acid analysis of bacteriophage T-3 with the aim of comparing the composition of strains of known genetic relationship in extension of the work of Stanley and Knight on plant virus strains. The method chosen was starch chromatography because of the apparent greater accuracy of the assay technique. This method has, however, been found to be extremely complicated and at times capricious" ("Semi-Annual Report of Progress," 11; see citation in note 136 above).

138. "Progress Report—July 1, 1953 to December 31, 1953," Stanley papers, carton 16, folder NFIP Progress Reports, 2.

in turn was developed as a model for poliovirus (more closely related on account of its possessing DNA like poliovirus, and unlike TMV's RNA).

In the early 1950s, Stanley's coworkers gave further consideration to the role of the nucleic acid in virus mutations. Knight analyzed the base content of the RNAs from several strains of TMV in order to determine whether they differed in chemical constitution. As he explained, "Since these strains of TMV are nucleoproteins, it is obvious that mutation in the virus might be associated with changes in the nucleic acid composition as well as with changes in the protein composition."[139] However, only in the strains most distantly related to TMV, cucumber viruses 3 and 4, could significant differences be detected; these viral RNAs contained less adenine and more uracil than TMV RNA. Knight did not find this result sufficiently compelling to change his earlier view that TMV mutations were registered in the protein, not the nucleic acid.[140] Thus when protein chemist Heinz Fraenkel-Conrat joined the group in 1952, he attempted to generate mutants of TMV by adding amino acids to the N-terminus of the protein.[141] As he observes, "The analytical consequences of this reaction were ambiguous, and, not surprisingly, no mutants were detected."[142]

Viruses and Plasmagenes

In the late 1930s, the characterization of TMV as a discrete particle had been achieved, one might even say operationalized, through the ultracentrifuge (see chapter 4); viruses were exogenous insofar as they could be separated from the host as discrete and autonomous units of infection. Yet the powerful tools which seemed to offer such a clear-cut differentiation between TMV and its host—there was nothing else in tobacco cells of such large dimension as the plant virus—failed to differentiate clearly endogenous from exogenous entities when used to retrieve viruses from

139. Knight, "Nucleic Acids of Some Strains," 241.

140. "It is proposed that mutation of tobacco mosaic virus is accompanied by a change in composition of the protein, and, conversely, that the nucleic acid remains essentially constant in composition through many mutations.... On the basis of this hypothesis, it is necessary to exclude CV3 and CV4 from classification as strains of TMV, since their nucleic acids differ from those of the other strains" (ibid., 248).

141. "The usefulness of N-carboxyamino acid anhydrides for the purpose of attaching amino acids to proteins by peptide linkages has recently been pointed out. Similar studies had been independently initiated in this laboratory, with the particular ultimate aim of producing self-duplicating virus modifications (mutations) by such means" (Fraenkel-Conrat, "Reaction of Proteins," 180).

142. Fraenkel-Conrat, "Protein Chemists Encounter Viruses," 310. Fraenkel-Conrat's work on TMV in the Virus Laboratory is discussed more fully in chapter 7.

animal cells. Even in uninfected tissue, the ultracentrifuge and electron microscope revealed a multitude of host particles that were strikingly similar to viruses in composition and size.[143] These host particles—for example, mitochondria, microsomes, and chloroplasts—were nucleoproteins, like viruses, and most seemed to reproduce independently of the nucleus. The newly apparent similarities, both structural and functional, between viruses and cellular constituents raised questions about whether viruses were entirely distinct from normal cellular components, as well as whether the capacity to self-reproduce extended to macromolecules beyond genes and viruses.

In fact, the "same" particle could apparently be classified as indigenous or infectious, as illustrated by the case of Rous sarcoma virus. F. Peyton Rous, using classic bacteriological techniques as a pathologist at the Rockefeller Institute, first isolated chicken tumor agent as a filterable virus in 1911.[144] Passing into the hands of his colleague James Murphy, who favored a biophysical approach, the entity was reclassified as an endogenous mitochondrial particle via the ultracentrifuge in the 1930s, and subsequently identified as a microsome, a normal cellular constituent.[145] In 1947, a chicken tumor virus was again sited in the cytoplasm via electron microscopy.[146] As Peter Medawar reflected at the time,

> [Albert] Claude's sedimentation from normal tissues of particles physically similar to those of the Rous sarcoma virus gives point to the suggestion that the Rous virus is an endogenous but prodigally aberrant member of the family of cytoplasmic self-reproducing particles. It is now becoming clear that these microsomal ingredients of cells are endowed with, and may even be the particular vehicles of, immunological specificity, and it is therefore not surprising to find that the purified Rous agent has an immunological "chicken-specific" component.[147]

143. Torbjörn Caspersson and Jack Schultz had pointed out in 1938 ("Nucleic Acid Metabolism") that all self-reproducing particles had been shown to contain nucleic acid, and some cytoplasmic particles shared with viruses the character of self-reproduction. On the growth of cell biology during this period and its particular reliance on biophysical tools, see Rasmussen, "Mitochondrial Structure" and *Picture Control;* Rheinberger, *Toward a History of Epistemic Things.*

144. Rous, "Transmission of a Malignant New Growth." See also Helvoort, "Century of Research."

145. Rheinberger, "From Microsomes to Ribosomes."

146. Ibid., 60, 76–77. As Rheinberger has noted of the case of Murphy's ultracentrifuged particle, "By the late 1940s, Claude [in Murphy's laboratory] had tended to look at microsomes as 'plasmagenes,' which 'might be endowed, at least temporarily, with a sort of genetic continuity'; consequently, 'the chromosomes might not constitute the sole reservoir of genetic characters' " (79). Rheinberger is quoting from Claude, "Studies on Cells," 156.

147. Medawar, "Cellular Inheritance and Transformation," 369.

Not only Rous sarcoma virus, but also fowl leucosis virus, influenza virus, and mouse pneumonitis virus were found to contain host structures or antigens when purified. As Seymour Cohen and Thomas Anderson observed, "These situations must be expected at some phase of virus synthesis, since it is highly improbable that metabolism of the host modified by the virus can occur without a close relationship of the virus particle and the organized metabolizing structures of the host."[148] For animal viruses, then, it could be difficult to differentiate virus from host in either macromolecular or immunological terms, and it seemed plausible to presume, like Medawar, that viruses might be classified with other kinds of self-reproducing subcellular particles.

A similar shift was occurring within bacteriophage research. The limitations of Delbrück's insistence on the autonomous nature of bacterial viruses, used to contest Northrop's precursor theory in the 1930s, were apparent by the late 1940s. By 1946, the problem of bacteriophage multiplication (Delbrück's "riddle of life") remained unanswered, and mounting observations of lysogeny made it increasingly difficult to argue that bacteriophage never arose from uninfected cells.[149] After confronting the restricted American approach to phage research at the 1946 Cold Spring Harbor Symposium, Jacques Monod and André Lwoff made it their mission to persuade Delbrück, high priest of the American phage church, that lysogeny was orthodoxy, not apostasy.[150] By virtue of Lwoff's painstaking labor with the micromanipulator and *Bacillus megaterium,* they succeeded in producing evidence that even Delbrück accepted.[151] Thus the long-standing emphasis on the autonomy of viruses, which had been used to justify viruses as unique models of self-duplication, was giving way, on the basis of diverse experimental evidence, to an interest in the similarities between viruses and other genetic particles.

A resurgence of interest in cytoplasmic inheritance further unsettled the grounds for the analogy between viruses and genes. As Sewell Wright noted in 1941, "It is doubtful whether a sharp line can be drawn between

148. Cohen and Anderson, "Chemical Studies on Host-Virus Interactions," 512.

149. On the "riddle of life," see Delbrück, "Experiments with Bacterial Viruses," 161–62. On the possible endogenous origin of bacteriophage, see Helvoort, "Construction of Bacteriophage." I have not focused in this chapter on the many studies of lysogenic bacteria in the 1920s and 1930s, but the investigations of F. Bordet at the Pasteur Institute and F. M. Burnet at the Walter and Eliza Hall Institute of Medical Research in Melbourne, Australia, were particularly important; see Galperin, "Le bactériophage."

150. Monod himself used religious figures in describing the American phage group: "Du microbe à l'homme," 7.

151. On the Pastorian contributions to understanding lysogeny in the context of the reception of Mendelian genetics in France, see Burian, Gayon, and Zallen, Singular Fate of Genetics.

non-Mendelian heredity and virus infection."[152] Embryologists had long ascribed the irreversibility of cell differentiation to cytoplasmic determinants, and the attribution of non-Mendelian inheritance of chlorophyll variegation in plants to cytoplasmic plastids (chloroplasts and leucoplasts) received additional corroboration in the 1930s.[153] During the 1940s, new and compelling research on non-nuclear inheritance implicated other cytoplasmic agents. Philippe L'Héritier and his collaborators in Paris provided evidence for cytoplasmic inheritance of sensitivity to carbon dioxide in *Drosophila*.[154] They considered the heritable factor to be a "genoid," called sigma, and showed that it was both mutable and capable of self-reproduction.

Genetic research on microbes provided other instances of extranuclear heredity. Tracy Sonneborn's study of the inheritance of the "killer" trait in strains of *Paramecium* was the most dramatic demonstration of non-nuclear inheritance. This trait enabled *Paramecia* possessing it to kill competing strains without the trait. Sonneborn ascribed the inheritance of this trait to a cytoplasmic genetic entity he called kappa. The prevalence of kappa in *Paramecia* was highly responsive to environmental conditions. Although kappa depended on a dominant nuclear gene, "K," for long-term maintenance, it was inherited cytoplasmically and could also mutate.[155] As Sonneborn pointed out, kappa's genelike properties also made it similar to a virus.[156] Wright referred to such "cytoplasmic proteins which possess the essential genic property" as *plasmagenes*, in contrast with nuclear (chromosomal) genes.[157]

In 1944, C. D. Darlington elaborated the analogy between viruses and plasmagenes in a widely cited review (see fig. 6.9).[158] He argued for classifying the general phenomena into two systems beyond Mendelian nuclear inheritance: the plastid system in plants (which he regarded as "corpuscular," since the bodies could be seen), and maternal inheritance (which Darlington argued was "molecular," since it could not be tied to specific visible particles). He classified plasmagenes as part of the "molecular" system and argued that these cytoplasmic agents provided the missing

152. Wright, "Physiology of the Gene," 496.

153. Sapp, *Beyond the Gene*. Sapp discusses plastids in chapter 1 (26) and the attractiveness of cytoplasmic inheritance to embryologists in chapter 2. The work of Otto Renner provided an important confirmation that plant plastids were inherited cytoplasmically: "Zur Kenntnis der nichtmendelnden Buntheit der Laubblätter."

154. L'Héritier, "Sensitivity to CO_2 in *Drosophila*."

155. Sonneborn, "Gene and Cytoplasm. Parts 1 and 2"; Sapp, *Beyond the Gene*, 98.

156. Sonneborn, "Beyond the Gene," 50.

157. Wright, "Physiology of the Gene," 501.

158. As Sapp has pointed out (*Beyond the Gene*), the term *plasmatischem Gene* had already been introduced in German: Winkler, "Über die Rolle von Kern und Protoplasma."

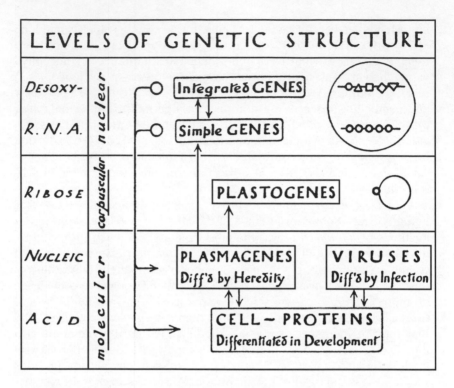

FIGURE 6.9 Diagram showing Darlington's postulated levels of genetic structure. (Reprinted, with permission, from Darlington, "Heredity, Development, and Infection," *Nature* 154 [1944]: 168. Copyright 1944 Macmillan Magazines Limited.)

link between embryology and genetics, the mutual control of "transmission in heredity and expression in development."[159] Darlington specified four shared properties of viruses and plasmagenes: (1) Both are nucleoproteins, or "proteins reproducing with the help of ribose nucleic acid"; (2) both are subject to developmental control of the host, "being excluded from certain tissues and reduced in others"; (3) both are subject to their host's nuclear control, "being suppressed by some host genotypes and permitted by others, either within limits or, pathologically, without limits"; and (4) both are capable of suppressing the expression of mutants (cf. cross-protection in plant viruses and interference in bacteriophage).[160] As he concluded: "The ultimate distinction between plasmagene and virus therefore seems to be the accidental one of transmission by heredity or by

159. Darlington, "Heredity, Development and Infection," 166.
160. Ibid., 167.

infection.... The plasmagene is a protein which can be made outside the nucleus and comes to be inherited through the egg. The virus is a similar protein which is capable of being acquired later."[161]

Darlington pointed to recent work on viruses to indicate that they, like plasmagenes, might originate from nuclear genes. In the first place, the postulation of plasmagenes seemed to provide a possible explanation for lysogeny: the bacteriophage could lose its infectivity, remain as an uninfectious nucleoprotein particle, and be replicated and passed on to daughter cells like an organelle. Second, work on cancer viruses during this period, like research on bacteriophage lysogeny, undermined any neat differentiation between exogenous and endogenous agents, between infection and heredity. For example, papilloma virus of rabbits was one of the best-characterized tumor viruses, having been isolated by Ralph Wyckoff and Joseph Beard in the ultracentrifuge as a homogeneous macromolecule.[162] However, it exhibited distressingly erratic patterns of transmission, being undetectable in some papilloma growths and often failing to induce disease when inoculated into domestic rabbits. Richard Shope hypothesized that the virus could exist in a "masked" state in the cell, perhaps indistinguishable from the host's genetic units and transmissible via inheritance.[163] Taken in conjunction with Murphy's identification of Rous sarcoma virus as being, or containing, a normal cell constituent, tumor viruses seemed peculiarly capable of arising from within the host cell rather than infecting it from without. As Stanley had noted in 1939, on the basis of the tumor virus research of his Rockefeller Institute colleagues,

> There is, however, a strong and growing tendency to consider that viruses or similar factors may, upon provocation of the cell, originate endogenously and give rise to tumors or cancers. Whether such viruses are actually derived from normal cell constituents or are formed by the mutation of a "masked" virus normally or usually carried within the cell, but in reality alien to the cell, is not known.[164]

Darlington suggested that the cell might produce viruses endogenously, just as it could produce plasmagenes.

Examples drawn from cancer research provided the focus of a second 1944 review of cytoplasmic inheritance in *Nature*, published a week

161. Ibid., 168.
162. Beard and Wyckoff, "Isolation of a Homogeneous Heavy Protein"; Beard, Bryan, and Wyckoff, "Isolation of the Rabbit Papilloma Virus."
163. Shope, "Immunization of Rabbits." For a more detailed historical account of "masked" viruses, see Creager and Gaudillière, "Experimental Arrangements."
164. Stanley, "Architecture of Viruses," 550.

after Darlington's. Alexander Haddow compared John Bittner's mouse mammary tumor agent, Rous sarcoma virus, and Claude's microsomes in an effort to cast light on the "nature of the cytoplasmic entities: plasto-genes, plasmagenes and viruses."[165] He reiterated the observation that, like plasmagenes and plant viruses, microsomes (from both normal and tumor cells) had been found to contain ribonucleic acid. Furthermore, the Green-Laidlaw hypothesis, that viruses were products of retrograde evolution from bacteria, could be invoked to account for the origins of all three sorts of particles:

> With the suggestion of an intrinsic origin for the avian tumour viruses may be related the view that many of the plant viruses are au-tocatalytic proteins of ultimate host-cell origin; and both possibilities should be compared with . . . hypotheses which envisage many viruses arising by a process of retrograde evolution, that is, by a progressive loss of enzyme systems and synthetic functions and an increasing degree of dependence upon the cellular host.[166]

Both differentiation and unregulated cell growth (tumors) could, in the view of Haddow, be accounted for by differential segregation or induc-tion of such cytoplasmic determinants. In particular, he suggested that the mechanism for unlimited cell growth might be traced to protein synthesis in the cytoplasm of rapidly divided cells, which depended on the presence of "considerable quantities of ribose nucleic acids."[167] The difference be-tween patterns of protein synthesis in differentiation and unlimited cell growth remained uncertain, however, and the role of RNA unclear.

Plasmagenes Questioned and Analogies Redefined

By classifying together a range of particles being actively investigated—mitochondria and microsomes in animal (and yeast) cells, plastids in plants (chloroplasts and leucoplasts), and viruses in a variety of hosts, this overarching framework of cytoplasmic genetic particles, or plasma-genes, extended the aim of isolating the gene as a material unit to new candidates beyond viruses. However, the relationship of plasmagenes to nuclear genes was a matter of not only uncertainty but contestation. As George Beadle put it in a 1945 review,

165. Haddow, "Transformation of Cells and Viruses," 196. Note that "plastogene" had been introduced by Yoshitaka Imai to refer to the invisible genelike component of the plastid responsible for its behavior of cytoplasmic inheritance. For an insightful historical account of Bittner's mouse mammary tumor agent, see Gaudillière, "Circulating Mice and Viruses."

166. Haddow, "Transformation of Cells and Viruses," 196.

167. Ibid., 198.

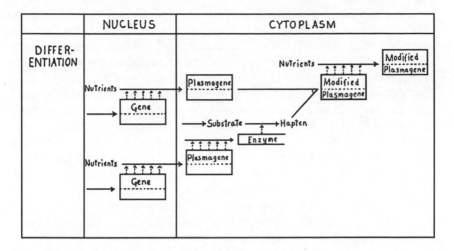

FIGURE 6.10 Sewell Wright's diagram showing the postulated role of plasmagenes in enzyme synthesis. (From Wright, "Genes as Physiological Agents," *American Naturalist* 79 [1945]: 299, © The University of Chicago. Reprinted by permission.)

> Plasmagenes are postulated self-duplicating cytoplasmic units which, like genes and unlike viruses, are normal cell constituents. Because it has not heretofore been possible to demonstrate the existence of these units in a manner as unequivocal as those used in the demonstration of genes and viruses, most geneticists and other biologists have been skeptical as to whether such units, comparable in autonomy to genes, really exist. The evidence is now sufficient, however, to justify serious attention to the possibility of gene-like units in the cytoplasm. Actually, of course, the distinction between viruses and plasmagenes on the basis of one being normal and the other an abnormal cell constituent may be quite artificial.[168]

The utility of plasmagenes was not restricted to accounting for cytoplasmic inheritance. In 1945, Wright offered a model for the role of plasmagenes in enzyme synthesis (see fig. 6.10).[169] In Wright's model, plasmagenes were derived from nuclear genes but could be altered (through "controlled mutation") and reproduce in the cytoplasm, particularly in response to chemical factors ("haptens"), either from within or from outside the cell.[170] Guido Pontecorvo also suggested that plasmagenes could account for patterns of protein synthesis, particularly as observed in the

168. Beadle, "Biochemical Genetics," 84–85.
169. Wright, "Genes as Physiological Agents."
170. Ibid., 299.

widely cited cases of enzyme adaptation, in which enzymes were produced in response to the presence of substrates in the culture medium.[171] Monod's studies of adaptation in bacteria and Sol Spiegelman's investigations of yeast adaptation provided the most important referents for adaptation.[172]

Spiegelman also began to use plasmagenes to explain enzyme formation in yeast. According to his scheme, the cytoplasmic gene was formed from the nuclear gene and was subsequently replicated without assistance from the nucleus.[173] Rather, the presence of the enzyme substrate stabilized and so maintained the plasmagene in the cytoplasm over time. Spiegelman argued that patterns of non-Mendelian inheritance surfaced when the rate of self-replication of plasmagenes was greater than the constant production of plasmagenes from nuclear genes. Thus he offered one way to reconcile observations of enzyme synthesis with those of cytoplasmic inheritance.

Spiegelman presented plasmagenes as generating adaptive enzymes (or, more generally, proteins), but unlike the case of viruses, the protein synthesis of plasmagenes could not be considered autocatalysis or self-reproduction. In this sense the old analogy between genes, enzymes, and viruses was being reconfigured, particularly once Beadle and Tatum's results with *Neurospora* indicated that genes and enzymes were correlated but not necessarily identical.[174] As Spiegelman explained (with coauthor Martin Kamen),

> It seems reasonable to adopt the tentative working hypothesis that the cytoplasmic self-duplicating entities previously found [by his laboratory] to be involved in enzyme formation are nucleoproteins rather than the enzyme itself. Such an hypothesis would be in harmony with the findings that all accepted self-duplicating units (with but one or two isolated exceptions) have been found to be inseparably linked with nucleic acid-containing compounds. Among such entities may be mentioned genes, plastogenes, viruses, and the pneumococcus 'transforming principle.'[175]

Monod provided an alternative model for adaptive enzyme synthesis in 1947, in which enzyme precursors were produced by genes and

171. Pontecorvo, "Microbiology, Biochemistry, and the Genetics."

172. See Gaudillière, "J. Monod, S. Spiegelman et l'adaptation enzymatique." Spiegelman's research on adaptive enzymes built on Carl and Gertrude Lindegren's work establishing yeast as a genetic system.

173. As Brock has described Spiegelman's model, "Once the plasmagene had formed, the gene was no longer needed" (*Emergence of Bacterial Genetics,* 272).

174. See Kay, "Selling Pure Science in Wartime"; and Kohler, "Systems of Production."

175. Spiegelman and Kamen, "Genes and Nucleoproteins," 583.

stabilized by enzyme substrates, but these were not self-duplicating units.[176]

Plasmagenes provided a focus of intense discussion at one of the first postwar European meetings on genetics, a 1948 symposium organized by Lwoff under the theme "Unités biologiques douées de continuité génétique." The importance given cytoplasmic inheritance on the program reflected the greater value placed on this topic by European biologists.[177] Virus genetics was also well represented. F. C. Bawden offered a critical summary of Stanley and Knight's chemical characterization of TMV mutants, Delbrück discussed recent experiments with bacteriophage, Darlington and Jean Brachet both contributed papers on plasmagenes, Marcus Rhoades offered an analysis of plastid inheritance in plants, and Geoffrey Beale presented a paper jointly authored by himself and Sonneborn on their current understanding of cytoplasmic factors in *Paramecium*.[178] Their paper, which became a touchstone for debates about plasmagenes, drew on the many years Sonneborn had invested studying the inheritance of eight serological types, or antigenic characters, in *Paramecium*. These observations, like his work on kappa, raised questions about the adequacy of chromosomal genes in accounting for patterns seen in microbial genetics.

The *Paramecium* serotype inheritance, as Sonneborn observed, exhibited a complex mixture of Mendelian and non-Mendelian patterns. Specifically, the inheritance of antigenic *potentialities* was determined by nuclear genes, but the antigenic type *realized* appeared to be environmentally determined and cytoplasmically inherited.[179] At the meeting in Paris, Beale explained the pattern of inheritance in terms of plasmagene theory. The plasmagenes he and Sonneborn hypothesized, unlike Wright's and Darlington's, were not derived continually from nuclear genes but had a more autonomous existence, relying on the nucleus solely for their long-term maintenance. As Sonneborn summarized in another lecture the same year, "The gene was specified by its action, the control of hereditary traits; its properties, self-duplication and mutability; and its location, in the nuclear chromosomes. . . The effects and basic properties of plasmagenes are identical with those of classical nuclear genes. Plasmagenes differ from

176. Brock, *Emergence of Bacterial Genetics,* 273; Monod, "Phenomenon of Enzymatic Adaptation." For a detailed and insightful analysis of the Monod-Spiegelman debate, see Gaudillière, "J. Monod, S. Spiegelman et l'adaptation enzymatique."

177. Brock, *Emergence of Bacterial Genetics,* 280; Sapp, *Beyond the Gene.*

178. Sonneborn and Beale, "Influence des gènes, des plasmagènes et du milieu." Sapp (*Beyond the Gene,* 140) notes that neither Sonneborn nor H. J. Muller were able to attend the meeting, although they were invited.

179. Sonneborn, "Beyond the Gene," 45–46. For a lucid analysis of these transitions, see Abraham, "From Plasmagenes to Steady States."

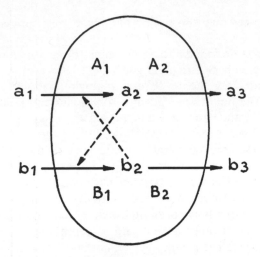

FIGURE 6.11 Max Delbrück's "steady-state" model for gene expression. (From Delbrück, "Discussion" [1949]: 34. Reprinted by permission of CNRS Editions.)

nuclear genes, so far as now known, only in their location in the cytoplasm instead of in the nucleus."[180]

Delbrück responded to Sonneborn and Beale's paper by presenting an alternative "steady-state" explanation of their data (see fig. 6.11).[181] He argued that if there were several alternative reaction pathways whose products inhibited each other, a pattern of flux equilibrium could develop (induced, for example, by transitory changes in the environment) in which the concentration of key metabolites maintained a distinct protein expression pattern, irrespective of any genetic changes. Accordingly, the first reaction to be activated would achieve dominance by inhibiting the other potential phenotypic states.[182] Boris Ephrussi considered this explanation "alluring," but felt it was ultimately inadequate to account for the long-standing expression patterns observed in cellular differentiation.[183] Nonetheless, Delbrück's theoretical model offered a way to conceptualize inducible protein synthesis apart from the phenomenon of cytoplasmic inheritance.[184]

180. Sonneborn, "Beyond the Gene," 57.
181. Delbrück, "Discussion."
182. As Sonneborn put it, "The essence of his idea is that the dominance of one reaction in a cell among two or more mutually exclusive ones may favor one line of development and inhibit the others, thus maintaining the favored trait" ("Partner of the Genes," 38).
183. Ephrussi, "Remarks on Cell Heredity," 243.
184. As Sonneborn appraised Delbrück's contribution, "It provides a totally different and fruitful way of envisaging the basis of cytoplasmic inheritance in those cases where

These debates over the nature and mechanism of protein synthesis tended to skew the meaning of plasmagenes away from reproduction and toward adaptation, and with the decline of Spiegelman's theory of adaptive enzyme induction, plasmagenes lost some of their explanatory power. Other developments also undercut the attraction of plasmagenes. For one thing, cytoplasmic inheritance threatened to erode the ascendant, neo-Darwinian synthesis with examples of non-Mendelian inheritance. The stakes extended beyond disciplinary realignments. As part of the reaction in the West to the brutal suppression of genetics in the U.S.S.R. by Trofim Lysenko, who offered his own theory of environment-induced "heredity," biologists looked more critically on cytoplasmic inheritance for its Lamarckian overtones.[185] Plasmagenes were politicized. Lastly, Edgar Altenburg had proposed that plasmagenes were viruses that had become cellular symbionts.[186] This interpretation made plasmagenes less interesting, rather than more interesting, to most geneticists, who viewed symbiosis as largely irrelevant to the predominant mechanisms of heredity.[187] To the degree that biologists doubted the scientific legitimacy of plasmagenes, continuing assertions of the virus-plasmagene equivalency threatened rather than strengthened the older analogy between viruses and genes.

Given the controversy surrounding plasmagenes, was the category of cytoplasmic genetic units still conceptually useful for investigating and contemplating the mechanism of self-reproduction? On this issue, many biologists remained skeptical, hesitant about generalizing genetic behavior from units that did not display Mendelian inheritance. Muller, like several other prominent geneticists, attacked the peculiarity of plasmagenes, arguing that the best-documented cytoplasmic agents were infectious and, hence, exogenous to the cell.[188] In this respect he invoked

---

cytoplasmic genetic particles are unknown, unlikely, or incapable of demonstration. . . . In the latter case [of self-perpetuating steady states], cytoplasmic control would be operating by self-perpetuating reactions, presumably on a chemical level" ("Beyond the Gene—Two Years Later," 197, 198).

185. See Sapp, *Beyond the Gene,* ch. 6. Cytoplasmic inheritance was Lamarckian insofar as it could involve the inheritance of acquired characteristics (or particles). As Sapp shows, because Lysenko's advocates used evidence of cytoplasmic inheritance to bolster their own critiques of Mendelian genetics, researchers such as Sonneborn had to defend themselves against their work being used by Soviet propagandists. See also Lindegren, *Cold War in Biology.*

186. Altenburg, "Viroid Theory," and "Symbiont Theory."

187. Sapp, *Evolution by Association,* esp. ch. 10. One notable exception to this generalization about geneticists is Joshua Lederberg who, as Sapp shows, both took seriously the many examples of cytoplasmic inheritance and sought to include hereditary symbiosis in a unified concept of heredity. See Lederberg, "Cell Genetics and Hereditary Symbiosis."

188. Sapp, *Beyond the Gene,* 117–18.

the classical notion of viruses as autonomous, using the analogy between viruses and plasmagenes to "pathologize" agents of heredity beyond the nucleus. Muller was particularly skeptical of Sonneborn's claim that the kappa factor he had identified in *Paramecium* was exemplary of a class of "self-duplicating, mutable, cytoplasmic particles—the plasmagenes—which depend on the nuclear genes for their maintenance or normal functioning, but not for their origin or for their specificity."[189] Indeed, even Sonneborn revised his plasmagene model between 1948 and 1950, reflecting new experimental evidence that suggested greater involvement of nuclear genes in *Paramecium* serotype inheritance.[190]

Nonetheless, the belief that viruses as well as other self-duplicating entities might shed light on general genetic mechanisms was not entirely abandoned. Muller still expressed his conviction that the "real core of gene theory, . . . its ability to cause the synthesis of another structure like itself" might best be illuminated by analyzing the chemical parallels between viruses and genetic particles, both nuclear and cytoplasmic:

> At least a part of the answer must somehow be buried in what is already known of the gene's chemical constitution. I refer to the fact that all genes—that is, all particles capable, in unlimited measure, of reproducing their variations, whether they be chromosomal genes, plastogenes, kappa particles, viruses, or the "transforming agents" of bacteria—have been found, whenever their composition could be approximately determined, to consist of or to contain polymerized nucleic acid, and usually, if not always, some protein as well. But as yet no one has been able to correlate these features of chemical structure with the gene's peculiar property of self-reproduction. . . . But whatever the secret of the gene's ability to reproduce itself and its mutations may consist in, it seems today clearer than ever, especially in light of modern knowledge of microorganisms and viruses, that this is also the most fundamental secret of life itself.[191]

Yet these examples of genetic units (see also table 6.1) were not equally valued as candidates for illuminating this "secret of life." As Jack Schultz surmised in a review of the 1948 Paris symposium volume,

> It is evident, with this resumé, that the only definite cases of self-perpetuating elements in the cytoplasm that require the presence of

189. Sonneborn, "Cytoplasm in Heredity," 31; also quoted in Sapp, *Beyond the Gene*, 107. Sonneborn and Muller were colleagues at the University of Indiana, and, as Sapp has pointed out (113–21), the Rockefeller Foundation's expectation that the two might work together was never realized, due in part to their profound disagreement over cytoplasmic inheritance.

190. See Sonneborn, "Beyond the Gene—Two Years Later," 194–95; Schloegel, "From Anomaly to Unification"; Sapp, *Beyond the Gene*.

191. Muller, "Development of the Gene Theory," 97–98.

identical preexisting elements for their formation, and that perpetuate specific changes in their own structure (mutants), are plastids,
viruses, and the "central apparatus" in plants and animals. The
existence of subsidiary levels of self-perpetuation remains in [a]
questionable state.[192]

Some examples of cytoplasmic inheritance remained more dubious than
others, whereas viruses retained their credibility as good models for self-
duplication (see table 6.1).

Joshua Lederberg urged that the various genetic units outside the
nucleus be considered along a spectrum of "infective heredity," with
"deleterious parasitic viruses at one extreme, and integrated cytoplasmic
genes like plastids, at the other." He continued,

> Within this interval, we find a host of transition forms; kappa, lyso-
> genic bacteriophages, genoids, tumor-viroids, male-sterility factors,
> Ephrussi's yeast granules, etc. . . . The objection has been voiced that
> this viewpoint is an attempt to relegate plasmagenes to pathology. I
> rather think that it may broaden our genetic insights if we consider the
> likenesses as well as the dissimilarities between pathogenic viruses
> and plasmagenes.[193]

He contended that lysogenic bacteriophage provided "the best material" for investigating the genetic character of such cytoplasmic genes,
"especially as infection with such a virus is formally indistinguishable
from events such as pneumococcus transformations."[194]

Viruses and Transforming Factors

The meeting entitled "Viruses 1950," held at Caltech, helped to consolidate the disparate information that had been gained about virus reproduction in bacteria, animals, and plants. In his contribution to the conference,
Luria reassessed the basic scientific problem:

> In many minds the terms "reproduction" and "self-reproduction"
> are connected with the idea of increase in size followed by division.
> Closer scrutiny reveals that increase in size followed by division is
> bound to be an epiphenomenon of some critical event of repro-
> duction, which must involve point-to-point replication of some ele-
> mentary structures responsible for the conservation of specificity
> from generation to generation. Thus, in dealing with cell growth
> and division we trace the critical event to gene and chromosome
> duplication. . . . In a repeat, crystal-like structure, such as has been
> suggested for rod-shaped particles of plant viruses, the elementary

192. Schultz, "Question of Plasmagenes," 407.
193. Lederberg, "Genetic Studies with Bacteria," 286.
194. Ibid.

Table 6.1 Characteristics of Viruses and Self-perpetuating Cytoplasmic Entities, circa 1950

	Nucleic Acid?	Protein?	Inherited?	Infective?	Particle Size
TMV	RNA (6%)	Yes	Yes	Yes	150 × 3000 Å
Bacteriophages[a]	DNA (T2: 37%)	Yes	When lysogenic	When lytic	T2: Head 600 × 800 Å Tail 1000 × 200 Å
Kappa factor[b]	DNA	No	Yes	Only in lab	2000–5000-Å diameter
Pneumococcus transforming factor[c]	DNA	No	Yes	Yes	Unknown
Claude's microsomes[d]	RNA (~5%)	Yes	Yes	No	500–2000-Å-diameter spheres
Rous sarcoma virus[e]	DNA	Yes		Yes	700-Å-diameter spheres
Rabbit papilloma virus (Shope)[f]			Yes	Yes (in wild jackrabbits)	440-Å- or 660-Å-diameter spheres
Mitochondria[g]	RNA	Yes	Yes	No	10,000–40,000 Å × 3,000–7,000 Å

Component	Nucleic acid				Size
Chloroplasts[h]	RNA	Yes	Yes	No	No	50,000Å × 20–30,000 Å
Enzyme-forming particles in yeast[i]	RNA	Yes	Yes (Spiegelman) No (Monod)	No	No	
Chromosomes[j]	DNA	Yes	Yes	No	1250 × 200 Å

[a] On DNA content: Cohen and Anderson, "Chemical Studies on Host-Virus Interactions." On particle size: Anderson, "Morphological and Chemical Relations."

[b] On nucleic acid content and size: Preer, "Killer Cytoplasmic Factor Kappa."

[c] Avery, MacLeod, and McCarty, "Studies on the Chemical Nature."

[d] Claude, "Particulate Components of Cytoplasm," 264.

[e] On particle size: Stanley and Knight, "Chemical Composition of Strains."

[f] On particle size: Sharp et al., "Study of the Papillomatosis Virus"; Sharp et al., "Density and Size of the Rabbit Papilloma Virus"; Sharp et al., "Rabbit Papilloma and Vaccinia Viruses and T2 Bacteriophage." On patterns of transmission: Shope, "Masking, Transformation, and Interepidemic Survival."

[g] On mitochondria as ribonucleoproteins: Claude, "Distribution of Nucleic Acids"; Lazarow, "Chemical Structure of Cytoplasm." On size: Palade, "Electron Microscope Study."

[h] On chloroplasts as ribonucleoproteins: Dubuy and Woods, "Evidence for the Evolution." On chloroplast size: Granick, "Plastic Structure, Development, and Inheritance," 513.

[i] Spiegelman and Kamen, "Genes and Nucleoproteins"; Monod, "Phenomenon of Enzymatic Adaptation."

[j] Stanley and Knight, "Chemical Composition of Strains."

repeated unit must be replicated. In other words, all growth and reproduction should ultimately be traceable to *replication of specific
chemical configurations by an essentially discontinuous appearance
of discrete replicas.*[195]

Yet the actual process of virus replication remained obscure: "The indispensable presence of the initial model (gene, virus) indicates that this
model plays a role in replication, but this role is by no means an obvious
one."[196]

As Luria observed, scientists tended to imagine three scenarios. One
was that the virus brought with it the enzymes "needed for its own synthesis from specific building blocks." According to a second possibility, the
virus "might act as a directive pattern for synthesis, according to which
building blocks are assembled by synthetic enzymes" not contained within
the virus. As Luria noted, this kind of replication by host-induced enzymes
might "require a two-dimensional unfolding of the [virus] model, to allow point-to-point replication followed by separation of the newly formed
unit." Lastly, the virus "might function as a directive pattern for folding a
pluripotential macromolecule into a specific tridimensional replica, possibly with the intervention of a negative template, by analogy with Pauling's
theory of antibody formation."[197]

The third prospect was particularly attractive to those interested in the
endogenous origin of viruses, for the precursor material could be a host
component. But recovering precursors, either through the use of mutant
phage strains or through biochemical means, had proven difficult. New investigations shed light on the reasons. Biochemical experiments in which
bacteriophage synthesis was traced with radioactive nitrogen, carbon, and
phosphorus revealed that new viruses were synthesized from small compounds derived from the bacterial medium.[198] Using genetic methods,
Gus Doermann demonstrated that bacteriophage underwent a profound
alteration after infecting cells and before reproduction, a period that was
termed the "eclipse" because of the disappearance of infective phage.[199]
Delbrück pointed out the surprising nature of this result—it meant that
not only was the virus in a different state when it was multiplying, but it
was completely uninfective.[200]

If the puzzle of virus multiplication remained unsolved in the early
1950s, confidence was growing that the "transforming factor" observed

195. Luria, "Bacteriophage: An Essay on Virus Reproduction," 7; emphasis in original.
196. Ibid., 8.
197. Ibid., 8 (all three quotations).
198. Kozloff, Putnam, and Evans, "Precursors."
199. See Doermann, "Vegetative State."
200. Delbrück, "Introductory Remarks," 1.

by Oswald Avery, Colin MacLeod, and Maclyn McCarty in 1944 in pneumococcus strains represented true genetic material, even if it was not a nucleoprotein.[201] Observations in microbial genetics that previously appeared to be the "inheritance of acquired characteristics" might now be accounted for through the transmission of acquired, autonomous hereditary units, or, as Peter Medawar phrased it, "the casual leakage of self-reproducing particles from one cell into neighbors."[202] Although this general line of analysis had been advocated before—for example, as early as 1938 Stanley had argued that pneumococcal transformation was a kind of "infection" and that the transforming agent should be further studied "because of its virus-like nature"—belief in the genetic equivalence of infection and transformation had gained substantial ground in the meantime.[203]

The perceived similarity between the processes of bacteriophage infection and pneumococcus transformation was reinforced by Roger Herriott's observations in 1951 of protein "ghosts" on the outside of phage-infected bacteria (as visualized in electron micrographs).[204] As he wrote Hershey, "I've been thinking—and perhaps you have, too—that the virus may act like a little hypodermic needle full of transforming principles; that the virus as such never enters the cell; that only the tail contacts the host and perhaps enzymatically cuts a small hole through the outer membrane and then the nucleic acid of the virus head flows into the cell."[205] This analogy, however, raised the issue that chemical differences between self-perpetuating particles might be highly consequential. As F. C. Bawden noted, "If only the nucleic acid from phages enters bacteria, chemically the phages seem to come more in line with transforming factors than with the plant viruses, for these are nucleoproteins, and in many of them the nucleic acid is, quantitatively, only a very minor component."[206] Hubert Chantrenne also noted that the two types of nucleic acid in "self-duplicating particles" were not identical in their cellular functions: "Chromosomes contain both [ribonucleic acid (RNA), and deoxyribonucleic acid (DNA)], chloroplasts contain RNA, and there are reasons to think that the self-duplicating particles responsible for enzyme production also contain ribonucleic acid. RNA is most probably involved

201. Avery, MacLeod, and McCarty, "Studies on the Chemical Nature."
202. Medawar, "Cellular Inheritance and Transformation," 371.
203. Stanley, "Biochemistry and Biophysics of Viruses," 491–92; Brock, *Emergence of Bacterial Genetics*, 231.
204. Herriott, "Nucleic Acid-Free T2 Virus 'Ghosts.'"
205. Letter from Roger Herriott to Hershey, 16 Nov 1951, as quoted in Hershey, "Injection of DNA into Cells," 102.
206. Bawden, "Discussion," 52; also quoted in Helvoort, "Construction of Bacteriophage," 112.

not only in the self-duplicating processes, but also in protein synthesis in general."[207] If DNA and RNA served different purposes in the cell, then the observation that various self-duplicating particles all contained some form of nucleic acid with protein might not be functionally important. Moreover, because bacteriophages contained DNA, whereas most viruses, including TMV, contained RNA, emphasizing that the two nucleic acid types had divergent biological roles seemed to weaken the grounds for grouping together all viruses.

At another level, experiments and theorizing by Lwoff on lysogenic phage in the early 1950s both evoked and reworked the earlier analogy between bacteriophage and plasmagenes. Lwoff and coworker A. Gutmann introduced the term "probacteriophage" in 1950 to describe the noninfectious hereditary structure necessary for the production of phage from lysogenic bacteria.[208] This term was shortened to "prophage" or, more generally, "provirus," and it was easiest to imagine this latent phage as the bacterial equivalent of an uninfectious cytoplasmic particle, such as a microsome.[209] Lwoff's discovery during that same year that exposure to UV radiation caused entire lysogenic cultures to lyse, a phenomenon termed *induction,* further solidified the status of lysogeny and made it more accessible to experimental study.[210] The fact that radiation could also serve as an inducer of cancer made the relationship between lysogeny and tumor viruses seem even more compelling. In both cases, one could imagine the asymptomatic cell harboring the potentially infective agent as a cytoplasmic particle.

In 1951, Esther Lederberg observed lysogeny in *E. coli* K-12, the strain that Joshua Lederberg had domesticated for genetic experimentation.[211] Given this system, a lysogenic bacteriophage could now be mapped genetically.[212] Joshua Lederberg expected that the lysogenic phage, λ, would reside as a genetic particle in the cytoplasm, in which case it would not exhibit linkage to chromosomal markers. However, the Lederbergs' experiment revealed that λ was genetically linked to *E. coli* chromosomal markers.[213] As Elie Wollman and François Jacob commented,

207. Chantrenne, "Problems of Protein Synthesis," 2. Chromosomes and chloroplasts are now known to contain DNA.

208. Lwoff and Gutmann, "Recherches sur un *Bacillus megatherium* lysogène."

209. Prophage was assumed to be a cytoplasmic particle; see Helvoort, "Construction of Bacteriophage," 128.

210. Lwoff, Siminovitch, and Kjeldgaard, "Induction de la production de bactériophages." See also Brock, *Emergence of Bacterial Genetics,* 176–79, for a historical account of induction.

211. Lederberg, "Lysogenicity in *E. coli* K-12."

212. This experiment was possible by virtue of Joshua Lederberg's work on a mating system in *E. coli,* enabling genetic crosses to be done (Lederberg, "Gene Recombination").

213. Lederberg and Lederberg, "Genetic Studies of Lysogenicity"; Brock, *Emergence*

> When first envisaged, this conclusion appeared somewhat surprising, since it would seem a priori that the noninfective structure which, in lysogenic bacteria, carries the genetic information of a virus, the bacteriophage, should be cytoplasmic rather than chromosomal. Although this problem was never seriously considered before 1950, that is, until the investigations of Lwoff, the hypothesis of a cytoplasmic determination of lysogeny had been accepted implicitly.[214]

The discovery that λ resided on the *E. coli* chromosome dispelled any remaining belief that virus genetics might be illuminated by reference to plasmagenes. However, under the new unifying framework of "infectious heredity," there remained a need for some way to refer to non-nuclear genetic units. Drawing on the similarities between *Pneumococcus* transformation, bacterial transduction, and viral infection, Joshua Lederberg proposed in 1952 that *plasmid* serve "as a generic term for any extra-chromosomal hereditary determinant."[215]

That same year, Alfred Hershey and Martha Chase performed their landmark experiment demonstrating that the reproductive capacity of bacteriophage was attributable specifically to the viral nucleic acid, not to the nucleoprotein. Hershey and Chase infected bacteria with ^{35}S-labeled bacteriophage and found that this radioactive label, only incorporated into protein, remained largely on the surface of the cell in the phage "ghosts." By contrast, when bacteria were infected with ^{32}P-labeled phage, in which the nucleic acid portion was radioactive, the label concentrated in the cell, and subsequently in the phage progeny.[216]

Thus, in the end, the mechanism of virus multiplication appeared to inhere in the *loss* of integrity by the "self-reproducing" agents, not in their duplication as integral units. As Martin Pollock has put it, "Typical viruses... replicate in a manner such that during the period of nucleic acid synthesis their original identity is lost, and the elements necessary for their reproduction—the new enzymes induced by their nucleic acids and the altered metabolic steps involved—are scattered throughout the cell and ultimately form no part of the emerging virion itself."[217]

The 1953 Cold Spring Harbor Symposium, "Viruses," is often cited as a historic occasion for molecular biology, in large part because it provided the forum for the first public presentation of the Watson-Crick model

of Bacterial Genetics, 180. One could only observe the genetic linkage between λ and *E. coli* markers when the phage was in the lysogenic state.

214. Jacob and Wollman, *Sexuality and the Genetics of Bacteria,* 90; also quoted in Brock, *Emergence of Bacterial Genetics,* 182.

215. Lederberg, "Cell Genetics and Hereditary Symbiosis," 403; see also Sapp, *Beyond the Gene,* 122; and Lederberg, "Plasmid (1952–1997)."

216. Hershey and Chase, "Independent Functions."

217. Pollock, "Changing Concept of Organism," 281.

for the double helical structure of DNA. The novel organization of the program around Lwoff's notion of a "provirus" and the diversity of virologists attending the conference further contributed to the significance of the event. While the important status of this meeting for these reasons is assuredly deserved, the conference also reflected the new emphasis on virus multiplication as intimately related to cell metabolism, superseding the earlier analogies between viruses, plasmagenes, organelles, and enzyme-forming centers. This conceptual reshuffling did not result in a restoration of the older notion of viruses as autonomous genetic agents, but established viruses as the defining representatives of "infective heredity." William Hayes's proposal that mating in *E. coli* strain K-12 was controlled by an agent analogous to a virus, the F (fertility) factor, provided yet another example of this new class of transmissible agents.[218]

As Delbrück noted, the predominating interest in the nature of the "vegetative" virus privileged bacteriophage research over that of plant viruses, for which the virus had been more successfully understood as an extracellular, infective entity.[219] Biophysicist and TMV researcher Robley Williams admitted that physicochemical studies of viruses had not revealed intermediates of viral replication but rather contributed to the "over-all understanding of their growth and multiplication primarily by establishing certain tangible specifications for the properties which the finished, mature particles must have."[220] Given the new significance accorded the study of viruses in the vegetative phase, some researchers attempted to create an experimental system for investigating the life cycle of plant viruses. For example, in 1953 phage biologist George Streisinger was attempting to establish TMV in a cell culture system, in order to perform quantitative experiments on infection like those typifying studies of bacteriophage.[221] However, the technical problems proved discouraging.[222]

On the other hand, work on animal viruses was expanding rapidly, due largely to the liberal funding provided by the National Foundation for

218. For a summary of the work on *E. coli* mating, which also drew on key experimental contributions from the Lederbergs and Luca Cavalli, see Hayes, "Mechanism of Genetic Recombination."

219. Delbrück, "Introductory Remarks"; Adams makes this same point in "Cold Spring Harbor Symposium."

220. Williams, "Shapes and Sizes of Purified Viruses," 185.

221. See George Streisinger to Stanley, 14 Dec 1953, Stanley papers, carton 14, folder California Institute of Technology.

222. Streisinger did get the local lesion assay established at Caltech, however—a resource Watson was counting on as he made plans for research on TMV RNA there (James D. Watson to Delbrück, 25 Mar 1954, Delbrück papers, folder 23.23).

Infantile Paralysis (NFIP).[223] Here phage provided an advantageous model system for studying virus multiplication. Delbrück voiced his optimism about the prospects for animal virology at the 1953 Cold Spring Harbor meeting (which, as it happens, was being funded by the NFIP):

> The paper on the last day of our program is devoted to a technique for the bio-assay of animal viruses which puts them on par with the phages, and wonder of wonders, polio virus appears to be a choice virus for this technique. I think it may safely be predicted that the next few years will see a fierce race between flu [for which genetic recombination had been demonstrated] and polio for the distinction of being called the "Drosophila of Animal Viruses."[224]

In the area of tumor virology, where latent infection was a characteristic feature, the new emphasis on infective heredity was particularly apt.[225] Here, both bacteriophage and TMV provided distinct and recognizable exemplars for the visualization and study of tumor viruses. Interest in the macromolecular nature of cancer-causing viruses was evident in studies of papilloma virus and chicken leucosis virus.[226] Researchers relied increasingly on the electron microscope to structurally identify putative tumor viruses—as particles—culminating in the late 1950s with the National Cancer Institute's massive screening programs for human cancer viruses.[227] Moreover, the ultracentrifuge remained a principal tool for purifying animal viruses, including tumor viruses.[228] This was the legacy of TMV. On the other hand, just as phage workers at Caltech managed to "plaque" poliomyelitis virus in 1953, making it into a system for quantitative experimentation, Howard Temin and Harry Rubin originated a

223. On the role of the NFIP in stimulating the work on animal viruses, see Dulbecco, "From Lysogeny to Animal Viruses," as well as chapter 5 of this book.

224. Delbrück, "Introductory Remarks," 2.

225. On tumor viruses and latent infection, see Pinkerton, "Pathogenesis and Pathology"; and Koprowski, "Latent or Dormant Viral Infections." On the history of research on tumor viruses, see Levine, "Origins of the Small DNA Tumor Viruses."

226. On papilloma virus, see Sharp et al., "Study of the Papillomatosis Virus Protein"; Sharp, Taylor, and Beard, "Density and Size of the Rabbit Papilloma Virus"; Sharp et al., "Rabbit Papilloma and Vaccinia Viruses and T2 Bacteriophage"; Schachman, "Physical Chemical Studies." On chicken leucosis, see, for example, Eckert, Beard, and Beard, "Dose-Response Relations." A more detailed account of investigations of these tumor viruses as macromolecules is given in Creager and Gaudillière, "Experimental Arrangements."

227. For a critical review of the electron microscope as a means to identify tumor viruses, see Porter and Kallman, "Significance of Cell Particulates." On the National Cancer Institute's screening program, see Gaudillière, "Molecularization of Cancer Etiology."

228. On the significance of the ultracentrifuge, see Beard, "Review: Purified Animal Viruses."

"plaque"-inspired transformation assay for Rous sarcoma virus in 1958.[229] This achievement significantly extended the reach of molecular genetics to tumor viruses, a growth industry in the 1960s and 1970s. Indeed, it was through the molecular genetic investigations of cancer viruses that "infective heredity" was reinterpreted in terms of DNA, for tumor viruses (like lysogenic phage) were found capable of integrating into the host genome.[230]

Conclusions

André Lwoff brought closure to the question of defining viruses in his Marjory Stephenson lecture of 1957, with his famous declaration, "*Viruses are viruses.*"[231] More specifically, Lwoff contended that viruses are "*strictly intracellular and potentially pathogenic entities with an infectious phase,* and (1) possessing only one type of nucleic acid, (2) multiplying in the form of their genetic material, (3) unable to grow and to undergo binary fission, (4) devoid of a Lipmann system [enzymes able to transfer energy from foodstuffs to high-energy bonds]."[232] Notable for historical reasons is what Lwoff's definition of a virus excluded—the virus was no longer considered to be a nucleoprotein particle capable of self-reproduction or growth. The revised depiction of the "virus" derived not only from new findings, but also from the changing understanding of how viruses related to genes and genetic particles.

From the late 1930s until the early 1950s viruses had been seen as material models for investigating the genetic property of self-multiplication. It was not that genetic research on viruses was experimentally unified; as we have seen, the laboratory practices employed in attempts to coax information about the material gene from bacteriophage and TMV were radically different. In effect, each system was constructed to follow a different aspect of Muller's definition of the material gene: self-duplicating and mutable. For Delbrück, the autonomous self-replication of bacteriophage was its key genetic property, and phage was visualized through the appearance of plaques on a lawn of bacterial cells. In the case of TMV, which could be chemically purified, Stanley and others expected that studies of TMV strains would illuminate reproduction and mutation of genes as material—that is, chemical—entities. In addition to the importance of TMV mutants, Stanley foregrounded structural similarities, including simple size comparisons, between TMV and chromosomal genes.

229. Dulbecco and Vogt, "Plaque Formation and Isolation"; Temin and Rubin, "Characteristics of an Assay."

230. For an overview of this development, see Kevles, "Pursuing the Unpopular."

231. Lwoff, "Concept of Virus," 240; emphasis in original.

232. Ibid., 246; emphasis in original.

During the 1940s, neither Delbrück nor Stanley gained the experimental access they sought through viruses to the gene as a physical entity. However, perceived similarities between their viruses and a range of other entities that appeared to self-duplicate reinforced expectations that one could investigate the gene as a material—a nucleoprotein. Many biologists became interested in plasmagenes as a class of self-perpetuating cytoplasmic particles, of which viruses might be an infectious representative. Although this line of reasoning did not ultimately produce a general mechanism for self-reproduction, it may well have affected the reception of the important work on transforming factors published by Avery, MacLeod, and McCarty. Scientists and historians have offered various reasons for the failure of scientists in the 1940s to register the compelling evidence that genes are made of DNA.[233] Attending to the analogies that researchers were drawing between viruses and cytoplasmic particles provides another perspective. The tendency to view bacterial transforming agents in conjunction with viruses, cell organelles, enzyme-forming centers, and other possible plasmagenes led scientists to underestimate the distinctiveness of transforming factors. They were, after all, the only known "self-perpetuating" agent that was not a nucleoprotein, but only a nucleic acid, mere DNA.

This account of why biologists did not appreciate the experimental evidence for DNA's specific role in heredity stresses the importance of analogies between different experimental systems in guiding scientific interpretation. Philosopher of science Mary Hesse provided the classic argument on behalf of the utility of analogies in scientific theorizing.[234] In her view, scientists use analogies creatively to extend explanations in one domain to another so as to prompt theory development in the new area.[235] Similarity relations between preexisting models and (new) observed properties can be classed in terms of positive analogies, negative analogies, and neutral analogies. One might well translate the information in table 6.1, which summarizes the properties of several representative genetic objects, into these terms. Yet my purpose has been somewhat different: to highlight the shifting grounds for these analogies, which did not necessarily grow out of any formalized theory, as in Hesse's analysis, but rather hinged on rapidly changing experimental developments. The

233. See Stanley, "'Undiscovered' Discovery"; Stent, "Prematurity and Uniqueness in Scientific Discovery"; and a response to Stent by Lederberg, "Greetings." See also Lederberg, "Forty Years" and "Transformation of Genetics by DNA." This issue has also been treated by Olby, *Path to the Double Helix,* and by Portugal and Cohen, *Century of DNA.*

234. Hesse, *Models and Analogies.*

235. For a valuable summary and extension of Hesse's approach, see Morgan, "Technology of Analogical Models."

fundamental analogy between viruses as research objects and the gene as a concept served as a focus for this larger set of comparisons between viruses and other particles. Yet the gene never attained precision as a theoretical concept; indeed its multifacetedness, even vagueness, may well have contributed to its wide scientific usefulness.[236]

By the early 1950s, confidence in plasmagenes had abated, replaced, I have argued, by a new emphasis on agents of infective heredity, including viruses, transforming factors, mating factors, and peculiar entities such as kappa. The fruitfulness of bacteriophage for genetics, and especially for the molecular mechanism of recombination, had continued to advance the gene as an *instrumental* entity and subsequently transformed its meaning. By the late 1950s, Seymour Benzer's pathbreaking mapping of the fine structure of the rII gene of phage T4 shattered the gene as a unitary conceptual entity into hundreds of mutable and recombinable (and thus mappable) sites.[237] At the same time, acknowledging the specificity of nucleic acids in transformation and heredity gave the *material* gene a new chemical meaning, now associated with the sequence of bases in polymers of DNA (or, in many viruses, RNA). The next chapter details how this emphasis on the chemical specificity of infectivity, in conjunction with efforts to establish the relationship between nucleic acid sequences and encoded proteins, reinvigorated TMV as an experimental tool for molecular biology in the 1950s and early 1960s.

236. See Rheinberger, "Gene Concepts." As he argues, the gene might best be considered a "boundary object," or "boundary concept," whose organizing power derived from its flexible meaning, even fuzziness. On boundary objects, see Star and Griesemer, "Institutional Ecology"; on the boundary concept, see Löwy, "Strength of Loose Concepts."

237. Benzer, "Elementary Units of Heredity"; Brock, *Emergence of Bacterial Genetics*, 139–43; Holmes, "Seymour Benzer." Falk has argued that "the ill-defined gene concept" that emerged from molecular biology of the late 1950s and early 1960s has been a "blessing" to subsequent researchers ("What Is a Gene?" 161). On the more recent erosion of the gene as a coherent conceptual entity, see also Rheinberger, "Gene Concepts"; and Keller, *Century of the Gene.*

Taking TMV to Pieces

As we have just learned, tobacco mosaic virus contains ribonucleic acid (RNA) apparently endowed with biological activity, embedded or encapsulated in a high, stiff collar of specific protein that is sufficiently constraining to impel the RNA to keep its family reputation intact, and not to dissipate its substance in low-molecular activities.

Rollin D. Hotchkiss, 1957

In the mid-1950s, TMV gained currency among molecular biologists as it became a leading prototype for working on two central problems: the genetic code and the assembly of viruses from their component pieces. The predominant experimental focus on TMV as an extracellular, biochemical entity that had proved so limiting once the topic of virus reproduction shifted to investigations of the virus in its "vegetative," intracellular phase (see chapter 6), proved enormously productive for understanding genetic specificity in chemical terms. New analogies between the RNA of TMV and other "informational" nucleic acids, and between the subunits of TMV and other well-studied proteins such as insulin, placed TMV in the center of the language and conceptualization of macromolecules associated with the emerging field of molecular biology. TMV continued to be a virus of "firsts," with the 1956 demonstrations that its nucleic acid alone was infectious, and the 1960 publications of its 158-amino-acid polypeptide sequence, the longest protein and first viral protein sequenced to date.[1] The kind of exemplar TMV provided for understanding other viruses shifted with these achievements, inspiring the search for the "infectious" nucleic acid portion of animal viruses and providing an important referent in the search for general principles of virus self-assembly.

1. Fraenkel-Conrat, "Role of the Nucleic Acid"; Gierer and Schramm, "Infectivity of Ribonucleic Acid"; Anderer et al., "Primary Structure"; Tsugita et al., "Complete Amino Acid Sequence."

The development of research on TMV from the late 1940s to the early 1960s may be viewed at one level as a continuation of the trajectory of Stanley's chemical work on viruses since 1931. While the instruments for representing the virus as a macromolecule became increasingly more sophisticated in the postwar period, the determination of TMV's precise protein structure by 1960 was the culmination of a quarter-century of chemical research. Yet the meanings associated with TMV as a macromolecule shifted decisively over this period. Stanley's emphasis on the rod-shaped virus particle as a unitary chemical object was slowly undermined by the increasing attention to the functional pieces of the virus, both protein and nucleic acid. Moreover, the demonstrations that the viral RNA specifically determined the viral protein provided a new way to envision the relationship between viruses and genes. The relationship between nucleic acids and proteins was increasingly organized around the notion of a mediating code, which could be established independently of the cellular mechanisms for genetic duplication.[2] TMV, with its readily accessible protein and nucleic acid components and a host of mutant strains, provided a promising resource for biochemically elucidating which triplet codons specified the twenty possible amino acids. (An alternative genetic strategy was being developed by Sydney Brenner and Francis Crick by incorporating mutagenesis into the system Seymour Benzer had developed for the fine-mapping of the bacteriophage T2 rII gene.)[3] Only with the surprising development by Marshall Nirenberg and Heinrich Matthaei of a cell-free protein synthesis system that could "translate" synthetic RNAs did a less laborious means become available to crack the genetic code.[4] Even then, TMV nucleic acid was used as a "template" for cell-free translation and a source for evidence corroborating the genetic code emerging from *in vitro* translation experiments.

This chapter focuses on TMV research in Stanley's Virus Laboratory in the 1950s and early 1960s, but includes analysis of developments in other laboratories as well. Pnina Abir-Am has argued that molecular biology was fundamentally a transnational field in the postwar period, reflecting the expanded international reach of politics and communication.[5] The TMV story provides a case in point for the increased importance of international collaboration and competition. The Virus Laboratory, even

2. The emphasis on the coding relationship between protein and nucleic acid was most clearly articulated by Crick in 1958 in his "sequence hypothesis," which asserted that "the specificity of a piece of nucleic acid is expressed solely by the sequence of its bases, and that this sequence is a (simple) code for the amino acid sequence of a particular protein" (Crick, "On Protein Synthesis," 152).

3. See Judson, *Eighth Day of Creation*, 443–47, 465–68.

4. See ibid., 452 ff.; Rheinberger, *Toward a History of Epistemic Things*, ch. 13; Kay, *Who Wrote the Book of Life?* ch. 6.

5. Abir-Am, "From Multidisciplinary Collaboration to Transnational Objectivity."

with its extensive instrumentation and large staff, did not encompass all of the important biochemical and biophysical techniques being used in TMV research. In particular, the X-ray crystallographic studies of TMV, led by Rosalind Franklin and a host of collaborators in the late 1950s, were critical to the emerging picture of TMV structure, and Stanley's only role in these investigations was to provide material for the studies. As the amount and diversity of research on TMV grew rapidly in the postwar period, researchers relied on personal communication and collaboration to keep abreast of the field. More important, the understanding of TMV as a model system was a composite of results and representations emerging from many different approaches to the virus as an epistemic object, from the diversity of experimental systems involving TMV.[6] This well-studied plant virus faced competition from a variety of other model systems in the 1960s, yet the emphasis on the relationship between structure and function, which had been so central to investigations of TMV during the previous decade, remained a central concern of molecular biology, now an established discipline.[7]

The Postwar Transition and the German Challenge

At the new Virus Laboratory in Berkeley in the late 1940s, researchers tended to treat TMV as a single functional unit, whose infectivity and capacity for self-reproduction depended on the integrity of the nucleoprotein. This viewpoint drew on Stanley's long-standing emphasis on TMV as a homogeneous macromolecule both *in vivo* and *in vitro*. In this respect, the debate at the end of the 1930s between colloidalists such as Vernon Frampton and molecularists such as Stanley over the nature of the virus as a macromolecule cast a long shadow over physicochemical research on TMV. As a result, the evidence presented in 1941 by Bernal and Fankuchen from their X-ray crystallographic studies of TMV that there were smaller repeating units in the virus structure, while acknowledged by Stanley, did not motivate further investigation in his laboratory.[8] As

6. On the significance of epistemic objects in research, see Rheinberger, *Toward a History of Epistemic Things*.

7. On the emergence of a Department of Molecular Biology at Berkeley, see Creager, "Stanley's Dream." Many of the other historical studies of molecular biology as a discipline have focused on other national contexts: Gaudillière, "Molecular Biology in the French Tradition?"; Uchida, "Building a Science in Japan"; de Chadarevian, "Sequences, Conformation, Information," and *Designs for Life;* Santesmases and Muñoz, "Scientific Organizations in Spain." The important exception to this generality is Lily Kay's analysis of developments at Caltech, *Molecular Vision of Life*, although her concerns are broader than those associated with disciplinary history.

8. Bernal and Fankuchen, "X-Ray and Crystallographic Studies." See chapter 4 for the debates over virus protein aggregation at the end of the 1930s. In a 1938 review, Stanley stated that the breakdown of the virus into smaller parts through treatment with proteolytic enzymes, or denaturation with acid, alkali, heat, dodecyl sulfate, or urea was

discussed in chapter 4, F. C. Bawden claimed that ultracentrifugation and other purification procedures caused TMV to aggregate into long rods, and he suggested that the biologically active virus was a much smaller particle, perhaps associated with other molecules that could not withstand purification.[9] Stanley vigorously opposed Bawden's suspicion that the 3,000-Å-long, rod-shaped particle resulting from his purification procedures was an artifact.[10]

The dispute between Stanley and Bawden over TMV aggregation informed their representations of TMV in publication well into the 1950s.[11] But during the war a new contributor to chemical research on viruses, not party to this old feud, had taken up the question of aggregation and regular repeats. In 1941, a new workshop branch ("Arbeitsstätte") for virus research was established in Berlin-Dahlem, Germany, by the Kaiser Wilhelm Institutes (KWI) for Biochemistry (headed by Adolf Butenandt) and Biology (headed by Alfred Kühn and Fritz von Wettstein).[12] Gerhard Schramm, who had been working as an organic chemist with Butenandt on sterols, was the principal biochemist associated with this new initiative.[13] Two years earlier, Georg Melchers of the KWI for Biology laid the groundwork for this collaboration by obtaining a sample of TMV and seeds for the test plant (*Phaseolus vulgaris*) from Stanley to begin "experiments about questions of physiological development on a genetical basis connected with the virus problem."[14] In fact, the strain that Stanley

---

always associated with a loss of biological activity. Thus he concluded that the small molecular weight materials produced through such procedures (and detectable in the analytical ultracentrifuge) were not of biological significance. Rather, he focused research attention on the agents of inactivation that did not cause the virus to break into pieces, such as chemical modification and exposure to ultraviolet light. He went so far as to state that "Frampton and Saum reported that, although low molecular weight material was found [in TMV disintegrated by altered salt and pH], there was no change in virus activity, a result that appears very questionable and one which it has not been found possible to confirm in the writer's laboratory" ("Architecture of Viruses," 534).

9. Bawden, "Virus Diseases of Plants."

10. In 1946, Stanley asserted, "The single virus particles are about 280 mμ in length and 15 mμ in diameter. Tobacco mosaic virus activity has never been demonstrated to be associated with smaller particles. However, there is good evidence that a single virus particle is built up from similar subunits fitted together in a hexagonal lattice to yield the final structure which possesses virus activity.... Because of the repeat pattern within a single virus particle, it can be regarded as a submicroscopic crystal" ("Viruses," 15).

11. See, for example, Bawden and Pirie, "Virus Multiplication," esp. 31–32.

12. A valuable institutional history of virology in Germany is given in Butenandt, "Historical Development of Modern Virus Research in Germany"; see also Deichman, *Biologists under Hitler,* 210 ff.; Macrakis, *Surviving the Swastika,* 119; and Rheinberger, "Virus Research at the Kaiser Wilhelm Institutes."

13. See, for example, Butenandt and Schramm, "Über die Bromierung"; Butenandt et al., "Einige Bemerkungen." For more on Butenandt's career, see Macrakis, "Adolf Butenandt."

14. Letter from Georg Melchers, Kaiser Wilhelm-Institut für Biology, to Stanley, 23 Mar 1938, Stanley papers, carton 10, folder Melchers, Johann Georg Friedrich.

sent provided the source material for all of the group's subsequent studies on TMV.[15] In comparison to Melchers, Schramm took the research
in a more chemical direction. One of his first efforts in virus research
involved constructing his own air-driven ultracentrifuge, based on the
design of Beams and Pickels.[16]

Schramm investigated TMV using quantitative chemical techniques
as well as physical-chemical instruments. In 1940, Schramm published on
the configuration of amino acids in tobacco mosaic virus, based on a crystalline sample.[17] The following year, he reported splitting the ribonucleic
acid from tobacco mosaic virus with an enzyme (nucleotidase).[18] Stanley's
laboratory paid particular attention to this publication. In October 1941,
Stanley wrote Schramm that his workers had been attempting in vain
to reproduce this result, and asked if Schramm would send a sample of
the enzyme he used as well as some nucleic-acid-free TMV.[19] According
to Georg Melchers, Schramm already knew that his report was incorrect when he received the letter from Stanley, but he was not eager to
admit the error, preferring to "write a footnote about it sometime."[20] The
subsequent involvement of the United States in the war against Germany
following the Pearl Harbor attack made an exchange of information
and materials more difficult. Stanley's coworkers published papers dis-
agreeing with the accuracy and interpretation of Schramm's enzymatic
and chemical denaturation of TMV.[21] But beyond his skepticism about

15. Schramm, "Die Struktur des Tabakmosaikvirus," 473.

16. Schramm, "Die luftgetriebene Ultrazentrifuge." Schramm also spent time in the
laboratories of The Svedberg and Arne Tiselius in 1938 (Deichman, *Biologists under Hitler,*
211). On the development of the air-driven ultracentrifuge, see the text and references in
chapter 4.

17. Schramm and Müller, "Über die Konfiguration der im Tabakmosaikvirus enthaltenen
Aminosäuren."

18. Schramm, "Über die enzymatische Abspaltung der Nucleinsäure aus dem Tabakmosaikvirus." Schramm's investigation followed up Edgar Pfankuch's cleavage of TMV into
nucleic acid and protein pieces with alkali; see Pfankuch, "Über die Spaltung von Virusproteinen."

19. Stanley to Gerhard Schramm, 13 Oct 1941, Stanley papers, carton 12, folder
Schramm, Gerhard.

20. Georg Melchers, personal communication to Ute Deichman, 21 May 1992, quoted in
Biologists under Hitler, 314. In a review several years after the original publication, Schramm
simply stated, "The observation of Schramm (1941) that by a glycerin extract from small
intestine mucous membrane the nucleic acid could be split off from TMV proved to be
mistaken. Cohen and Stanley (1942) showed that by such an enzyme preparation the nucleic
acid is not removed from the molecule, and also no inactivation occurs" (Schramm, "Die
Struktur des Tabakmosaikvirus," 461; author's translation).

21. Cohen and Stanley, "Action of Intestinal Nucleophosphatase"; Lauffer and Stanley,
"Denaturation of Tobacco Mosaic Virus by Urea." In an exchange of letters late in 1946,
Schramm agreed with Stanley's assessment that the early claim to have separated the
viral nucleic acid from protein with calf intestine phosphatase was mistaken, and he drew
Stanley's attention to his published correction and reinterpretation in his review, "Über die

Schramm's result, Stanley's work on the question of aggregation continued to be directed principally at Bawden's criticism of the long virus rods as purification-generated artifacts.[22] More significant, the Princeton laboratory had shifted its focus from TMV to influenza virus as part of the mobilization for war.[23] Schramm, however, continued publishing steadily on TMV structure and mutants.

In 1943, Schramm reported that exposing TMV to alkaline solution resulted in three products: normal TMV, a small molecular weight nucleoprotein, and a small, nucleic-acid-free protein.[24] This last species was found to be homogeneous in the ultracentrifuge and electrophoresis apparatus; based on the sedimentation and diffusion constants, a molecular weight of 360,000 daltons was assigned. More surprising, a drop in pH caused this protein to form a large, rod-shaped macromolecule similar in its size and properties to TMV.[25] This protein, however, was not infective, and so had lost the property of self-reproduction. On the basis of these striking findings, Schramm concluded that the size and shape of TMV were due to the characteristics of its subunits but maintained that the nucleic acid played no role. In a 1944 review article, Schramm argued that TMV was composed of building blocks, homogeneous protein subunits with a molecular weight of 370,000 daltons and an approximate dimension of 67×87 Å. These dimensions corresponded closely to the repeating units seen in Bernal and Fankuchen's X-ray crystallographic studies.[26] Schramm ventured that each subunit might contain one molecule of nucleic acid, although his TMV dissociation experiments always yielded some nucleic-acid-free virus fragments. Schramm also drew an analogy between TMV and the large protein hemocyanin, whose dissociation into smaller homogeneous protein subunits had been well documented.[27]

Konstitution des Tabakmosaikvirus"; see Schramm to Stanley, 26 Nov 1946, Stanley papers, carton 12, folder Schramm, Gerhard.

22. See, for example, Scientific Reports, RAC RU 439, box 6, vol. 34, 1945–46, 262.

23. This is how Schramm explained the five-year delay in his answer to Stanley, 2 Aug 1946, Stanley papers, carton 12, folder Schramm, Gerhard. On the influenza research in Stanley's laboratory during World War II, see chapter 4.

24. Schramm, "Über die Spaltung des Tabakmosaikvirus in niedermolekulare Proteine"; see also Pfankuch and Piedenbrock, "Zur Spaltung von Virusproteinen."

25. This macromolecule produced through reassociating virus fragments was further characterized by Schramm in "Über die Spaltung des Tabakmosaikvirus und die Wiedervereinigung.... Part 2."

26. Schramm, "Über die Konstitution des Tabakmosaikvirus," esp. 112; Bernal and Fankuchen, "X-Ray and Crystallographic Studies." Schramm was at the same time working on the structure of native and regenerated cellulose; perhaps this provided the basis for an analogy with virus regenerated from "denatured" pieces (Schramm, "Über periodische Fällungen").

27. Schramm, "Über die Spaltung des Tabakmosaikvirus in niedermolekulare Proteine," 96.

Further studies of the hydrodynamic properties of the alkali-produced protein fragments revealed that they were the same size, but not the same shape, as the repeating units found by Bernal and Fankuchen. Specifically, most of the variously sized fragments Schramm was producing seemed to be long and asymmetric like the intact TMV (see fig. 7.1).[28] These fragments could be assembled into a macromolecule very similar in size and shape to TMV, but the resulting entity was not infectious, despite repeated attempts to preserve and measure low-level activity.[29]

Schramm's portrayal of TMV as composed of discrete subunits attracted surprisingly little attention from biologists in the United States and Great Britain. As James D. Watson reflected later in *The Double Helix,*

> There already existed biochemical evidence for protein building blocks. Experiments of the German Gerhard Schramm, first published in 1944, reported that TMV particles in mild alkali fell apart into free RNA and a large number of similar, if not identical, protein molecules. Virtually no one outside Germany, however, thought that Schramm's story was right. This was because of the war. It was inconceivable to most people that the German beasts would have permitted the extensive experiments underlying his claims to be routinely carried out during the last years of a war they were so badly losing. It was all too easy to imagine that the work had direct Nazi support and that his experiments were incorrectly analyzed.[30]

Watson voiced a commonplace American opinion that all German science during the war must have been corrupted by Nazism.

Schramm clearly tolerated the National Socialist regime—he was, in fact, a party member.[31] Yet, remarkably enough, his research at the KWI remained entirely unrelated to the war effort or government activities. It is difficult to find specific evidence, moreover, that his political affiliation was the reason for neglect of his work by Stanley and other American biologists, beyond Max Delbrück's pronounced (and perhaps influential) antipathy to Nazi biologists.[32] Rather, as a German colleague concluded,

28. Schramm, "Über die Spaltung des Tabakmosaikvirus und die Wiedervereinigung. ...Part 1."

29. For the continued efforts to measure biological activity in the reassociated virus fragments, see Schramm, "Über die Spaltung des Tabakmosaikvirus und die Wiedervereinigung. ... Part 2."

30. Watson, *Double Helix,* 112.

31. "Schramm had been a member of the NSDAP as well as the SS" (Deichman, *Biologists under Hitler,* 313). Deichman also discusses Max Delbrück's opposition to German biologists who had been Nazis (420, n. 32).

32. Bear in mind that among scientists at German universities and the Kaiser Wilhelm Institutes, Schramm was very frequently cited from 1945 to 1954, as assessed quantitatively by Deichman, *Biologists under Hitler,* 99–102.

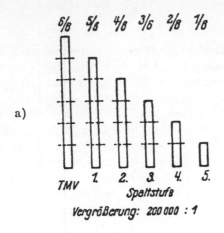

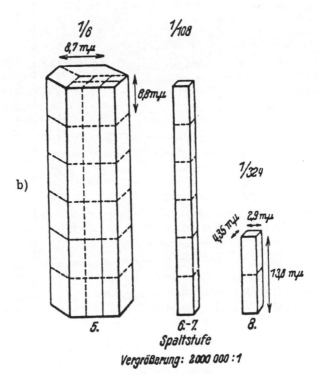

FIGURE 7.1 Schramm's schematic diagram of the splitting of the TMV particle. *a:* The splitting of a transverse piece of the TMV rod by sixths along its length, as seemed to be evident in electron micrographs of the fragments. *b:* A drawing to show how these long segments of the rod might fit together to make the elementary unit cells determined by Bernal and Fankuchen in their X-ray crystallographic analysis of TMV. ("Schema der Spaltung," from Gerhard Schramm, "Über die Spaltung des Tabakmosaikvirus und die Wiedervereinigung.... Part 1" [1947]: 118. Reprinted by permission of *Zeitschrift für Naturforschung.*)

"Schramm's reputation in the United States suffered considerably from his scientific carelessness and his failure to react to the refutation of his results."[33] Even so, Schramm and Stanley exchanged a set of reprints in 1947, and Schramm offered comments on Stanley's work on influenza. By this time, the KWI for Biology and the KWI for Biochemistry had been moved to Tübingen.[34] Stanley wrote Delbrück, "The paper [by the group in Tübingen] published in the *Biologisches Zentralblatt* in 1946 is very interesting and is similar in many respects to the results which we have obtained and published from this laboratory."[35] Yet Stanley showed little interest in Schramm's dissociation experiments.[36]

In an effort to resolve differences in the literature on the size and shape of TMV, Stanley, Gerald Oster, and C. Arthur Knight published a comparison in 1947 of their laboratory's TMV stock with the strains being studied by Bawden and Pirie in Rothamsted and by Schramm, Melchers, and Friedrich-Freksa in Tübingen, Germany.[37] A variety of lengths for TMV rods had been reported by these three groups, ranging from 190 mμ to 300 mμ (1 mμ = 10^{-9} m = 10 Å).[38] Stanley's group as well as Schramm and Friedrich-Freksa had suggested that these inconsistencies might derive from actual genetic differences between the stocks that the groups were investigating.[39] By contrast, Bawden attributed the differences to aggregation of "smaller, possibly spherical, biologically active

33. Deichman, *Biologists under Hitler*, 313 14. Deichman here is summarizing the view of Georg Melchers, whom she cites.

34. These institutes in Tübingen were renamed the Max Planck Institutes for Biology and Biochemistry in 1949. A new Max Planck Institute for Virus Research was established in Tübingen in 1954. I am indebted to Christina Brandt and her work on TMV research in Germany for clarification of the institutional history.

35. Stanley to Max Delbrück, 12 Dec 1947, Stanley papers, carton 2, folder Dec 1947. The paper to which he referred was Friedrich-Freksa, Melchers, and Schramm, "Biologischer, chemischer und serologischer Vergleich."

36. Strikingly, in the 1948 review Stanley wrote with Max Lauffer on physicochemical approaches to virus research ("Chemical and Physical Procedures"), they only cited one paper by Schramm, and none on his work on TMV subunits: see Schramm, Born, and Lang, "Versuch über den Phosphorastausch zwischen radiophosphorhaltigen Tabakmosaikvirus und Natriumphosphat."

37. Oster, Knight, and Stanley, "Electron Microscope Studies." Stanley thanked Schramm for the sample of TMV he sent and reported early results from the comparison in a letter, 25 Feb 1947, Stanley papers, carton 12, folder Schramm, Gerhard.

38. The original electron micrographs published by Kausche, Pfankuch, and Ruska ("Die Sichtbarmachung von pflanzlichem Virus im Übermikroskop") showed a variety of lengths from 150 to 300 mμ. The other German group reported a length of 190 mμ in micrographs of TMV purified using ammonium sulphate precipitation. By contrast, Stanley's group found a length of 280–300 mμ in electron micrographs of their ultracentrifuge-purified TMV.

39. Stanley and Anderson, "Study of Purified Viruses"; Schramm and Friedrich-Freksa, "Die Präcipitinreaktion des Tabakmosaikvirus."

units" through the process of isolation.[40] Indeed, as late as 1946 he stated that the size of TMV "remains a matter of speculation and controversy," and that the 280-mμ particles observed in electron micrographs were probably aggregates.[41] In order to counter Bawden's contention, Oster, Knight, and Stanley visualized TMV following a minimal amount of laboratory preparation. In electron micrographs of virus in crushed tobacco leaf hairs infected with the three TMV samples, they observed identically sized virus rods, with very similar particle-size distributions. The median particle length in all three cases was 280 $\pm$ 8.6 mμ, agreeing with the Princeton value they had obtained earlier. Schramm himself largely confirmed these results in his own analysis of the range of sizes of TMV observed in the ultracentrifuge and electron microscope under different conditions.[42] Whatever differences were arising from the laboratories appeared to inhere not in their materials, but their methods of visualizing the virus.

Oster was also interested in investigating viruses using X-ray diffraction techniques, and Stanley assisted him in arranging for a year's training abroad in this specialty.[43] Indeed, Stanley intended his new Virus Laboratory in Berkeley to include an active program of X-ray crystallography. Inquiring about a Rockefeller Foundation fellowship for Oster, he stated, "As our work develops, I should like to make use of X-ray diffraction, since I believe that such studies will yield valuable information concerning the structure of viruses which can be obtained in no other way."[44] Oster set out for Bernal's laboratory in the summer of 1948, planning to use X-ray

40. Quotation from Oster, Knight, and Stanley, "Electron Microscope Studies," 280, summarizing Bawden's views in *Plant Viruses and Virus Diseases,* 2d edition.

41. Bawden, "Virus Diseases of Plants," 165. On the following page he asserts, "It seems probable that, as produced in the plant, the virus particles may not be significantly rod-like, but when exposed to conditions in expressed sap and the processes of purification, these small particles aggregate linearly to produce the elongated particles responsible for the characteristic properties of the isolated virus. The precise nature of the change leading to this aggregation is still uncertain, but the most likely explanation is that in the plant, the particles are chemically more complex than the nucleoprotein of the purified preparations. In the plant, the nucleoprotein is probably combined with some other substance or substances, presumably inessential for infectivity, whose removal sets free groups in the virus particles which can then combine one with another. Thus, aggregation to form elongated particles may be an inescapable result of purification, and the length of the particles may be a direct measure of the chemical purity of a preparation" (166).

42. Schramm and Wiedemann, "Größenverteilung des Tabakmosaikvirus."

43. The itinerary Stanley envisioned for Oster included nine months in J. D. Bernal's laboratory and two months of visits with biophysicists in Europe (Paris, Amsterdam, Copenhagen, and Stockholm); see Stanley to H. Marshall Chadwell, 7 Apr 1948, Stanley papers, carton 2, folder Apr 1948.

44. Stanley to H. Marshall Chadwell, 18 Feb 1948, Stanley papers, carton 11, folder Oster, Gerald.

crystallographic techniques to study the structure of TMV nucleic acid (although he rapidly shifted his focus to proteins). But he decided in 1949 that he would rather accept a position in Strassbourg than join Stanley's laboratory in California. This defection caused a good deal of consternation on the part of both Stanley and Warren Weaver.[45] Thus, Stanley did not set up the crystallography program he had anticipated during the first few years of the Virus Laboratory. Subsequently, when Stanley's second Rockefeller Foundation equipment grant was cut from $100,000 to $50,000, the nearly $20,000 of X-ray equipment in this proposal was eliminated. Linus Pauling inquired in 1952 whether Stanley might be interested in hiring his former student Jerry Donohue, who was completing a postdoctoral fellowship at the Medical Research Council (MRC) laboratory in Cambridge. Stanley was uncertain of his ability to provide adequate space and funding and demurred.[46] Instead, the biophysical approach at the Virus Laboratory was increasingly shaped by electron microscopy.

Stanley also intended that electron microscopy would be a central technology for his new Virus Laboratory, and first attempted to woo James Hillier from RCA Laboratories to develop this approach.[47] However, when Hillier declined the invitation to move to Berkeley, Stanley sought to recruit Robley Williams from Michigan, and succeeded. Williams joined the biochemistry department in 1950 as a professor of Biophysics and began a program of electron microscopy of viruses (see fig. 7.2).

When he arrived in Berkeley, Williams was already a leader in techniques of electron microscopy, to which he turned his attention after establishing his scientific reputation in astrophysics at the University of Michigan.[48] There he became acquainted with Ralph Wyckoff, whose collaboration with Stanley on ultracentrifugation of viruses proved so consequential for virus research generally.[49] Wyckoff was serving as a

45. See Warren Weaver to Gerald Oster, 27 Jan 1950, Stanley to Oster, 31 Jan 1950, and Oster to Weaver, 11 Feb 1949, Stanley papers, carton 11, folder Oster, Gerald.

46. Linus Pauling to Stanley, 13 May 1952; Stanley to Pauling, 15 May 1952; Pauling to Stanley, 19 Nov 1952; Stanley to Pauling, 24 Nov 1952, all in Stanley papers, carton 14, folder California Institute of Technology. Stanley wrote Donohue himself in these terms: "I also indicated to Dr. Pauling that he should proceed with recommendations on your behalf to other good laboratories because, frankly, there is a considerable amount of indecision here as to whether or not we can afford to go into x-ray work in the protein field" (Letter, 4 Dec 1952, Stanley papers, carton 3, folder Dec 1952).

47. Letter from Stanley to H. M. Weaver, 17 Jun 1948, Stanley papers, carton 13, folder We misc., 4.

48. For a valuable account of Williams's contributions to biophysics and his program of research in the Virus Laboratory, see Rasmussen, *Picture Control,* ch. 5, "Wendell Stanley, Robley Williams, and the Land of the Virus."

49. On Wyckoff and Stanley's collaboration, see chapter 4.

FIGURE 7.2 Robley Williams at the console of his RCA electron microscope in the Virus Laboratory. Undated, circa 1955. (Reprinted courtesy of the Bancroft Library, University of California, Berkeley.)

research associate at the School of Public Health during World War II, and he collaborated with Williams to develop a new method for visualizing macromolecular particles, involving "shadow-casting" specimens with metals.[50] As Nicolas Rasmussen has observed, the rapid adoption of this method by other electron microscopists derived not only from its more accurate measurement of particle dimensions, but also from the way in which the casting revealed hitherto unseen topologies and textures by concentrating the electron-scattering on the surface of the particles.[51] One

50. Williams and Wyckoff, "Thickness of Electron Microscopic Objects."
51. Rasmussen, *Picture Control,* 200.

of Williams's first publications from the Berkeley Virus Lab reported on his use of the electron microscope to measure more carefully the length of TMV. As he summarized in a subsequent paper, his observations showed "that the particles in a TMV suspension can exist almost entirely in precisely monomeric lengths, or in multiples of the monomeric length. This result is of considerable importance in its bearing on the problem of the particle weight of the virus, since it implies that one can think of a unit TMV particle in suspension as a unique entity with a definite and determinable weight."[52]

The other principal biophysicist of the group was Howard Schachman, who had worked with Stanley at Princeton and mastered the ultracentrifuge under the tutelage of Max Lauffer there (see fig. 7.3). Like Knight, who also came with Stanley to Berkeley, Schachman was given a faculty appointment in the new Department of Biochemistry as well as a staff position in the Virus Laboratory. Schachman achieved wide recognition for his technical ingenuity in extending the range of questions that could be answered using the analytical ultracentrifuge. For instance, his invention of the synthetic boundary cell, in collaboration with Pickels (who subsequently commercialized the innovation at Spinco), enabled the study of the sedimentation of smaller molecular weight materials (e.g., 300–10,000 daltons) than had been previously possible.[53] Similarly, Schachman developed methods for studying viscosity in the ultracentrifuge, an approach that became increasingly significant for assessing the length and behavior of nucleic acids.[54] His charismatic personality and gift for clearly communicating complex technical issues made him a popular teacher of physical biochemistry.[55]

Schachman also provided a reason for Stanley's first serious involvement in academic politics at Berkeley. As Schachman recounts,

> A prominent state senator in California, who was ahead of his time in anticipating McCarthyism, initiated a campaign to find and fire all Communists employed by the university. Anticipating potential difficulties from that source, the President of the university proposed an addition to the standard loyalty oath required of all state employees.... In the addendum, individuals were required to affirm

52. Williams, Backus, and Steere, "Macromolecular Weights," 2063. The earlier paper to which he referred was Williams and Steere, "Electron Microscopic Observations."

53. Pickels, Harrington, and Schachman, "Ultracentrifuge Cell for Producing Boundaries"; Schachman and Harrington, "Ultracentrifuge Studies with a Synthetic Boundary Cell. Part 1."

54. Schachman and Harrington, "Viscosity Measurement in the Ultracentrifuge."

55. For a summary of Schachman's early achievements at Berkeley, see the letter from H. O. L. Fischer and Stanley to Dean A. R. Davis, 12 Jan 1954, Stanley papers, carton 4, folder Jan 1954.

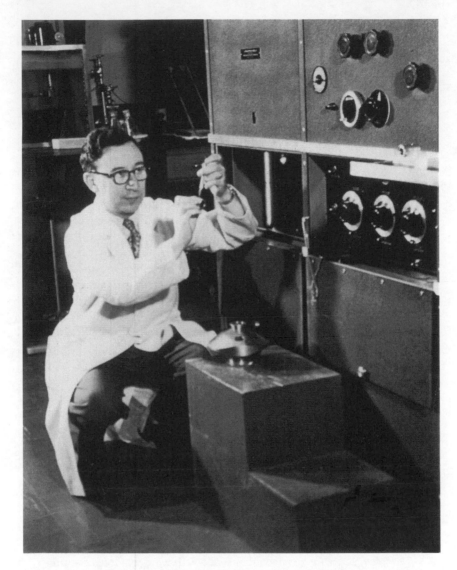

FIGURE 7.3 Howard Schachman in front of a Model E ultracentrifuge, in the Virus Laboratory. Undated, circa 1955. (Reprinted courtesy of the Bancroft Library, University of California, Berkeley.)

"that (I) do not believe in, and am not a member of, nor do (I) support any party or organization that believes in, advocates, or teaches the overthrow of the United States Government, by force or by any illegal or unconstitutional methods."[56]

56. Schachman, "Still Looking for the Ivory Tower," 7.

Along with about two hundred others, Schachman protested this infringement on political freedom by refusing to sign the loyalty oath. Although Stanley himself had promptly signed the oath, he decided to defend those who had refused to sign. Stanley became one of a four-person faculty Committee on Conference with the Regents in the midst of intense and unproductive negotiations over the loyalty oath through the spring of 1950. In addition to his role in the dispute on campus, he also wrote a resolution that he urged the National Academy of Sciences to adopt, and solicited journalists to cover the dispute in popular magazines.[57]

Stanley's activism on this issue was motivated by a strong desire to prevent the opposition to the oath from becoming an outright conflict between the faculty and the Regents.[58] He wrote Williams in the letter offering him a professorship in April of 1950, "The question of the special oath is at white heat here just now, and the majority of the faculty will be fired or will resign if the special oath is not rescinded."[59] He encouraged Williams to make his acceptance of the job contingent on the elimination of the oath. At the end of that month, an alumni-brokered compromise was reached between the faculty and the Regents.[60]

Stanley's line-up of researchers for the Virus Laboratory by the end of 1950 included Knight; Williams; Schachman; Carlton Schwerdt, who (as we saw in chapter 5) became Stanley's lead researcher on poliovirus; Arthur Pardee, a graduate of Linus Pauling's laboratory who used bacteriophage as a means of studying cellular regulation of nucleic acid and protein synthesis; and Dean Fraser, a phage biochemist who had trained with Delbrück. Another of Delbrück's postdoctoral fellows, Gunther Stent, joined the staff in 1952. The mission of the group, as Stanley articulated it in these early years, was wide-ranging even as it focused on the particular experimental promise of viruses:

> The purpose of the Virus Laboratory is to foster investigations on the essential nature of viruses and on the interrelationships between viruses and host tissues, including the phenomena of replication and mutation. Other basic problems in molecular biology are also under investigation since the viruses provide excellent systems for such studies. Various approaches are employed involving, for example, biological, physical, organic, and analytical chemistry, biophysics, tracer techniques, and genetics.[61]

57. See draft of "Statement by Members of the National Academy of Sciences," Stanley papers, carton 21, folder UC Loyalty Oath controversy, and letter from Stanley to T. S. Matthews, *Time*, 3 Apr 1950, carton 3, folder Apr 1950.

58. Gardner, *California Oath Controversy,* 87.

59. Stanley to Robley C. Williams, 6 Apr 1950, Stanley papers, carton 3, folder Apr 1950.

60. Gardner, *California Oath Controversy,* 154–59.

61. Stanley papers, undated, carton 21, folder UC Loyalty Oath controversy.

FIGURE 7.4 Heinz Fraenkel-Conrat (*left*) and Wendell Stanley in front of some tobacco plants in the greenhouse on the fifth floor of the Biochemistry and Virus Laboratory building. Undated, circa 1955. (Reprinted courtesy of the Bancroft Library, University of California, Berkeley.)

Although Stanley included genetics among the approaches being taken in his laboratory, the research programs reflected the priority given biochemistry and biophysics.

Another recruit to the Virus Laboratory, who reinforced the biochemical bent of research there, was Heinz Fraenkel-Conrat (see fig. 7.4). Fraenkel-Conrat had grown up the son of a prominent Jewish obstetrician in Breslau, Germany.[62] He followed his father's example by receiving a medical degree in 1933, but then earned a doctorate in biochemistry at the University of Edinburgh in 1936. His subsequent research career in protein chemistry, while distinguished, was peripatetic, including stints at the Rockefeller Institute, the São Paulo Institute of Brazil, and the Institute of Experimental Biology at Berkeley.[63] In 1942 he assumed a post

62. Unpublished biographical sketch of Heinz Fraenkel-Conrat, Stanley papers, carton 14, folder California Museum of Science and Industry, enclosed with letter from Stanley to Burrill, 27 May 1958.

63. Fraenkel-Conrat worked with protein chemist Max Bergmann at the Rockefeller Institute. He then spent one year at the São Paulo Institute of Brazil, where he was the

as a protein chemist for the Western Regional Laboratory of the U.S. Department of Agriculture located in Albany, California.

Fraenkel-Conrat first wrote Stanley in 1946, telling him of his own studies of formaldehyde modification of proteins, similar to the inactivation studies with formaldehyde that Stanley published in 1938.[64] Stanley sent a sample of TMV for further protein modification studies, which Fraenkel-Conrat used to determine the likely mode of chemical inactivation of TMV by formaldehyde.[65] However, Fraenkel-Conrat felt that his employer's highly directed research mission would prevent his publishing these results. As he wrote Stanley, "The restrictions and limitations imposed upon us by the rigid system of bureaucracy make me wish ever so often that I might be associated with a free research institution, similar to the Rockefeller, or with a University. If you should know of any openings of that nature, and paying not too much less than what I am getting now ($5,000), would you be good enough to think of me?"[66] Within several months, Stanley sent Fraenkel-Conrat word that he would likely be relocating to California, and that he hoped to explore their mutual interest further.[67] Stanley did in fact offer Fraenkel-Conrat a staff research position in the Berkeley Virus Laboratory three years later.[68]

In 1951, Stanley helped Fraenkel-Conrat obtain a Rockefeller Foundation Fellowship to spend a year with Frederick Sanger in Cambridge and with Kaj Linderstrøm-Lang in Copenhagen "so as to become thoroughly acquainted with all of the significant facts on the work on the organic reactions of proteins."[69] Drawing on the Rockefeller Institute's traditions

---

first to crystallize an animal neurotoxin, that of rattlesnake venom, crotoxin (Slotta and Fraenkel-Conrat, "Schlangengifte. Part 3"). He investigated pituitary hormones (which, like crotoxin, are small polypeptides) upon returning to the United States in 1938 to work with H. M. Evans and C. H. Li at the University of California's Institute of Experimental Biology. See the biographical sketch of Fraenkel-Conrat, in Srinivasan, Fruton, and Edsall, *Origins of Modern Biochemistry,* 308.

64. Ross and Stanley, "Partial Reactivation."

65. Heinz Fraenkel-Conrat to Stanley, 15 Jan 1946, Stanley papers, carton 8, folder Fraenkel-Conrat, Heinz; Stanley to Fraenkel-Conrat, 26 Feb 1946, Stanley papers, carton 2, folder Feb 1946; Fraenkel-Conrat to Stanley, 4 Jul 1947, Stanley papers, carton 8, folder Fraenkel-Conrat, Heinz. See the biographical sketch of Fraenkel-Conrat in Stanley's application for U.S.P.H. graduate training grant, Stanley papers, carton 21, folder Department of Biochemistry, history.

66. Fraenkel-Conrat to Stanley, 4 Jul 1947, Stanley papers, carton 8, folder Fraenkel-Conrat, Heinz.

67. Stanley to Fraenkel-Conrat, 23 Oct 1947, Stanley papers, carton 2, folder Oct 1947.

68. Stanley explained that there would be a delay of at least two years in his letter to Fraenkel-Conrat, 13 Nov 1947, carton 2, folder Nov 1947; see also Stanley to Fraenkel-Conrat, 15 May 1947, carton 3, folder May 1951.

69. Stanley to Weaver, 15 Jan 1951, Stanley papers, carton 3, folder Jan 1951. See Stanley to Fraenkel-Conrat, 15 May 1951 (Stanley papers, carton 3, folder May 1951) on his response to Fraenkel-Conrat's consideration of a possible position at Krebs's department at Sheffield.

of scientific education, Stanley had a practice of sending his "men" off for study in other laboratories in preparation for their work with him. In preparation for the arrival of Williams's student Russell Steere at the Virus Laboratory, Stanley helped arrange for him to spend the year at the Rockefeller Institute "to broaden his knowledge of viruses."[70]

Getting these researchers to return to Berkeley sometimes proved difficult, as illustrated in the disappointing situation with Gerald Oster. Fraenkel-Conrat's time away raised similar concerns: while Fraenkel-Conrat was in England, Hans Krebs attempted to recruit him to his department at Sheffield. Stanley sent a carefully worded letter to Fraenkel-Conrat, assuring that he would do his best to turn his research associate position at Berkeley into a regular faculty appointment as soon as he could.[71] In fact, it would be seven more years before Fraenkel-Conrat was made a professor.[72] In retrospect, the tacit expectations of staff loyalty Stanley held for the Virus Laboratory researchers provide an uneasy contrast to his highly public opposition during these years to the loyalty oath that the Regents of the University of California sought to impose upon faculty.

TMV from Structure to Sequence

Fraenkel-Conrat began working in the Virus Lab shortly after a serendipitous observation had pointed to the promise of protein chemistry for elucidating the fine structural details of TMV. In 1951, J. Ieuan Harris was investigating adrenocorticotropic hormone (ACTH) as a postdoctoral fellow in the laboratory of C. H. Li, a well-known hormone biochemist whom Stanley had attempted (unsuccessfully) to assimilate into his department of biochemistry. Li's group was already using state-of-the-art techniques of protein chemistry to purify and characterize hormones, and Harris was focusing on the amino acids in the polypeptide ACTH. Searching for a control for his carboxypeptidase digest of the hormone,[73] he came to Stanley's laboratory to procure some TMV, which, if composed of a single polypeptide, was presumed to be 350,000 amino acids

70. Stanley to Weaver, 15 Jan 1951, Stanley papers, carton 3, folder Jan 1951.

71. Stanley to Fraenkel-Conrat, 15 May 1951, Stanley papers, carton 3, folder May 1951.

72. The wait that Fraenkel-Conrat had to endure before finally receiving his faculty appointment in 1958 caused a good deal of dissatisfaction in the Virus Laboratory. Gunther Stent complained about this to Max Delbrück in 1957: "[Fraenkel-Conrat] has really been getting a raw deal for the past six years. Stanley has promised him a professorship year after year, without doing anything much about it.... Stanley's attitude seems to be that there is no hurry; after all, Heinz has waited already for so many years, why can't he wait a little longer?" (Letter dated 30 Oct 1957, Delbrück papers, folder 20.20; see also Stanley to Fraenkel-Conrat, 9 Jul 1954, Stanley papers, carton 4, folder Jul 1954; Fraenkel-Conrat to Stanley, 7 Feb 1957, Stanley papers, carton 21, folder Department of Biochemistry, history).

73. Harris and Li, "Biological Activity of Enzymatic Digests."

long.[74] But rather than yielding one mole of end-product amino acid from each mole of TMV, as Harris expected, the carboxypeptidase split off more than 3,000 molecules of threonine per virus![75]

Harris and Knight found the same large amount of threonine released from the Holmes ribgrass strain of TMV. Hoping that the "dethreonized" TMV would itself serve as a genetic variant, they inoculated plants with the modified virus (which retained its infectivity). However, they were disappointed to find that all progeny contained the C-terminal threonine in the same amount as wild-type virus.

As Stanley reflected on the unexpected finding of the carboxy-terminal threonines, "The fact that the virus can be dissociated to small molecular weight subunits suggests that these chains are held together only by secondary and ionic valence forces."[76] Indeed, on the basis of his recent chemical analysis of the N-terminal amino acids of the virus protein, Gerhard Schramm had already been arguing that the TMV protein fragments he had observed previously were probably made up of smaller polypeptides. Following the example of Sanger's work on insulin, Schramm and Gerhard Braunitzer used fluorodinitrophenol (FDNP) to identify amino-terminal residues, turning up 3,000 N-terminal prolines.[77] From the number of prolines generated by the reaction with DNP, Schramm calculated that TMV was made up of 1,500–2,300 polypeptides, each of a molecular weight between 17,000 and 27,000 daltons.[78] He noted the striking resemblance between the number of amino end-groups generated by his method and the number of carboxy end-groups detected by Harris and Knight. In addition, given that various strains of TMV showed distinct alterations in the amino acid composition, he argued that the polypeptides should reflect these changes in the order of amino acids.[79]

Despite the apparent agreement between the number of amino-terminal residues identified in Tübingen and the number of carboxy-terminal residues found in Berkeley, Stanley and his coworkers were skeptical of the German result, which Fraenkel-Conrat was unable to

74. Fraenkel-Conrat, "Protein Chemists Encounter Viruses," 311.

75. Harris and Knight, "Action of Carboxypeptidase." For an account of the surprising result of this experiment, see Fraenkel-Conrat, "History of Tobacco Mosaic Virus," 8–9. Harris and Knight initially calculated 3,400 threonine residues released from each TMV nucleoprotein, based on a virus molecular weight of 50 million daltons; as the molecular weight was revised downward to 40 million, the number of C-terminal threonines was readjusted to 3,000.

76. Stanley, "NFIP Progress Report—January 1, 1953 to June 30, 1953," attached to letter to H. M. Weaver, carton 16, folder NFIP Progress Reports, 2–3.

77. Sanger, "Free Amino Groups of Insulin"; Schramm and Braunitzer, "Prolin als Endgruppe des Tabakmosaikvirus."

78. Schramm and Braunitzer, "Prolin als Endgruppe des Tabakmosaikvirus," 63.

79. Ibid., 64.

duplicate. Fraenkel-Conrat and Bea Singer reported that Schramm's acidic digestion of TMV had actually hydrolysed the protein, exposing prolines that were internal to the subunits.[80] As Stanley commented in a 1953 report to the NFIP, "It appears that the end-groups observed by Schramm represent artifacts due to hydrolysis of the most labile peptide bonds. The numerical similarity of these bonds with the C-terminal threonine was proven to be purely coincidental when this preparation was subjected to the action of carboxypeptidase."[81]

As Stanley explained, carboxypeptidase split off up to ten times as many residues from virus prepared according to the German method as it did from the Berkeley preparation of native TMV. This provided evidence that the German technique for preparing the sample was itself producing many shorter polypeptides. In accounting for the absence of detectable N-terminal residues, Stanley was open to considering nonlinear polypeptide structures. For example, he suggested that the discrepancy between reliable estimates of N-terminal residues and C-terminal residues might be attributed to "cyclical or 6-shaped [chains], with an unusual peptide linkage involving the α-carboxyl group of one aspartic or glutamic acid residue."[82]

Even the published disagreement over the end-groups between the virologists in Berkeley and Tübingen exhibited a combative tone. Fraenkel-Conrat and Singer commented on the discrepancies between the results coming from the two laboratories in a 1954 publication:

> When the paper by Schramm and Braunitzer became available to us, the main difference in our experiments became immediately evident. These authors had used only one type of preparation for their study, a TMV protein sample obtained by treating the virus with 5% trichloroacetic acid for 30 min. at 100°C. It appeared possible that spurious "end groups" had been generated through hydrolysis of peptide bonds under these conditions. This *suspicion* [emphasis added] was confirmed when we studied preparations made according to Schramm and Braunitzer.[83]

Schramm rapidly published a paper defending his claim.[84] Responding to Fraenkel-Conrat's interpretation, he asserted, "We think that this opinion is incorrect, and believe that we are able to account partially for the sources of error."[85] The following year, in a short paper confirming the

80. Fraenkel-Conrat and Singer, "Peptide Chains."

81. Stanley, "NFIP Progress Report–January 1, 1953 to June 30, 1953," attached to letter to H. M. Weaver, carton 16, folder NFIP Progress Reports, 3.

82. Ibid.

83. Fraenkel-Conrat and Singer, "Peptide Chains," 180.

84. Schramm, Braunitzer, and Schneider, "Zur Bestimmung der Amino-Endgruppe."

85. Ibid., 300 (author's translation).

carboxypeptidase finding of threonine as the carboxy-terminal residue, Fraenkel-Conrat wrote, "During the preparation of this manuscript, a paper by G. Braunitzer [and Gerhard Schramm] ... came to our attention, which also demonstrates the C-terminal position of threonine in TMV protein by hydrazinolysis. Other conclusions contained in that paper appear not valid."[86]

But underlying this rather strident debate over the identification of TMV's terminal groups was a new consensus concerning the conceptualization of viruses as assemblies of identical polypeptides and nucleic acid. This was a marked shift in the Virus Laboratory, where the emphasis since the late 1930s had been on the indivisible macromolecular nature of the active virus. The analytical ultracentrifuge was now trained on a different manifestation of TMV, as Schachman and his student William Harrington characterized the dissociation via sodium dodecyl sulfate of TMV into subunits, and found a homogeneous subunit species of molecular weight 10,000–20,000 daltons.[87] Schramm's early reporting of subunits of TMV was thus correct, in essence if not in detail. (The early details of virus disassembly, as the Berkeley group saw it, were indeed mistaken: Harrington and Schachman argued that many of the fragments Schramm had identified were reaggregation products, not degradation products as the virus particle split gradually by sixths.)

More to the point, Stanley's workers were no longer seeking to refute Bawden and Pirie's skepticism about whether the rod-shaped macromolecule was TMV. They were competing with Schramm to characterize the TMV subunit in the same manner that Sanger was characterizing insulin: as a sequence.[88] This rivalry between Berkeley and Tübingen would dominate the work on TMV in both laboratories over the next decade.[89] In his 1954 paper defending the identification of the amino-terminal end-group as proline, Schramm introduced a term, "A-protein," to describe the nucleic-acid-free TMV subunits of molecular weight about 120,000 daltons.[90] A-protein could be further dissociated into peptides of molecular weight 17,400, or about 150 amino acids long, of which there were over 2,000 in the entire virus particle.

86. Niu and Fraenkel-Conrat, "C-Terminal Amino-Acid Sequence," 598.

87. Harrington and Schachman, "Studies on the Alkaline Degradation of Tobacco Mosaic Virus. Part 1. Ultracentrifugal Analysis." Fraenkel-Conrat cited this result in 1954, well in advance of its publication. As Schachman has commented since, "When Bill Harrington and I were studying the degradation of TMV (he used alkaline solutions and I used SDS), I found that the sedimentation coefficient of the virus decreased from about 185 S to 1 S. This was strong evidence that the virus was composed of subunits of molecular weight about 10^4. I didn't publish it then because, at the time, we did not know how to deal with multicomponent systems on the ultracentrifuge" (letter to the author, 16 Sep 1999).

88. de Chadarevian, "Sequences, Conformation, Information."

89. See Zaitlin, "Tobacco Mosaic Virus," 678–79.

90. Schramm, Braunitzer, and Schneider, "Zur Bestimmung der Amino-Endgruppe."

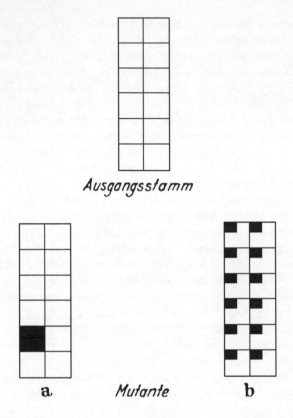

Ausgangsslamm

a   *Mulanle*   b

FIGURE 7.5 Schramm's schematic diagram of two possible locations of a TMV mutation. At the top is the original (nonmutant) TMV, with the small blocks depicting protein subunits. *a:* The structural alteration (indicated in black) is located in one virus subunit. *b:* The alteration is present in every virus subunit. Schramm favored the interpretation in *b*. (From Gerhard Schramm, "Die Struktur des Tabakmosaikvirus und seiner Mutanten," *Advances in Enzymology* 15 [1954]: 481. Copyright © 1954 John Wiley & Sons, Inc. Reprinted by permission.)

Increasingly both the virus protein and the nucleic acid were being viewed in terms of linear sequence, of amino acids or nucleotide bases. As Schramm argued, if the virus polypeptides had identical carboxy-terminal and amino-terminal endgroups, then most likely all of the other amino acids in the polypeptide were also identical.[91] Similarly, a mutation might involve an identical change in every subunit rather than a single change in the virus macromolecule (see fig. 7.5).[92] Indeed,

91. Schramm, "Die Struktur des Tabakmosaikvirus," 464.
92. Ibid., 481. But at this stage, Schramm argued that a mutation was not an alteration of the nucleic acid or protein composition, but a structural realignment; see 479–80.

serological evidence supported this view: The Tübingen workers had observed that isolated nucleic-acid-free subunits of TMV could be precipitated with both group-specific antiserum and strain-specific antiserum, leading them to conclude that every subunit must contain both antigenic markers.[93]

An observation by William Takahashi and Mamoru Ishii was particularly suggestive of the role of these subunits in assembly of the virus, as Schramm pointed out.[94] The two plant pathologists at Berkeley isolated a protein specific to TMV-infected plants of molecular weight 120,000. This nucleic-acid-free protein, which the researchers called "X-protein," could be assembled into TMV-resembling rods *in vitro*, and might represent an intermediate in virus synthesis or a product of virus degradation.[95] In the Virus Laboratory, Roger Hart obtained electron micrographs of short rods of X-protein, which appeared as little disks. Hart called attention to the apparent holes or pits at the axis of each disk. Equally striking were his micrographs of partially depolymerized TMV rods, from which protruded nucleic acid, thread-like in appearance (see fig. 7.6). If these same structures were exposed to ribonuclease, the filamentous structures disappeared.[96]

Fraenkel-Conrat continued using the latest techniques of protein chemistry to study the virus, drawing on what he had learned in his year abroad before joining the Virus Lab, when he worked on N-terminal analysis techniques with Frederick Sanger, R. R. Porter, and K. Linderstrøm-Lang. Under Fraenkel-Conrat's supervision, Ching-I. Niu employed protein degradation techniques to determine the amino acid composition beyond the carboxy-terminal unit. Through hydrazinolysis, the last three amino acids were identified as –proline–alanine–threonine.[97] In less than six months, three more amino acids were identified, giving a carboxy-terminal sequence of –threonine–serine–glycine–proline–alanine–threonine.[98] As Fraenkel-Conrat later put it, "The era of protein sequencing was upon us."[99] Soon, two students of Shiro Akabori, the Japanese protein chemist who had developed the important technique of hydrazinolysis

93. Ibid., 481; Friedrich-Freksa, Melchers, and Schramm, "Biologischer, chemischer, und serologischer Vergleich," 216–18.

94. Schramm, "Die Struktur des Tabakmosaikvirus," 470.

95. Takahashi and Ishii, "Abnormal Protein," and "Macromolecular Protein." Similar work was being carried out at laboratories in St. Louis and in Brussels: see Commoner et al., "Proteins Synthesized in Tissue"; Jeener, Lemoine, and Lavand'Homme, "Détection et propriétés." On the work of Jeener and coworkers in Brussels, see Thieffry, "Contributions of the 'Rouge-Cloître Group.'"

96. Hart, "Electron-Microscopic Evidence."

97. Niu and Fraenkel-Conrat, "C-Terminal Amino-Acid Sequence."

98. Niu and Fraenkel-Conrat, "Determination of C-Terminal Amino Acids."

99. Fraenkel-Conrat, "Protein Chemists Encounter Viruses," 311; see also Fraenkel-Conrat, Harris, and Levy, "Recent Developments."

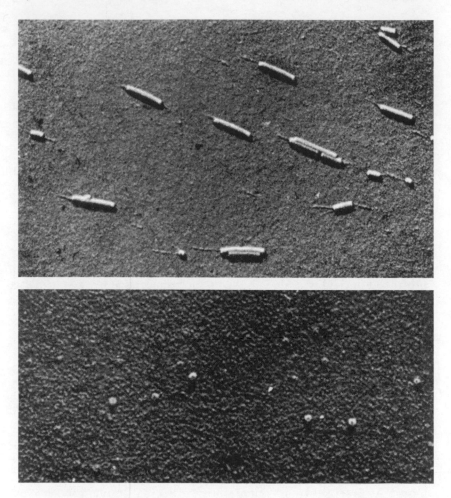

FIGURE 7.6 Electron micrographs of treated TMV. *Top:* "TMV particles after treatment with a hot detergent solution. The narrow fibers seen projecting from the ends of the rods consist of ribonucleic acid, which presumably is localized along the axis of the virus rod. ×60,000." *Bottom:* "Short rods of X-protein seen in an end-on view. Note that each rod appears to have a depression or pit at its axis. ×80,000." (Images and captions from Hart, "Electron-Microscopic Evidence" [1955]: 262, 263. Reprinted courtesy of Mrs. Dorothy M. Hart.)

for identifying C-terminal residues, arrived at the Virus Laboratory to work on TMV. An additional material resource proved particularly advantageous to these efforts—when Stanley recruited carbohydrate chemist H. O. L. Fischer as a professor of biochemistry for his new department, he brought along the historic peptide collection of his father,

Emil Fischer, who had synthesized the compounds in his work establishing the chemical nature of the linkage between amino acids. These samples provided a very useful set of controls for identifying the peptide fragments of the virus protein.[100]

Fraenkel-Conrat also drew on his expertise in the techniques of chemical modification to analyze the relationship between structure and function in TMV. In 1955, he showed that iodine could be added stoichiometrically to the sulfhydryl groups of the polypeptides, producing a stable and bright yellow derivative.[101] In addition, he showed that formaldehyde inactivated TMV not by reacting with the protein, as had been assumed for a decade, but by modifying the nucleic acid.[102] However, this result did nothing to revise the tepid reaction of Virus Laboratory members to the significance of the Hershey-Chase experiment and other evidence for the role of nucleic acids in virus genetics. As Knight argued in a 1954 review, "One should be cautious in ascribing genetic significance to the differences in nucleic acid composition observed for different viral species. . . . It is possible that the major function of viral nucleic acid is to hold the protein in a specific configuration in which its several biological properties are made manifest."[103] Likewise, Gerhard Schramm showed no interest in nucleic acid's role in virus reproduction. In fact, he was equally unconvinced by Knight's evidence that strain differences were correlated with changed amino acid composition, arguing instead that virus mutations must consist only in relatively small structural changes independent of chemical alteration.[104]

Electron Microscopic and X-Ray Diffraction Approaches to the Subunit Problem

Research on bacteriophage structure in Berkeley's Virus Laboratory manifested the same shift toward conceptualizing viruses in terms of their functional components. Fraser and Williams teamed up to use Williams's new freeze-drying technique for the electron microscope to study the morphology of all of the T-odd and T-even bacteriophages. Their new and more vivid images revealed many inaccuracies in earlier visualizations. All of the bacteriophages were found to possess a head-and-tail structure (the short tails of T3 and T7 phage had not been previously observed),

100. Fraenkel-Conrat, "Protein Chemists Encounter Viruses."

101. Fraenkel-Conrat, "Reaction of Tobacco Mosaic Virus with Iodine."

102. Fraenkel-Conrat, "Reaction of Nucleic Acid with Formaldehyde"; "Progress Report for Jan. 1–June 30, 1954," Stanley papers, carton 16, folder NFIP Progress Reports, 3–4.

103. Knight, "Chemical Constitution of Viruses," 179.

104. Schramm, "Die Struktur des Tabakmosaikvirus," 480–81.

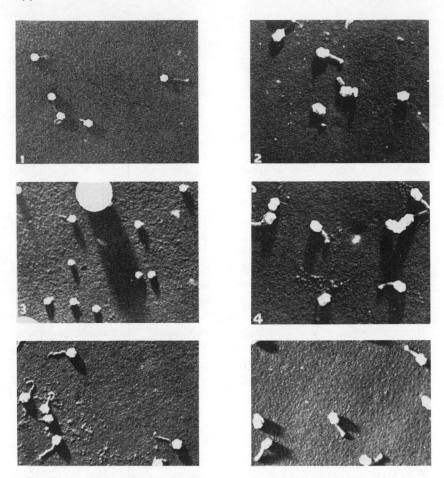

FIGURE 7.7 Electron micrographs of six of the seven T-phages prepared by freeze-drying. Uranium shadowed; ×50,000 for all figures. *Panels 1–6* depict T1 to T6." (Images and text from Williams and Fraser, "Morphology of the Seven T-Bacteriophages" (1953): 459. Reprinted by permission of the American Society for Microbiology.)

and the heads were found to be spheroid, ellipsoid, or hexagonal, depending on the strain (see fig. 7.7).[105] The sharp polyhedra characteristic of the bacteriophage heads stimulated structural consideration of what kinds of subunit arrangements might produce these three-dimensional configurations.

105. Williams, "Shapes and Sizes of Purified Viruses"; Williams and Fraser, "Morphology of the Seven T-Bacteriophages"; Fraser and Williams, "Electron Microscopy of the Nucleic Acid"; Fraser and Williams, "Details of Frozen-Dried T3 and T7 Bacteriophage."

Even more strikingly, Williams and Fraser showed that upon osmotic shock, the bacteriophage head spilled forth the encapsulated DNA in long strands (see fig. 7.8). As Stanley stated, "Since this is the first time that unequivocal genetic material has been made available for observation with virtually no likelihood of preparative distortion, it is of great interest to learn as much about its structure as present electron microscope technique will permit."[106] Just prior to the Cold Spring Harbor Symposium that summer on viruses, at which Williams planned to show the stunning pictures of the DNA spilling out like ringlets from the bacteriophage, Delbrück wrote Stanley, "I would not be surprised if this picture, and the DNA structure of Watson and Crick, would steal the show at Cold Spring Harbor."[107]

Advances in X-ray crystallography, particularly in fiber diffraction, were also important in the elucidation of finer structural details of viruses. In 1952, the young American phage geneticist James Watson received a postdoctoral fellowship from the NFIP to undertake diffraction studies of TMV in Lawrence Bragg's Cambridge laboratory. A novice to crystallography, Watson learned a great deal about the techniques and interpretation from Francis Crick, as recorded in his famous account, *The Double Helix*.[108] Watson and Crick's rivalry with Linus Pauling at Caltech and their uneasy mixture of collaboration and competition with X-ray crystallographers Rosalind Franklin and Maurice Wilkins at King's College reflected a larger tension between the approaches of model-building and of fiber diffraction analysis, the theoretical and experimental approaches to structural determination.

In 1954, Watson published the results of his X-ray diffraction work on TMV, conducted while he was a fellow with Roy Markham at the Molteno Institute at Cambridge (see fig. 7.9). In fact, the original draft of this paper was submitted to *Biochimica et Biophysica Acta* the week before his famous note with Crick proposing the double-stranded model of DNA was published in *Nature*.[109] His analysis of TMV diffraction corroborated the earlier finding of Bernal and Fankuchen that the virus was composed of many equivalent subunits (estimated to number between 900 and 1,500). Building on Crick's predictions of the pattern expected from a discontinuous helix, Watson argued that TMV was a helix

106. Stanley, "Progress Report—January 1, 1953 to June 30, 1953," Stanley papers, carton 16, folder NFIP Progress Reports, 6.

107. Max Delbrück to Stanley, 5 May 1953, Stanley papers, carton 7, folder Delbrück, Max.

108. Watson, *Double Helix*.

109. Watson and Crick, "Structure for Deoxyribose Nucleic Acid." This paper was submitted on 2 April and published on 25 April 1953; Watson's paper on TMV diffraction was submitted on 16 April of the same spring.

FIGURE 7.8 "Electron micrograph of a broken T6 bacteriophage particle. The empty membrane of the head and the intact tail are shown at the bottom of the picture. The fine strands of deoxyribonucleic acid are seen spread out in an approximately circular array over most of the micrograph. Shadowed with uranium. ×94,000." (Image and caption from Fraser and Williams, "Electron Microscopy of the Nucleic Acid" [1953]: 752. Reprinted courtesy of Dr. Robley C. Williams.)

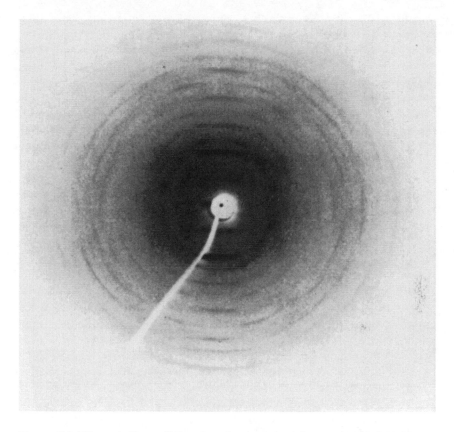

FIGURE 7.9 Watson's X-ray diffraction photograph of dry specimen of TMV.
(Reprinted from Watson, "The Structure of Tobacco Mosaic Virus. Part 1,"
Biochimica et Biophysica Acta 13 [1954]: 11, with permission from Elsevier
Science.)

repeating every three turns with a period of 68Å.[110] He also suggested that
RNA was in the center of this helix (analogous to the placement of nucleic
acid in the center of two spherical viruses—Turnip Yellow Mosaic Virus
and T2 bacteriophage), although his own diffraction data did not actually
allow for resolution of the nucleic acid in TMV.[111]

Rosalind Franklin also began work on the structure of TMV, having
joined J. D. Bernal's laboratory at Birkbeck in 1953 after leaving DNA
and Wilkins behind at King's College, London.[112] She contacted Stanley

110. Cochran, Crick, and Vand, "Structure of Synthetic Polypeptides. Part 1"; Watson,
"Structure of Tobacco Mosaic Virus. Part 1."
111. Watson, "Structure of Tobacco Mosaic Virus. Part 1," 18.
112. Piper, "Light on a Dark Lady." Birkbeck College is also in London.

in the spring of 1954, describing her work and inquiring if she might visit and give a lecture that fall, following her participation in a Gordon Conference. As she described her own work, "I have obtained X-ray diffraction diagrams which have better orientation and better resolution than previous ones, and show some 300 discrete maxima, and am now in the middle of a detailed analysis of the diagram for calculation of the Patterson function."[113] She visited the Virus Laboratory in October 1954 following a few days in Linus Pauling's laboratory in Pasadena.[114] While in Berkeley she arranged with Knight and Fraenkel-Conrat to obtain samples of TMV, including one with heavy metal derivatives, for her continuing crystallographic studies.[115] At the same time, she was also in contact with Gerhard Schramm in Tübingen; his method of preparing nucleic-acid-free A-protein from TMV became an important resource for her studies as well.

Drawing on her formidable skills in crystallography, Franklin was able to obtain much clearer X-ray diagrams of TMV than anyone else had previously (see fig. 7.10), which she published along with a more detailed structural picture in *Nature*.[116] Her data were in agreement with Watson's contention that each turn of the helix contained $3n + 1$ subunits; she calculated a value for n of 12, yielding units of molecular weight 29,000 daltons. She argued that these units were subdivided into two equivalent or near-equivalent subunits, and that these smaller units in turn correlated with the subunits detected through chemical methods. Her Patterson function analysis indicated external grooving of the virus structure along the helix. This groove between helical turns would give the virus an extensive surface area, which "may perhaps account for the surprising variety and extent of chemical modifications which it is possible to make in tobacco mosaic virus without breaking up the particle and, in some cases, without destroying its infectivity."[117]

Making use of Watson's suggestion that each protein subunit might accommodate two alpha-helical turns, Franklin offered a schematic representation of the subunit arrangement in TMV (see fig. 7.11). Her description of this representation, furthermore, acknowledged helpful conversations with Crick. In this respect, the picture of TMV emerging

113. Rosalind Franklin to Stanley, 7 May 1954, Stanley papers, carton 8, folder Franklin, Rosalind.

114. Rosalind Franklin to Stanley, 6 Jul 1954, Stanley papers, carton 8, folder Franklin, Rosalind.

115. Rosalind Franklin to Stanley, 14 Oct 1954, Stanley papers, carton 8, folder Franklin, Rosalind.

116. Franklin, "Structure of Tobacco Mosaic Virus."

117. Ibid., 381.

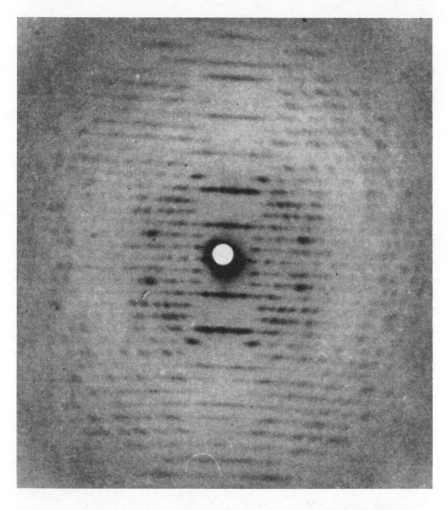

FIGURE 7.10 Franklin's X-ray diffraction photograph of TMV gel. Sample of
TMV obtained from N. W. Pirie, Rothamsted culture, and incubated with
proteolytic enzymes. (Reprinted, with permission, from Franklin, "Structure of
Tobacco Mosaic Virus," *Nature* 175 [1955]: 379. Copyright 1955 Macmillan
Magazines Limited.)

from crystallographic analysis in 1955 was strongly reminiscent of the
double-helical model of DNA, reflecting the same set of scientific contributors, including Wilkins as well as Watson and Crick.[118] But of this set,
it was Franklin that moved the resolution of TMV forward from 1955–58,

118. Wilkins, Stokes, Seeds, and Oster, "Tobacco Mosaic Virus Crystals."

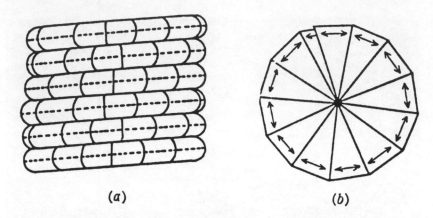

(a)   (b)

FIGURE 7.11 Franklin's schematic diagram of the arrangement of the protein
subunits in TMV. "(a) View of a short length of the virus particle, showing units
and subunits on six turns of the helix. (b) Transverse section of the virus rod,
showing twelve units in one turn of the helix. Arrows indicate principal direction
of protein chains in each unit. Ribonucleic acid not shown." (Diagrams and
captions reprinted, with permission, from Franklin, "Structure of Tobacco
Mosaic Virus," *Nature* 175 [1955]: 381. Copyright 1955 Macmillan Magazines
Limited.)

working alongside Aaron Klug at Birkbeck. In her 1955 *Nature* paper,
she stated that she had been able to calculate a Fourier projection of the
virus from the intensity distribution in the X-ray diagram. It revealed
a ring of high density at 55 Å from the center, and she proposed that
this was the nucleic acid.[119] Thus, the RNA was not at the center of the
virus—as suggested by the electron micrographs in which it appeared
like the wick of a candle—but was closely associated with the protein
subunits.[120]

Reconstituting TMV

In 1955, Fraenkel-Conrat and Williams published a startling result: purified viral protein and purified viral nucleic acid could be recombined to
yield infective nucleoprotein particles, possessing about 1 percent of the
activity of TMV.[121] This "reconstitution" experiment demonstrated that

---

119. Using a different technique, Donald Caspar calculated the same pattern for the
radial distribution; he and Franklin published their papers together in *Nature*. See Caspar,
"Radial Density Distribution"; Franklin, "Location of the Ribonucleic Acid."

120. Schramm himself used this metaphor of the candlewick to describe viral nucleic
acid (Schramm, "Neue Untersuchungen über die Struktur des Tabakmosaikvirus," 322).

121. Fraenkel-Conrat and Williams, "Reconstitution of Active Tobacco Mosaic Virus."
When the authors compared the particle count of reconstituted virus with control TMV

virus assembly from its components could take place in a test tube. Purified protein alone was not infectious, and neither was the purified RNA,
except at very high concentrations, a result attributed by Fraenkel-Conrat
and Williams to contamination with virus (in part because the same
result seemed to appear with high concentration of phosphate alone).[122]
Thus the biochemical reassembly resulted in the production of infectious
virus from apparently inactive components. Concerned that Schramm's
method for separating protein and RNA in alkaline conditions would
degrade the nucleic acid, Fraenkel-Conrat and Williams developed a detergent method to isolate the RNA. (The protein was prepared sepa-
rately using an alkaline sodium carbonate solution.) Ultracentrifuge runs
of the individual components (by lab mate Schachman) revealed them
to be small: the protein exhibited a sedimentation coefficient of 4.5 S,
and the nucleic acid a sedimentation constant of 8 S, resembling that
observed in Princeton by Cohen and Stanley for purified viral RNA.[123]
The protein and RNA were combined in a 10:1 ratio at pH 9.0, and the
pH was then adjusted to neutrality. The solution then began to take on
the opalescent quality characteristic of TMV solution. Reconstitution was
not instantaneous, requiring "about one hour at room temperature ... for
the formation of any active rods."[124] Strikingly, the reconstituted virus
appeared to increase in activity and become more resistant to ribonuclease and alkali upon cold storage.

Electron micrographs revealed the purified protein to consist of
disk-shaped particles with central holes; these perforated disks were
identical to those arising from Takahashi's polymerized X-protein.[125] The
reconstituted virus resembled a sample of ordinary TMV except for the
larger proportion of short rods (see fig. 7.12). As Fraenkel-Conrat and
Williams concluded their report, "The evidence thus seems reasonably
complete that, under the conditions described, TMV nucleic acid enters
into combination with TMV protein subunits and favors aggregation to
rods, some of which are of sufficient length and structural integration
to carry infectivity."[126] Within several months, Barry Commoner and

---

in the electron microscope, there were about one hundred times as many reconstituted
particles as native TMV in samples of the same infectivity (696). Thus, it could be concluded
that many of the reconstituted rods were not infective.

122. Ibid., 697, n. 9.

123. Cohen and Stanley, "Molecular Size and Shape."

124. Fraenkel-Conrat and Williams, "Reconstitution of Active Tobacco Mosaic Virus,"
693.

125. Polymerized X-protein had been visualized in the Virus Lab's electron microscope
by Hart ("Electron-Microscopic Evidence").

126. Fraenkel-Conrat and Williams, "Reconstitution of Active Tobacco Mosaic Virus,"
697.

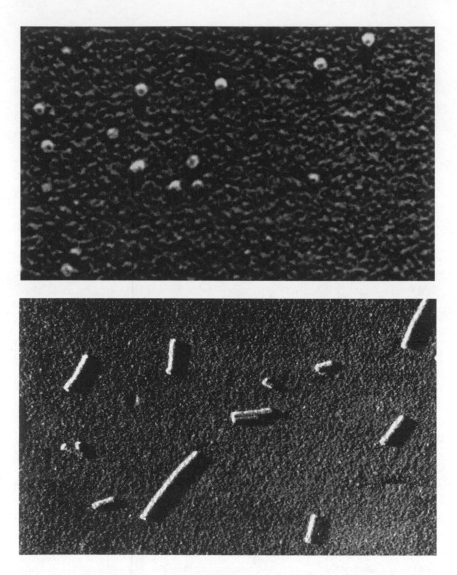

FIGURE 7.12 Electron micrographs of TMV protein disks and reconstituted TMV. *Top:* "Electron micrograph of the TMV protein used in the reconstitution experiments. The particles are characteristically disk-shaped, with central holes. ×120,000." *Bottom:* "Particles of the reconstituted TMV particles. Their morphology is identical with that of normal TMV, except for a greater proportion of short particles. The long rod in this field is 300 mμ long. ×60,000." (Images and captions from Fraenkel-Conrat and Williams, "Reconstitution of Active Tobacco Mosaic Virus" [1955]: 695. Reprinted by permission of Dr. Bea A. Singer and Dr. Robley C. Williams.)

coworkers at Washington University in St. Louis provided striking corroboration of the reconstitution results.[127]

This experiment, like the crystallization of poliovirus achieved in
the Virus Laboratory the same year, garnered a great deal of popular
attention. Newspapers carried the U.P.-syndicated story under the headline "Scientists Create Life in a Test Tube."[128] *Collier's* magazine informed
the public of these developments under the title, "Now—Man-Made
Virus—First Step in Controlling Heredity?"[129] Science Service, a science
popularization agency sponsored by the National Academy of Sciences
and the American Association for the Advancement of Science, reviewed the research accomplishments of 1955 under the headline: "Top
achievement of past year may be reconstitution of virus as first step in
conquest of such diseases."[130] And Waldemar Kaempffert, science editor
for the *New York Times,* included the reconstitution of TMV as one of
the four most important scientific events of the year.[131] Building on the
publicity, *Life* magazine prepared a photoessay on virus research, with
several pictures from the Virus Laboratory. Among the shots they hoped
to capture for the public eye were "examples of huge, complex, or expensive equipment which typify the newest look in virus tools."[132] Stanley,
always mindful of the possible medical benefits from his laboratory's
research, predicted that the reconstitution experiments would lead to
"better ways to produce immunity to diseases in 10 to 20 years" by
enabling production of "a vaccine composed only of the virus fraction
causing immunity."[133]

127. Commoner et al., "Reconstitution of Tobacco Mosaic Virus Components";
Lippincott and Commoner, "Reactivation of Tobacco Mosaic Virus."

128. See clipping from the *Detroit Free Press,* Friday, 28 Oct 1955, Stanley papers, carton
22, folder Clippings 1955.

129. Milton Silverman, "Now—Man-Made Virus—First Step in Controlling Heredity?"
Collier's Magazine, 31 Aug 1956, 74–77. The reconstitution experiment was subsequently
featured by Thomas E. Stimson Jr. in *Popular Mechanics:* "Science Probes the Secrets of
the Virus," 15 Aug 1956. See also Williams, *Virus Hunters,* 481.

130. Stanley to Warren Young, 6 Jan 1956, Stanley papers, carton 19, folder *Life.*

131. Ibid. See also Waldemar Kaempffert, "Reconstruction of Virus in Laboratory
Reopens the Question: What Is Life?" *New York Times,* Sunday, 30 Oct 1955, E9. Kaempffert
had previously written on a corroboration of the reconstitution experiment in the *New York
Times,* Sunday, 15 Sep 1955, E11. Clippings of both are found in Stanley papers, carton 22,
folder Clippings 1955.

132. Telegram from Warren Young, science editor for *Life,* Time, Inc., New York, to
Stanley, 3 Oct 1956, Stanley papers, carton 19, folder *Life.*

133. First quotation from Science Service, "Make Virus Hybrid by Crossing Viruses,"
enclosed in letter from Watson Davis to Stanley, 21 Mar 1956, Stanley papers, carton 19,
folder Science Service; second from University of California press release in association
with NFIP grant to Stanley, enclosed with letter from Daniel Wilkes to Stanley, 27 Jun 1956,
Stanley papers, carton 20, folder University of California—Public Information.

The ability to purify and recombine the components of the virus meant that Fraenkel-Conrat could make hybrids between the available strains of TMV. Fraenkel-Conrat and Singer constructed various combinations of nucleic acid from four strains and protein from three strains. At one level, these experiments confirmed the results of the simple reconstitution experiment, supplying "what appears to be incontrovertible evidence that the infectivity of the reconstituted virus is actually a property of the newly formed virus particles."[134] But even more important, in every case where the nucleic acid from one strain was combined with the protein from another strain, the progeny virus was identical to the "parent" strain from which the nucleic acid had been derived (see fig. 7.13).

A particularly striking example was achieved by mixing TMV protein with RNA from Holmes ribgrass (HR), a strain whose protein is quite distinct from garden variety TMV in both amino acid composition and antigenicity. The reconstituted hybrid virus showed none of the distinctive antigenic properties of HR, but when infected into tobacco it produced the virus progeny with protein indistinguishable from the HR strain. "Thus the ribonucleic acid seems to represent the main genetic determinant even for the progeny protein in the TMV strains."[135] Moreover, the residual infectivity of the isolated nucleic acid no longer appeared to be artifactual. As Fraenkel-Conrat explained,

> In attempts to free the original protein and nucleic acid fractions from any traces of undegraded virus, assay at high concentrations (up to 0.05%) in a very sensitive variety of *nicotiana* test plants indicated the presence of no more than about 0.0003% of active virus in either. Yet, when the nucleic acid was tested in a plant variety less sensitive to TMV (Holmes necrotic), sufficient lesions were sometimes produced to indicate the presence of about 0.1% of TMV. This residual infectivity differed from the virus in its relative activity on different hosts and at different levels, in its marked instability, and in its susceptibility to ribonuclease. Virus rods could not be found to account for this residual infectivity, and it is therefore now regarded as a characteristic property of the nucleic acid itself.[136]

In the longer paper on the reconstituted hybrids, Fraenkel-Conrat and Singer made explicit the similarity to results with phage: "These findings strongly suggest that the nucleic acid is the genetic determinant in TMV, and related strains, playing the same decisive role which DNA seems to play in the bacteriophages."[137]

134. Fraenkel-Conrat and Singer, "Virus Reconstitution. Part 2," 540.
135. Fraenkel-Conrat, "Role of the Nucleic Acid."
136. Ibid., 883.
137. Fraenkel-Conrat and Singer, "Virus Reconstitution. Part 2," 544.

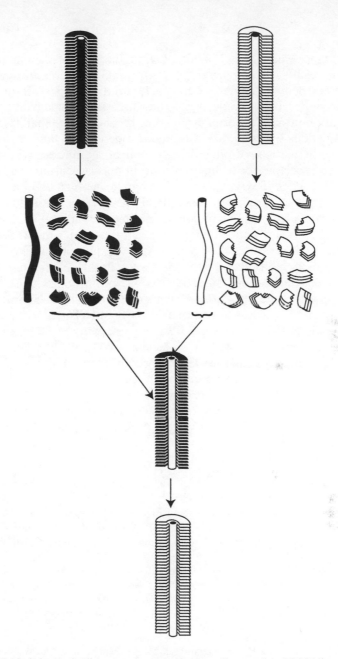

FIGURE 7.13 Schematic diagram showing the hereditary role of RNA in reconstituted hybrid TMV. Two strains of virus (*shortened cross-sections in top row*) were broken down into their constituent nucleic acids and protein subunits (*second row*), and the protein of one was allowed to recombine with the nucleic acid (RNA) of the other (*third row*). The progeny of this hybrid (*fourth row*) had the protein originally associated with their nucleic acid. (After Fraenkel-Conrat, "Rebuilding a Virus" [1956]: 47.)

Just three weeks after Fraenkel-Conrat filed his brief communication of these results to the *Journal of the American Chemical Society,* Alfred Gierer and Schramm independently submitted a letter to *Nature,* entitled "Infectivity of Ribonucleic Acid from Tobacco Mosaic Virus."[138] This publication provided a brief synopsis of their evidence that TMV RNA was infectious by itself, at a rate about 2 percent of that of the native virus. Their method of phenol extraction of the nucleic acid, adapted from the repertoire of techniques used in polysaccharide biochemistry, produced a larger RNA (12–18 S) than had been previously identified,[139] and the isolation procedure virtually eliminated the possibility that the infectivity could be carried by contaminating protein.[140] While the results from the Berkeley and Tübingen laboratories were similar, the RNAs they isolated were not: the phenol-extracted nucleic acid in Germany was higher in molecular weight, corresponding to 2 million daltons or about 6,000 nucleotides—indeed it could be imagined to stretch the entire length of the TMV rod. A single disruption of the complete RNA chain seemed to cause a complete loss of infectivity.[141] By contrast, the infectious RNA isolated in Berkeley appeared to be much smaller in the ultracentrifuge, with a molecular weight of 300,000, or about 1,000 nucleotides. Unlike the Berkeley counterpart, however, the larger nucleic acid did not seem to be able to be reconstituted into infectious viral rods with purified virus protein.[142]

Despite the mounting evidence, some scientists were skeptical of the central role being ascribed to TMV RNA. Rollin Hotchkiss held Fraenkel-Conrat's result to the same high standards that had been used to question the role of DNA in transformation:

> There are several difficult stages in the demonstration of an investigation of such a complex biological activity as initiation of an infection by isolated RNA: first, the isolation of adequately pure RNA; second, the isolation of that pure RNA in an adequately native or undamaged state; third, the subjection of this material to a suitable

138. Gierer and Schramm, "Infectivity of Ribonucleic Acid." This paper was submitted the same day Fraenkel-Conrat's short communication, "Role of the Nucleic Acid," was published in *Journal of the American Chemical Society* (10 Feb 1956); there is no indication that either group knew of the other's result. The longer version of the German result was also submitted just three days after their account in *Nature* (see Gierer and Schramm, "Die Infektiosität der Nucleinsäure").

139. In fact, when the Berkeley group measured the sedimentation coefficient of phenol-extracted RNA preparations, they obtained 20 to 30 S, an even larger value than the Tübingen group (Fraenkel-Conrat, "Infectivity of Tobacco Mosaic Virus Nucleic Acid," 219).

140. Mundry, "TMV in Tübingen," 155–60.

141. Gierer, "Structure and Biological Function."

142. Schramm, "Aufbau und Vermehrung phytopathogener Viren."

test of biological infectivity; and, fourth, the demonstration that any infectivity so observed is attributable to the native RNA present in the preparation tested. It seems to me that the first and third of these criteria still are not adequately fulfilled.[143]

Similarly, the Rothamsted group viewed the RNA infectivity results with skepticism as late as 1957. As Pirie commented, "There is no advantage in attributing the infectivity of nucleic acid preparations to any other component in them besides the nucleic acid. This is, however, far from sufficient reason for assuming that the infectivity is indeed carried by pure nucleic acid."[144] Bawden added, "There is no evidence to suggest that the protein alone is ever infective, and there is nothing to conflict with Schramm's conclusion that the nucleic acid *can* be infective. The farthest I shall go toward questioning his conclusions is to say that the absence of contrary evidence does not constitute proof."[144] Barry Commoner, who was also working on TMV infectivity, argued in 1958 that "although there is no question that virus RNA is essential for infectivity, there is considerable doubt that it is itself a sufficient cause for biological activity."[146]

Nevertheless, most biologists took these experiments as a decisive demonstration that the "nucleic acid is the main carrier of genetic information from parent to offspring,"[147] and that the TMV RNA specified the exact sequence and composition of the virus protein. As Crick put it for readers of *Scientific American* in 1957,

> The most convincing evidence that RNA is responsible for the specific construction of proteins has come from recent work on the tobacco mosaic virus, done mainly by Heinz Fraenkel-Conrat and his colleagues at the University of California and by Gerhard Schramm and co-workers at the University of Tübingen in Germany. They have separated the RNA from the proteins of the virus and used the RNA, separately and in combination with different proteins, to generate virus progeny. RNA alone, inoculated into a tobacco plant, has proved capable of reproducing the virus. In this case the progeny multiplying in the infected plant had the protein corresponding to the virus strain from which the RNA was taken, although the plant had never seen this protein before.[148]

Crick took the reconstitution experiment as a good illustration of his "Central Dogma," that "once information (meaning here the determination of a sequence of units) has been passed into a protein it cannot get out again, either to form a copy of the molecule, or to affect the

143. Hotchkiss, "Discussion," 226.   145. Ibid., 249.
144. Pirie, "Discussion," 247.   146. Commoner, "Biochemical Basis," 26.
147. Fraenkel-Conrat, "Rebuilding a Virus," 44.
148. Crick, "Nucleic Acids," 199.

blueprint of a nucleic acid."[149] The example of TMV, whose RNA could itself "reproduce the virus," handily conflated transmission between generations and translation from nucleic acid to protein. At the same time, the RNA experiments put the significance of the earlier reconstitution experiments in a new light. As Fraenkel-Conrat summarized, "The value of reconstitution both in opening up a new field of study and in supplying a powerful spade with which to till it appears thus evident. On the other hand, we can now gracefully retreat from a position which we have never held or expressed. Life was not here created in test tubes, since the nucleic acid alone shows 'signs of life' similar to those of the original virus."[150]

Not surprisingly, Fraenkel-Conrat and Schramm disagreed over which of their laboratories first offered clear demonstration of the infectivity of RNA, and their priority dispute still echoes.[151] The rivalry between the virus chemists in Tübingen and Berkeley was not attributable to political or personal differences in any obvious way, as Gunther Stent asserted in a letter to Delbrück about Fraenkel-Conrat:

> It is very true that he doesn't have much feeling for biology (that, I think, is one of the reasons why he tried such an intrinsically improbable experiment as putting RNA + protein back together) and that he keeps on insulting Schramm quite unnecessarily. His dislike of Schramm, by the way, does not originate from any emotional prejudices (I suppose you meant that F.-C., the German Jew, hates Schramm, the ex-Nazi; but F.-C. is not only completely unpolitical but also had no idea of Schramm's background until I told him about it not long ago) but from their fight about the N-terminal residue of the TMV protein subunit, from which F.-C. has concluded that Schramm is a crook. But F.-C. is strictly an Einzelgänger, rather difficult to get along with as a collaborator, but pleasant to have around if you don't happen to work on exactly the same thing.[152]

But given the degree of competitiveness in the TMV field, it remains striking that none of these experiments with viral RNA had ever been attempted prior to 1956. As Seymour Cohen has pointed out, all of the necessary components were available in the 1940s:

149. Ibid., 199–200.

150. Fraenkel-Conrat, Singer, and Williams, "Nature of the Progeny," 502. I follow Greer Williams (*Virus Hunters,* 481) in attributing this quote to Fraenkel-Conrat (rather than to all three coauthors).

151. Mundry, "TMV in Tübingen"; correspondence between Karen-Beth G. Scholthof and Heinz Fraenkel-Conrat concerning the selections for Scholthof, Shaw, and Zaitlin, *Tobacco Mosaic Virus,* personal communication, Karen-Beth Scholthof.

152. Gunther Stent to Max Delbrück, 7 Nov 1957, Delbrück papers, folder 20.20. (An Einzelgänger is a loner.)

> Despite the availability of appropriate viral RNA after 1936 and
> inactivating crystalline ribonuclease in 1940, and despite the demon-
> stration of DNA as pneumococcal transforming agent in 1944 and the
> apparent infectivity of phage DNA, accepted by the community of
> phage workers in 1953 following the discovery of the Watson-Crick
> model, the thought that the viral RNA might be the genetic element
> of this virus was not tested before 1956.[153]

The relative disinterest on the part of plant virologists in the hereditary
role of nucleic acid derived from a number of factors. For one thing, the
experimental cultures prevailing in bacteriophage genetics and structural
studies of TMV were quite divergent. Consequently, the Hershey-Chase
experiment did not inspire a parallel attempt to distinguish the role of
virus protein from nucleic acid in plant viruses. At a conceptual level, as
I argued in the last chapter, interest in analogous properties of various
self-duplicating nucleoproteins effectively drew attention away from the
unique and specific role of nucleic acid in transformation and in virus
replication. Furthermore, the growing sense that DNA and RNA fulfilled
different roles in the cell discouraged researchers from extrapolating from
the Watson-Crick model for DNA, with its implications for replication and
mutation, to the smaller, single-stranded RNAs of plant viruses. As David
Nanney argued in 1957, "Although recent work, particularly on *Pneumococcus,* has demonstrated beyond reasonable doubt that DNA can carry
genetic information, the conclusion that only DNA can do so *in vivo* is
unwarranted; in this connection one needs to mention only the fact that
DNA has not been found in tobacco mosaic virus."[154] The demonstration
that TMV RNA was infectious showed that the capacity to carry genetic
information was a general property of nucleic acids, not only of DNA.

Once the central role of TMV RNA was recognized, researchers in
both Berkeley and Tübingen rapidly reconfigured their experimental
systems to exploit the fact that one could work with a virus in terms of
its purified components to analyze chemically the role of nucleic acid in
heredity. Thus earlier analogies between macromolecular particles were
now replaced with analogies between types of chemical agents—TMV
RNA was understood to be similar to bacteriophage DNA and to chromosomal DNA in carrying or constituting genes, and TMV subunit polypeptides were being modeled after insulin and ribonuclease, proteins
whose sequence and ability to fold into a native protein were being actively pursued by Sanger and Anfinsen.[155] The visual similarities between

153. Cohen, "Wendell Meredith Stanley," 845.
154. Nanney, "Role of the Cytoplasm in Heredity," 140.
155. On Sanger, see de Chadarevian, "Sequences, Conformation, Information";
Anfinsen et al., "Studies on the Gross Structure."

microsomes and viruses, which had seemed so meaningful in the late 1940s, were displaced by the preeminence of similar chemical specificities and roles.[156] The redefined problematics of molecular biology centered on the relations, informational and biochemical, between different types of cellular constituents, especially DNA, RNA, and protein. TMV provided a good model system for this line of inquiry precisely because it could be understood in terms of its smaller functional pieces, the protein subunits and RNA that could be isolated in pure form.

In Berkeley, the ability to reconstitute TMV enabled particularly elegant biochemical studies of the virus. Fraenkel-Conrat and Singer extended their study of reconstituted viruses with protein from one strain and nucleic acid from another.[157] The most decisive results were obtained with the strain most divergent from common TMV, the Holmes ribgrass (hereafter HR) strain. Following the early indications reported in the *Journal of the American Chemical Society,* Fraenkel-Conrat and Singer published a more detailed account of their experiments on hybrid viruses. The best-studied of the mixed, reconstituted viruses were HR/TMV, combining HR nucleic acid with the usual TMV protein, and the reciprocal combination TMV/HR, containing normal-strain nucleic acid and HR protein. Mixed virus HR/TMV reacted with TMV antiserum, whereas the reciprocal TMV/HR was neutralized with anti-HR serum. At the same time, the progeny of mixed virus HR/TMV were true-to-type for Holmes ribgrass, yielding its characteristic lesions and viruses with HR protein, whereas mixed virus TMV/HR produced wild-type TMV upon infection (see fig. 7.14). Amino acid analyses of the progeny of the mixed viruses confirmed that the nucleic acid of the mixed virus determined the protein type of the progeny. As Stanley stated, "These results demonstrate that the ribonucleic acid core of the virus carries the genetic message."[158] This use of communications metaphors, so pervasive already in phage genetics, remained uncommon in publications from the Virus Laboratory, but TMV soon took on these meanings in the context of the emerging "coding problem."

156. George Palade pointed out that the size and shape of TMV differentiated it from the nucleoprotein particles associated with protein synthesis that had been isolated from a variety of organisms and tissues. See discussion following Fraenkel-Conrat, "Infectivity of Tobacco Mosaic Virus Nucleic Acid," 224–25.

157. Fraenkel-Conrat and Singer, "Virus Reconstitution. Part 2."

158. Stanley, "Virus Composition and Structure," 816. Similarly, Fraenkel-Conrat and Singer used this language when discussing the very slight differences observed between the amino acids of the original strains and those of the progeny: "These differences are too small to be regarded as more than a suggestion that the protein component may slightly influence the *genetic message* transferred by the nucleic acid [emphasis added]" ("Virus Reconstitution. Part 2," 546).

NATURE OF DISEASE PRODUCED WITH RECONSTITUTED VIRUS OR NUCLEIC ACID OF
DIFFERENT STRAINS OF TOBACCO MOSAIC VIRUS

Strain supplying Nucleic Acid	Protein	Disease Symptoms†
TMV	Masked	TMV
TMV	HR	TMV
Masked	TMV	Masked
YA	TMV	YA
YA	Masked	YA
HR	TMV	HR
HR	Masked	HR
TMV	TMV	TMV
HR	HR	HR
HR	—	HR
TMV	—	TMV

FIGURE 7.14 The following notes accompanied this table in the source. "TMV: common TMV; Masked: Masked strain; HR: Holmes ribgrass; YA: Yellow aucuba. Note that in every case the nucleic acid portion determined the disease symptomotology. †: The disease symptoms could be distinguished by infecting each combination into several strains of tobacco. In Turkish tobacco TMV, YA, and HR give systemic infection of different appearance; Masked gives no symptoms, but its presence can be demonstrated upon transfer to another host. In *Nicotiana sylvestris,* TMV and Masked give a systemic disease, while HR and YA give local lesions only. In *Nicotiana glutinosa* HR gives small local lesions with little necrosis and pigmentation, and is strikingly different from the spreading lesions produced by other strains." (Table and caption from Fraenkel-Conrat, Singer, and Williams, "Nature of the Progeny of Virus," 504. Copyright 1957 the Johns Hopkins University Press. Reprinted by permission.)

TMV and the RNA-Protein Relation

In their second coauthored article on the structure of DNA, Watson and Crick argued that "the precise sequence of the bases is the code which carries the genetical information."[159] As Lily Kay has pointed out, the implications of this coding language for the relationship between nucleic acid and protein structure were first explored not by biochemists but by physicist George Gamow and information theorists such as Henry Quastler.[160] Gamow published a note on the coding problem in *Nature* in response to Watson and Crick's double helical model for DNA, proposing a direct correspondence between combinations of the four

159. Watson and Crick, "Genetical Implications," 965.
160. Kay, *Who Wrote the Book of Life?* esp. chs. 3 and 4. On the development of the notion of the genetic code, see also Judson, *Eighth Day of Creation,* 236 ff.; and Morange, *History of Molecular Biology,* 120–38 ("Deciphering the Genetic Code").

nucleic acid bases and the twenty amino acids found in proteins.[161] In his schema, termed the "diamond code," amino acids matched up with the rhombic-shaped holes formed by four bases between the strands of DNA in a helix (see fig. 7.15). Crick and Pauling immediately called Gamow's attention to the problems with the diamond code, motivating Gamow to consider other possible solutions to the coding problem.[162] He turned to the newly published sequence of insulin for clues.[163]

Expanding his use of experimental data, Gamow drew on the available amino acid compositions and nucleic acid compositions of TMV and turnip yellow mosaic virus as resources for genetic decoding. This endeavor was part of a larger collaborative analysis, with Alex Rich and Martynas Yčas, on the current state of the coding problem. As they explained,

> If nucleic acids function as protein templates, a correlation should exist between their composition and that of proteins. The most promising biological material for the study of such correlations is presented by the viruses since, in this case, there is reason to believe that the protein forming the shell of the particle is synthesized according to the information carried by the nucleic acid constituting its interior.[164]

Differences in the RNA composition of the two viruses could then be expected to correlate with changes in the amino acid compositions:

> RNA composition of TYV [tobacco yellow mosaic virus] differs from that of TMV by a large increase of *Cy* [cytosine], a moderate decrease of *Ad* [adenine] and *Gu* [guanine], and a small decrease of *Ur* [uracil]. If the sequence of amino acids in the protein is determined by the sequence of bases in the nucleic acid, we should expect that TYV will have an increased amount of those amino acids which are determined by base configurations containing one or more molecules of *Cy*. On the other hand, amino acids determined exclusively by the other three bases must be decreased in their amount.[165]

161. Gamow, "Possible Relation." (Communicated to *Nature* on 22 Oct 1953.)

162. The complete, 121-amino-acid sequence of insulin had recently been published by Frederick Sanger, and Crick used the sequence to "disprove all possible versions of Gamow's code." Kay, *Who Wrote the Book of Life?* 138. Kay offers a fascinating and much more detailed account of Gamow's code-breaking efforts and of his collaboration with Martynas Yčas.

163. Gamow, "Possible Mathematical Relations."

164. Gamow, Rich, and Yčas, "Problem of Information Transfer," 55. As Lily Kay makes clear, the authors had been working on this paper since the fall of 1954, and even though many of the coding schemes they reviewed had been disproved by the time the article appeared, it was an important contribution to the literature (Kay, *Who Wrote the Book of Life?* 148).

165. Gamow, Rich, and Yčas, "Problem of Information Transfer," 55–56.

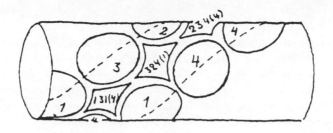

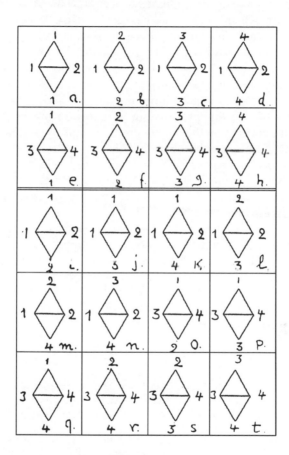

FIGURE 7.15 George Gamow's schematic diagram depicting his diamond code. The cylinder at top depicts the structure of the DNA molecule in accord with Watson and Crick's double-stranded helical model. The panel below shows the twenty different types of "holes" created by different neighboring bases between adjacent strands of the helix. Gamow postulated that each base combination around the diamond hole specified a particular amino acid. (Reprinted, with permission, from Gamow, "Possible Relation between Deoxyribonucleic Acid and Protein Structure," *Nature* 173 [1954]: 318. Copyright 1954 Macmillan Magazines Limited.)

The relative frequencies of amino acids and nucleic acids in the two viruses led them to conclude that the amino acids histidine, methionine, lysine, threonine, and cysteine were encoded by base combinations containing one or more molecules of cytosine, whereas four other amino acid residues—arginine, aspartic acid, proline, and glutamic acid—were specified mostly by the other bases.[166] Gamow further used the relative frequencies of amino acids and bases to propose which base triplets might correlate with which amino acids (see fig. 7.16). Unfortunately, "the data for TMV and TYV lead to different and contradictory assignments between the amino acids and the base triplets."[167] During the next two years, several more coding schemes were proposed based on mathematical and combinatorial approaches to the problem, but none found clear support from the amino acid composition and sequence data available.[168]

Gamow was a visiting professor in Berkeley during much of the time when he was working on the coding problem, and spent time at the Virus Laboratory.[169] Before publishing the article with Rich and Yčas, Gamow sent a copy of the analysis to Stanley to check if he was representing the TMV data accurately. Stanley offered only small corrections on the use of data but pointedly expressed skepticism of Gamow's emphasis on the Chargaff rule for RNA.[170] However, Stanley and his colleagues seem to have been unpersuaded (and were perhaps simply baffled) by his theoretical approach to the RNA-protein relation.[171] Their efforts in this direction were entirely experimental in spirit and were inspired by newly isolated nucleic-acid-synthesizing enzymes. In this respect, they reveal the biochemical preference for approaching the RNA-protein relationship as a problem of macromolecular synthesis rather than as an example of information transfer.

166. Ibid., 56.

167. Ibid.

168. Kay, *Who Wrote the Book of Life?* ch. 4. On the inadequacy of the various proposed coding schemes, see Brenner, "Impossibility of All Overlapping Triplet Codes"; Crick, Griffith, and Orgel, "Codes without Commas"; and Golomb, Gordon, and Welch, "Comma-Free Codes."

169. As Kay comments, "Gamow spent considerable time at the new Virus Lab in 1954, devoting one day a week to biology, as Stent informed Delbrück" (*Who Wrote the Book of Life?* 186).

170. Gamow to Stanley, 28 Mar 1955, including a preprint of "On Information Transfer from Nucleic Acids to Proteins"; and Stanley to Gamow, 1 Apr 1955, Stanley papers, carton 8, folder Gamow, George.

171. Kay, *Who Wrote the Book of Life?* 186. As Kay explains, despite the time Gamow spent around the Virus Laboratory, and the fact that he taught a course in the fall of 1956 on "Biology at the Molecular Level," his interest in genetic coding as a theoretical problem made little impression on Stanley's biochemists.

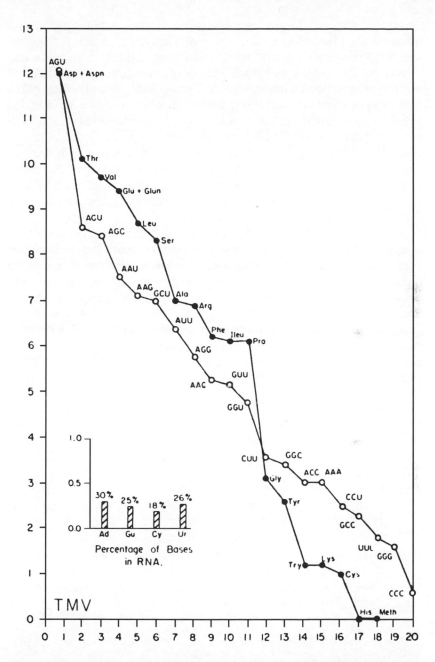

FIGURE 7.16 Graph showing the distribution of amino acids and base triads in TMV. (From Gamow, Rich, and Yčas, "The Problem of Information Transfer" [1956]: 57.)

In 1955, Severo Ochoa and Marianne Grunberg-Manago had announced the identification of a new enzyme system capable of synthesizing RNA in the test tube from simple nucleotides.[172] The principal enzyme in this system, polynucleotide phosphorylase, could be used to make synthetic RNA from nucleotide diphosphates, and members of the Virus Laboratory wondered if a synthetic RNA could be used in the reconstitution experiment. John Smith and Roger Hart in Stanley's laboratory prepared a synthetic polynucleotide by Ochoa's method and succeeded in combining it with tobacco mosaic protein into rods. Stanley reported on their results at a meeting of the Federation of American Societies of Experimental Biology (see fig. 7.17): "The result: a new, man-made, 'virus,' but one for which a susceptible host is yet to be found."[173]

Ochoa came to Berkeley in the summer of 1956 to collaborate with members of the Virus Laboratory. The work on synthetic nucleic acids with Ochoa led the TMV researchers in an unexpected direction. In particular, they began to seek an explanation for the wide and inexplicable variation they observed when assaying the infectivity of isolated viral nucleic acid and also of reconstituted virus rods. As Fraenkel-Conrat explained, "Recently we were led by a somewhat devious trail of research (started with S. Ochoa in 1956) to the realization that metal-chelating compounds show pronounced effects on the infectivity of nucleic acid."[174] In pyrophosphate buffer, Fraenkel-Conrat went on to explain, reconstitution rates of as high as 34 percent had been obtained. By contrast, divalent metal ions caused a marked loss in infectivity, although this inactivation could be prevented by adding metal-chelating compounds, which became standard practice in the laboratory.

Ochoa's intent was to use the TMV system to determine if the enzyme he had isolated might play a physiological role in virus reproduction. With a sample of Fraenkel-Conrat's TMV, he had already tried to determine whether infection caused polynucleotide phosphorylase activity in conjunction with (and presumably with responsibility for) the production of viral RNA.[175] Initial results were not encouraging. Moreover, while the activity of this enzyme might illuminate the mechanism of nucleic

172. Grunberg-Manago and Ochoa, "Enzymatic Synthesis"; Grande and Asensio, "Severo Ochoa"; Santesmases, "Severo Ochoa."

173. Draft of *Medical News* piece on the synthetic virus work, enclosed in letter from Elma T. Wadsworth to Stanley, 26 Apr 1956, Stanley papers, carton 20, folder World Wide Medical News Service, Inc. See also Robert K. Plumb, "Scientist Makes Synthetic Virus," *New York Times,* 18 Apr 1956, C33; and Waldemar Kaempffert, "Artificial Virus Needs a 'Host' So That Its Real Nature Can Be Determined," *New York Times,* 22 Apr 1956, E9.

174. Fraenkel-Conrat, "Infectivity of Tobacco Mosaic Virus Nucleic Acid," 220–21.

175. Ochoa to Stanley, 4 Jun 1956, Stanley papers, carton 11, folder Ochoa, Severo.

FIGURE 7.17 Stanley presenting the latest results from the Virus Laboratory at the May 1956 meeting of the Federation of American Societies of Experimental Biology. (From Williams, *Virus Hunters,* plate 4, 117.)

acid synthesis, it could not alone account for the sequence specificity of biologically active RNAs.[176] Virus Laboratory members undertook other biochemical studies of TMV nucleic acid. For example, investigation of viral RNA synthesis *in vivo* through labeling with ^{32}P demonstrated that the nucleic acid was synthesized from single 5′-nucleotides.[177]

But in the end, Fraenkel-Conrat recognized that obtaining base sequence data for TMV RNA would be more valuable than any other kind of information about viral nucleic acid. "Since this material alone can evoke and determine the synthesis in the plant not only of replicates of itself but also of a specific protein, the comparative sequential relationship of nucleotide and corresponding amino acid residues should be most illuminating. Unfortunately, the methodology for analysis of polynucleotide structure is as yet woefully inadequate for this task."[178] The sequencing of TMV subunit protein, in contrast to determining its RNA base sequence, was proceeding apace. Kozo Narita, working with Heinz Fraenkel-Conrat, showed that an acetylated dipeptide was likely to be the N-terminal fragment, providing the reason for the early discrepancies with Schramm over the N-terminus.[179] By late 1957, Stanley remarked that his laboratory had already "worked out about half of the amino acid sequence of a TMV subunit."[180] Fraenkel-Conrat forecast that once the primary protein structure of the virus was available, TMV "may then become the Rosetta

176. As Ochoa stated in 1957, "Even if polynucleotide phosphorylase or a similar enzyme proved to be responsible for the intracellular synthesis of RNA, it would be difficult to see how such an enzyme would lead to specific sequential arrangements of nucleotides in the polynucleotide chains without auxiliary mechanisms. Characteristic nucleotide sequences are considered to be the main basis for the specific biological activity exhibited by ribonucleic acids such as those of plant viruses. It is unlikely that the mechanism of biosynthesis of specific ribonucleic acids will be understood until the synthesis of a biologically active virus RNA can be obtained in the test tube" (Ochoa, "Biosynthesis of Ribonucleic Acid," 199).

177. "Progress Report, January 1, 1958–December 31, 1958," Stanley papers, carton 16, folder NFIP Progress Reports from Jul 1955. These experiments were done with turnip yellow mosaic virus as well as TMV.

178. Fraenkel-Conrat, "Tobacco Mosaic, a Molecular Infection," 208. Stanley also emphasized the value of getting the RNA sequence in a description he wrote (in third person) of his laboratory's goals: "More recently he [Stanley] has been directing another group of collaborators in the determination of the exact amino acid sequence of the protein subunits of which the tobacco mosaic virus is built up, in order that it may ultimately be possible to correlate that sequence with the specific succession of nucleotides in the ribonucleic acid, which, as one of Stanley's associates showed in 1955, is the carrier of the viral hereditary information" (unpublished sketch of scientific achievements, 19 Mar 1959, Stanley papers, carton 22, folder Biographical data).

179. Narita, "Isolation of Acetylpeptide."

180. Typescript of *Medical News* interview with Stanley, enclosed with letter from William H. White to Stanley, 12 Nov 1957, Stanley papers, carton 20, folder World Wide Medical News Service, Inc., 5.

stone of biochemical genetics, in supplying the bilingual record needed to decipher nucleic acid in terms of protein structure."[181]

TMV Subunit Organization

The accumulation of TMV protein sequence data reinforced the value of TMV as a model system for virus assembly. Fraenkel-Conrat had pointed out that the fact that reconstitution succeeded at all for quite divergent strains of TMV had important ramifications. Because the usual TMV protein and the HR protein exhibited many chemical differences, the fact that either one could combine with nucleic acid from each strain showed that the ability of the virus to assemble relied on relatively few structural signals. As he and Singer concluded, "An important functional property of the virus proteins probably resides in their specific tendency to aggregation in a superhelical array around, if they are present, nucleic acid strands. In view of the exchangeability of different virus proteins, as observed in the present experiments, one must conclude that this activity is dependent only upon a few suitably situated key sites surrounded by nonspecific areas."[182]

X-ray crystallography studies of TMV contributed greatly to the conceptualization of virus structure in terms of its subunit organization. In 1956, Franklin, in Bernal's Birkbeck laboratory, and Donald L. D. Caspar, working at Yale, provided independent evidence that the TMV rod was hollow rather than solid.[183] Watson found this work particularly important, as he reported to Stanley:

> Rosalind Franklin's X-ray analysis of TMV is becoming more and more exciting. The RNA seems now to be [at] a radius of 42 Å, not 24 Å as Caspar and I originally suspected. Also the subunit number has been revised. Using Fraenkel-Conrat's Hg^{++} substituted TMV, she now has good evidence that there are 49 subunits/3 turns, and so there is good reason for believing that only one chemical subunit is present per crystallographic unit. If so, the subunit number is ∼2,150, and the TMV molecular weight around 40 million.
>
> It is really quite fascinating how the facts are beginning to fit together.[184]

At the Ciba Foundation symposium on viruses in the summer of 1956, Franklin and her colleagues unveiled a more detailed proposal for TMV subunit structure (see fig. 7.18).[185]

181. Fraenkel-Conrat, "Degradation and Structure," 810.

182. Fraenkel-Conrat and Singer, "Virus Reconstitution. Part 2," 545.

183. Caspar, "Radial Density Distribution"; Franklin, "Location of the Ribonucleic Acid."

184. Jim Watson to Stanley, 27 Jan 1956, Stanley papers, carton 13, folder Wa misc.

185. Franklin, Klug, and Holmes, "X-ray Diffraction Studies."

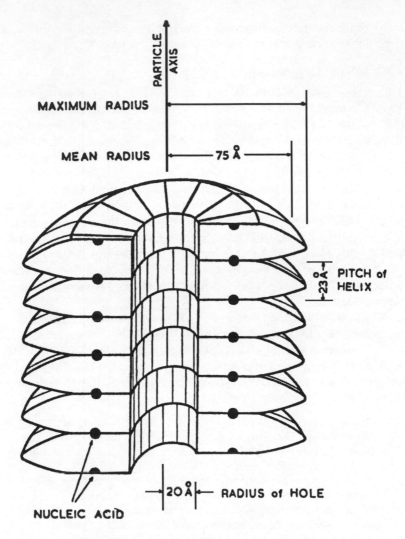

FIGURE 7.18 "Schematic representation of a short length of the virus particle cut in half along a plane through the particle axis, showing the helical arrangement of protein subunits (49 subunits on 3 turns of the helix), the helical groove and its accompanying helical ridge extending beyond the *mean* radius of the particle, and the hollow axial core. The serrated character of the helical ridge is not shown. The nucleic acid is shown at a radius of 40 Å." (Image and caption from Franklin, Klug, and Holmes, "X-ray Diffraction Studies" [1956]: 43. Reprinted by permission of Dr. Kenneth C. Holmes and Sir Aaron Klug.)

Franklin had made good use of her collaborations with TMV biochemists to obtain materials that would help her resolve the subunits through X-ray diffraction. In addition to the metal-substituted TMV she obtained from Fraenkel-Conrat (in which a mercury atom bound each protein subunit at its single cysteine residue), she used Schramm's A-protein to obtain repolymerized, nucleic-acid-free viral protein. Comparing the electron density of native TMV to nucleic-acid-free TMV revealed that the RNA was 55Å from the center of the helix. Her results also showed the high degree of regularity in the structural repeats of the TMV protein helix. Experimental evidence from these studies on TMV provided key evidence for Crick and Watson's contention that all viruses must be built up symmetrically from identical protein subunits that surround the nucleic acid.[186] The elegant simplicity of this observation prompted the witticism, attributed to Crick, that "Any child could make a virus."[187]

During the summer of 1956, Franklin spent three weeks in the Virus Laboratory "in order to get a little experience of the techniques of handling virus material."[188] She was particularly interested in working with Fraenkel-Conrat to prepare new, heavy-atom derivatives of TMV for diffraction analysis, and brought X-ray specimen tubes with her to take back the modified virus. Franklin also asked Stanley for "a fresh supply of your standard TMV, as this has now become my standard preparation, and if I have to change to Cambridge or any other preparation I should have to repeat a large amount of work on the basic measurements."[189]

The collaboration between Franklin and members of the Virus Laboratory extended beyond TMV. On 24 June 1957, Schwerdt, who had crystallized poliovirus with Schaffer, hand-delivered some polio crystals to Franklin so that she could begin X-ray diffraction work on them. Results were promising, and after Franklin's untimely death the next year, Aaron Klug continued the project, working on the structure of poliovirus alongside his continuing studies of TMV and tobacco yellow mosaic virus (TYMV).[190] Once again, TMV became the point of reference for similar work on other viruses.

186. Crick and Watson, "Structure of Small Viruses."

187. As quoted by Aaron Klug in his historical introduction to session 1, "Particle Structure," at Symposium on Tobacco Mosaic Virus: Pioneering Research for a Century, sponsored by the Royal Societies of Edinburgh and London, 7 Aug 1998, Edinburgh. Klug dates the quote to the late 1950s (personal communication, 17 Apr 2001).

188. Franklin to Stanley, 20 Apr 1956, Stanley papers, carton 8, folder Franklin, Rosalind. Franklin's thank-you note to Stanley, dated 30 Aug 1956 (same folder) makes clear that she remained in Berkeley for three weeks.

189. Franklin to Stanley, 4 Jul 1956, Stanley papers, carton 8, folder Franklin, Rosalind.

190. Klug to Stanley, 2 Oct 1959, Stanley papers, carton 9, folder Kl–Ku misc.

The productivity of focusing on the functional pieces of TMV also provided a model for efforts with other viruses. Fraenkel-Conrat noted in 1959 that methods "are being worked out for other viruses to separate chemically the two components and study their separate functions."[191] As a first step, scientists began to recognize that the plant viruses were not the only viruses to contain RNA rather than DNA. "The nucleic acid content of influenza virus may be taken as a case in point. Early reports indicated deoxyribonucleic acid (DNA) at a level of 3 percent, and the absence of ribonucleic acid (RNA). Subsequently both nucleic acids were considered to be present, whereas currently it is generally accepted that DNA is absent and RNA is present at a level of about 1 percent."[192] By 1959, RNA had been identified as the hereditary material in nearly a dozen other animal viruses, including ECHO virus, foot-and-mouth disease virus, western and eastern equine encephalitis viruses, and one of the mouse leukemia viruses.[193] Moreover, Frederick L. Schaffer and Carl F. T. Mattern of the Virus Laboratory demonstrated that the isolated nucleic acid of poliomyelitis virus and Coxsackie virus were sufficient alone to infect.[194] They prepared the poliovirus nucleic acid using Schramm and Gierer's phenol extraction method.[195] Thus the use of TMV as a model for poliovirus in this instance took its shape from experimental systems in both Berkeley and in Tübingen. Stanley was interested in extending the same model to the work on cancer research in the Virus Laboratory, initiating "attempts to secure a biologically active nucleic acid from tumor viruses."[196] As Fraenkel-Conrat explained, many of these isolated viral RNAs exhibited very low rates of infectivity, which might have been otherwise discounted if not for the example of TMV: "The low yield in infectivity has not unduly

191. Fraenkel-Conrat, "Tobacco Mosaic, a Molecular Infection," 210.

192. Schaffer, "Purification and Physicochemical Properties," 177; Schaffer and Schwerdt, "Nucleic Acid Composition."

193. "RNA Infectivity Shown," *Medical News* 5 (29 Apr 1959): 1; Williams, *Virus Hunters,* 483. A letter from Cutter Laboratory's Houlihan to Schwerdt, 6 May 1958, cc: Stanley, implies that Hattie Alexander first tried isolating infective nucleic acid from poliovirus.

194. Infectivity of poliovirus RNA and Coxsackie virus RNA was demonstrated in both tissue culture and in mice (Schaffer and Mattern, "Infectivity and Physicochemical Studies"). Schaffer had crystallized poliomyelitis virus in Stanley's laboratory (with Carlton Schwerdt; see chapter 5) in 1955, and Carl Mattern had crystallized Coxsackie virus. On the isolation of infectious RNA from other animal viruses, see Schramm, "Biosynthese des Tabakmo-saikvirus," 4.

195. "Progress Report, January 1 to December 31, 1955," Stanley papers, carton 16, folder NFIP Progress Reports from Jul 1955.

196. Stanley's entry for "Biblio-Director" of American tumor virus research, enclosed with letter from Elizabeth Koenig to Stanley, Aug 1961, Stanley papers, carton 17, folder Rockefeller Institute for Medical Research.

concerned these investigators since they could refer to the precedent of TMV."[197]

Producing Mutants: TMV and the Genetic Code Revisited

The Fourth International Congress for Biochemistry, held in Vienna during early September of 1958, provided an occasion for a symposium on the biochemistry of viruses. Eleven of the thirty papers concerned recent work with TMV. Three of the contributions were from the Tübingen laboratory, and one elicited particular excitement. Karl Mundry and Gierer reported their success with using the host Java tobacco in a genetic screen to pick up nitrous-acid-generated mutants of TMV.[198] This achievement built on the recent demonstration by Heinz Schuster and Schramm that the mutagen nitrous acid (HNO_2) could be used to inactivate purified TMV RNA (by inducing lethal mutations).[199] In this area they were considerably ahead of their Berkeley competitors. Mundry recalls that after the session in which he and Gierer presented these results, Fraenkel-Conrat "invited me for a cup of coffee, and I wrote for him the recipe for the incubation procedure on a paper napkin."[200] Nitrous acid caused the deamination of three bases, converting adenine to hypoxanthine, guanine to xanthine, and cytosine to uracil. Hypoxanthine and xanthine were not naturally occurring bases, and were likely be "read" as guanine, so that modification of adenine could result in a mutation to guanine. In addition, a cytosine-to-uracil change would register as a mutation, and could change the amino acid specified if it fell in a coding region of RNA.

In Tübingen, Heinz Günter Wittmann was already sequencing TMV polypeptides of the mutants in an attempt to find changes in the amino acid sequence associated with these mutations. (Wittmann had worked on the mutability of T phages with Wolfhard Weidel in Tübingen and then spent a year in the Virus Laboratory as a postdoctoral fellow with Knight.) The first mutant to be fully analyzed was disappointing—no amino acid changes were evident in the strain's protein.[201] In Berkeley, Tsugita and Fraenkel-Conrat, using the newly acquired German recipe, used the nitrous acid technique to generate a mutant whose effects did show up in

197. Fraenkel-Conrat and Singer, "Structural Basis of Activity," 9.
198. Mundry and Gierer, "Erzeugung von Mutanten des Tabakmosaikvirus"; Gierer and Mundry, "Production of Mutants." Not everyone believed this result; Bawden ("Effect of Nitrous Acid on Tobacco Mosaic Virus") claimed that the German genetic screening was selecting naturally occurring mutants, rather than producing them. Mundry defended his group's methods ("Effect of Nitrous Acid"); in the end, the specific mutagenicity of the nitrous acid was generally accepted.
199. Schuster and Schramm, "Bestimmung der biologisch wirksamen Einheit."
200. Mundry, "TMV in Tübingen," 157–58.
201. Wittmann, "Vergleich der Proteine."

the protein.[202] Three of the 158 amino acids had been altered: a proline was changed to leucine, aspartic acid to alanine, and threonine was replaced by serine. Shortly thereafter, Wittmann reported on nine amino acid changes in chemically generated mutants of TMV.[203]

As Lily Kay has pointed out in her analysis of the significance of these events for efforts at "cracking the genetic code," chemical methods for mutagenizing TMV RNA provided a powerful new probe for correlating the nucleic acid sequence with protein sequence. Moreover, during this period the textual metaphors ubiquitous in the discussions of the genetic code began to appear commonly in the publications of the Virus Laboratory.[204] For example, Stanley, well aware of the significance of the mutant work, informed a medical news reporter of the latest developments from the Virus Laboratory in these terms: "As soon as we have definite information concerning the end portion of the polynucleotide chain, namely, that portion which codes for the change from proline to leucine, we will have the first bit of information which can be regarded as the first break in the code of what might be called 'the language of life.' About a million more such breaks may prove to be necessary before the full language can be understood."[205] As he suggested, using single mutants of TMV to deduce the genetic code would be a very laborious process. Furthermore, as Wittmann had found, not every mutation in the RNA would result in an amino acid alteration. Thus Fraenkel-Conrat and Stanley referred to the 1960 publications on TMV mutants as "the first small steps taken to climb the Mt. Everest of molecular biology, the code that relates nucleotide sequence to amino acid sequence."[206]

That same year, the race for the amino acid sequence of TMV protein culminated in publications from both laboratories. The Tübingen group published their primary structure of the TMV protein in the 18 June 1960 issue of *Nature*.[207] It included 157 amino acids, although it did not assign the amide groups, leaving ambiguous the differentiation of glutamic acid from glutamine and aspartic acid from asparagine. For this reason, when the Berkeley Virus Laboratory published its 158-amino acid sequence (shown in fig. 7.19) in the 15 November issue of *Proceedings of the National Academy of Sciences,* the article stated that the list comprised "the first complete amino acid sequence for TMV protein."[208] Stanley was listed

202. Tsugita and Fraenkel-Conrat, "Amino Acid Composition."

203. Wittmann, "Comparison of the Tryptic Peptides."

204. Kay, *Who Wrote the Book of Life?* 187.

205. Stanley to Margaret Markham, 19 Apr 1960, Stanley papers, carton 20, folder World Wide Medical News Service, Inc.

206. Fraenkel-Conrat and Stanley, "Chemistry of Life. Part 2," 144.

207. Anderer et al., "Primary Structure."

208. Tsugita et al., "Complete Amino Acid Sequence," 1464.

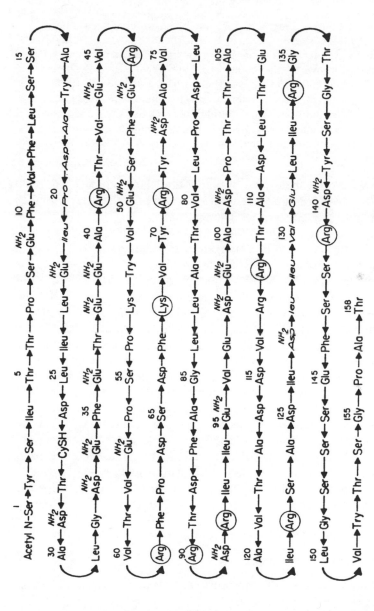

FIGURE 7.19 Sequence of the 158 amino acid residues in the protein subunit of TMV. The encircled residues indicate the points of splitting by trypsin to produce the shorter fragments for sequencing. (As published in Tsugita et al., "The Complete Amino Acid Sequence" [1960]: 1465. Reprinted courtesy of the Bancroft Library, University of California, Berkeley.)

as an author of this paper, in contrast to nearly all of the other research publications coming from the Berkeley laboratory. George Gamow wrote Stanley on RNA Tie Club stationery,

> Please accept my congratulations for the beautiful 158-jewel necklace. Some six or seven years ago, I would spend some sleepless nights attempting to decode it using an overlapping code. But now I am disillusioned in overlapping codes, and believe that each a. a. [amino acid] is determined by an independent triplet of bases in RNA. I added your data to the previously known protein sequences (which enlarged the statistical sample by almost a factor of two), and found that the randomness, as testified by Poisson formula, is even better than it was before. Hope that somebody will get the RNA sequence in TMV soon.[209]

Gamow's wish for a speedy resolution of the RNA sequence was highly optimistic—it was not published until 1982, and even then was among the first completed sequences of viral genomes.[210]

There were some differences between the amino acid sequences obtained in Berkeley and in Tübingen. Stanley tended to downplay the discrepancies: "This sequence is almost in complete agreement with one worked out independently by a German group in Tübingen, a rather remarkable result even though we supplied their original TMV over 20 years ago and it has been grown in Germany under obviously different conditions ever since."[211] Fraenkel-Conrat tended to attribute the relatively high degree of agreement to the competition between their groups. "TMV protein was the second large protein to be sequenced; in contrast to the first, pancreatic ribonuclease [124 amino acids], TMV sequencing was more a competitive than a cooperative effort of two laboratories, and possibly for that reason there were fewer or smaller errors in the first published structures of this then longest sequence of 158 amino acids."[212] However, even if the sequencing had not been cooperative, Stanley did correspond with Melchers and Wittmann to see if the groups might exchange samples of TMV. He was particularly eager to see if genetic differences might possibly account for sequence discrepancies, but there is no evidence that the preparations were exchanged.[213] Stanley

209. George Gamow to Stanley, 30 Nov 1960, Stanley papers, carton 8, folder Gamow, George.
210. Goelet et al., "Nucleotide Sequence."
211. Stanley, "Regulation and Transfer of Biological Information," 147.
212. Fraenkel-Conrat, "Protein Chemists Encounter Viruses," 311; see also Fraenkel-Conrat, "History of Tobacco Mosaic Virus," 9.
213. Stanley to G. Melchers, 4 Nov 1960, Stanley papers, carton 10, folder Me–Mi misc.; Stanley to H. G. Wittmann, 4 Nov 1960, Stanley papers, carton 13, folder Willis–Witton misc.

continued negotiating with Schramm over how to clear up remaining differences:

> There seem to be two areas of uncertainty, and I wonder if your work has proceeded to the point where it is possible for you to accept our sequence for 25–29. I also wonder whether or not you are now able to accept Leu[cine] at 76. I would gather that we still have differences with respect to the presence or absence of amides at 22, 64, 73, 97, 98, 106, and 125. We feel rather certain about the distribution in these positions and wonder whether or not you have any additional information.... When I was in Tübingen about a year ago, we had a discussion regarding the possibility of a final joint paper, and I wonder whether or not we have sufficient agreement at this time to plan such a paper. In view of the large amount of work on the strains of TMV, I think that it is highly important for us all to have a common foundation structure as a basis of comparison.[214]

As in his earlier correspondence with Bawden and Pirie, Stanley was eager to emphasize corroboration over competition.

In his survey of the state of the "coding problem" in 1959, Crick had been pessimistic about the lack of direct evidence supporting his contention that "the amino acid sequence of a particular protein is in some way determined by the sequence of the bases in some particular length of nucleic acid."[215] He pointed to work on the correlation between the RNA and protein of TMV as among the best-known experimental facts that would need to be explained by any plausible coding theory. With the complete amino acid sequence of TMV in hand, efforts to use chemical mutagens to correlate nucleotide substitutions with amino acid changes accelerated. Yčas showed how the logic of code-breaking would work as TMV results accumulated, and Carl Woese actually used the available information from virus mutations to predict the assignments for a triplet code, many of which were borne out by later experimental work.[216]

However, the use of chemically induced TMV mutants to crack the code was slow. Of the first 120 nitrous acid mutants analyzed by Wittmann

214. Stanley to Gerhard Schramm, 2 Mar 1962, Stanley papers, carton 12, folder Schramm, Gerhard. Stanley wrote again the next month asking Schramm to respond to the differences between their structures, but it seems Schramm was not forthcoming in acknowledging how many of the sequence discrepancies he would concede to the Berkeley group.

215. Crick, "Present Position of the Coding Problem," 35. On his sequence hypothesis, see Crick, "On Protein Synthesis."

216. Yčas, "Correlation of Viral Ribonucleic Acid"; Woese, "Coding Ratio"; Kay, *Who Wrote the Book of Life?* 191.

AMINO ACID EXCHANGES IN CHEMICAL MUTANTS OF COMMON TMV

Mutagen	Mutant No.	Amino acid exchanges	Location
HNO₂	273	Asp^b → Ser	Peptide #1
HNO₂	282	Asp^b → Ala	Peptide #10
		Thr → Ser	Peptide #1
HNO₂	171	Pro → Leu	Residue #156
		Asp^b → Ala	Peptide #10
		Thr → Ser	Peptide #1
HNO₂	332	Pro → Leu	Peptide #1
HNO₂	220	Ser → Phe	Residue #138
HNO₂	329	Ser → Phe	Residue #138
		Ser → Phe	Peptide #12
HNO₂	237	Arg → Gly	Residue #61
		Arg → Gly	Residue #134
HNO₂	321B^c	Arg → Gly	Residue #61
		Arg → Gly	Residue #134
		Arg → Gly	Residue #122
HNO₂	284	Glu → Gly	Residue #97
		Arg → Lys	
HNO₂	328	16 amino acids exchanged in many peptides; compo	
HNO₂	262	sition identical with G-TAMV (see Table I)	
NBSI	218	Pro → Leu	Peptide #1

FIGURE 7.20 Example of a table produced by Tsugita and Fraenkel-Conrat, showing amino acid exchanges in chemically induced mutants of TMV. (From Tsugita and Fraenkel-Conrat,"Contributions from TMV Studies" [1963]: 503.)

and the 56 chemically induced mutants analyzed by Fraenkel-Conrat and Tsugita (theirs derived using brominating and methylating agents as well as the deaminating nitrous acid), at least half of the mutants showed no amino acid changes.[217] Of those exhibiting changes, locating the exact alteration required painstaking protein sequencing (see fig. 7.20). Correlating these amino acid alterations with the RNA mutations relied in large part on deduction, as techniques for RNA sequencing were not available and chemical analysis of RNA base composition was not sufficiently precise to detect a single base change.

TMV was not the only system being employed in biochemical efforts to break the genetic code. As Lily Kay has pointed out, "the code" served

217. Tsugita and Fraenkel-Conrat, "Contributions from TMV Studies," 512–13. See also Wittmann, "Comparison of the Tryptic Peptides"; Tsugita and Fraenkel-Conrat, "Composition of Proteins."

NBSI	233[d]	Pro → Leu	Peptide #1
NBSI	207	Pro → Leu	Peptide #1
NBSI	235	Pro → Leu	Peptide #1
NBSI	187[d]	Arg → Gly Asp[b] → Ser	Residue #46 Peptide #1
NBSI	326	Asp[b] → Ser Asp[b] → Ser Asp[b] → Ser	Peptide #1 Peptide #1
NBSI	330	Ileu → Thr	
NBSI NBSI	223 206	17 exchanges, similar to G-TAMV, but for 1 Ala → Gly	
NBSI	331	7 exchanges, similar to Y-TAMV (Table I), but for 1 Leu → Ileu	
DMS	214	Pro → Leu	Peptide #1
DMS	278	Pro → Leu	Peptide #1
DMS	176	Pro → Leu	Peptide #1
DMS	215	Ser → Fhe	Residue #138
DMS	178	Arg → Gly	
P.O.[e]	249	16 amino acid exchanges in many peptides; composition identical with G-TAMV	

Figure 7.20 (continued)

primarily as a shorthand for "protein synthesis" circa 1960, which was a hot topic among a whole cadre of biochemists concerned with metabolism and bioenergetics.[218] Marshall Nirenberg, working at the National Institute of Arthritis and Metabolic Diseases at the intramural campus of the National Institutes of Health, was one of many researchers to develop a cell-free translation system, that is, an extract of cells (in his case ribosomes, nucleic acids, and enzymes from *E. coli*) that, when supplemented with amino acids and energy-providing ATP and GTP, would synthesize proteins in the test tube.[219] Nirenberg's work related to

218. Kay, *Who Wrote the Book of Life?* 240.
219. Kay (ibid.) mentions the efforts of many of Nirenberg's competitors in developing bacterial cell-free translation systems, such as Sol Spiegelman at the University of Illinois, Alfred Tissières at Harvard, and Paul Zamecnik at Massachusetts General Hospital. For more on Zamecnik's contributions, see Rheinberger, *Toward a History of Epistemic Things*. Rheinberger also offers a list of those using bacterial extracts in protein synthesis research (208).

ongoing concerns in basic biochemistry—such as the interdependence of nucleic acid synthesis and protein synthesis—as well as to the coding problem.

Just one floor above Nirenberg's research space was the laboratory of Leon Heppel, chief of the Institute, one of the leading centers for the synthesis of RNAs (as was Ochoa's laboratory in New York). This provided Nirenberg with an unusual resource as he sought to improve the results from his cell-free translation system (which at first, like most others, depended on endogenous nucleic acids from the cellular extracts). In the fall of 1960, Nirenberg was joined by the German plant physiologist Heinrich Matthaei. In the very competitive field of experimentation with protein-synthesizing systems, Nirenberg and Matthaei's approach focused on using purified RNAs to investigate template specificity. To this end, they had recourse to synthetic RNA polymers from Heppel's freezer as artificial templates for protein synthesis.[220] Their notion of a template was not identical to what crystallized as "messenger RNA" over the course of the next year; most biochemical researchers of protein synthesis tended to assume that the RNA directing protein synthesis was part of the ribosome. Thus Nirenberg and Matthaei sought to augment protein synthesis with a non-ribosomal template.[221] TMV RNA was an attractive material for this aim, since it could be obtained in pure form and was well characterized (as was the protein it encoded). Early attempts to use TMV RNA and TYMV RNA to stimulate protein synthesis *in vitro* were promising.[222]

In May of 1961, Nirenberg headed to Berkeley to collaborate with Fraenkel-Conrat, to see if TMV RNA would stimulate specific virus protein synthesis in the cell-free system. While Nirenberg was in California, Matthaei tested synthetic polymers from Heppel's laboratory, such as poly-A and poly-U, as templates for cell-free translation. The poly-A trial stimulated a surprisingly high level of protein synthesis. On 27 May, Matthaei conducted an experiment with poly-U as a template and phenylalanine as a radioactively labeled amino acid. The results were definitive that poly-U specified the synthesis of a polypeptide of pheny-lalanine residues. This provided a direct experimental approach for

220. A much more detailed account of Nirenberg and Matthaei's work is given by Kay, *Who Wrote the Book of Life?* ch. 6. See also Judson, *Eighth Day of Creation,* 452 ff.

221. Rheinberger, *Toward a History of Epistemic Things,* 212.

222. Kay, *Who Wrote the Book of Life?* 250. Kay cites Matthaei's notebooks from May 1961. Judson quotes Nirenberg's recollections of his first experiments with TMV RNA: "Then I got hold of some tobacco-mosaic virus RNA, some viral RNA. Which I thought would be a pure template. Pure message. And we added it to the extracts, here, and it was just *superb.* . . . We hadn't seen anything even approaching the activity. With respect to amino-acid incorporation. . . . I mean it was just *beautiful* . . . So, I called Fraenkel-Conrat" (*Eighth Day of Creation,* 475).

determining which nucleotide sequences encoded which amino acids—
in this case that the RNA codon UUU "spelled" phenylalanine to the
protein-synthesizing machinery. In the meantime, Nirenberg's experiments demonstrated that TMV RNA could serve as a template for protein
synthesis.[223] However, by the time he returned to Bethesda on 11 June,
the possible use of TMV to crack the genetic code in a cell-free translation
system had been surpassed, as Nirenberg eagerly took up and extended
Matthaei's success with synthetic RNAs.

Nirenberg presented their startling results using poly-U to synthesize
polyphenylalanine in August of 1961 at the Fifth International Congress
of Biochemistry in Moscow. Although the session was sparsely attended,
word spread rapidly through the research community that Matthaei and
Nirenberg had begun to crack the code.[224] By September, Ochoa had
announced that his laboratory of twenty biochemists would join the competition to decipher the other codons.[225] These developments changed the
status of TMV as a system for solving the coding problem. (The biochemical decoding engine of cell-free translation systems similarly left behind
the genetic competition of Sydney Brenner and Francis Crick's elegant
rII suppressor mutants.)[226] The accumulating data from TMV could still
serve important functions, such as differentiating among alternative coding possibilities, as Stanley sketched out in a letter responding to some of
Thomas Jukes's theoretical predictions (see fig. 7.21).[227] In addition, the
TMV mutant collection provided a source of confirmatory data for the
codons emerging from *in vitro* studies with synthetic RNAs. But Stanley
and Fraenkel-Conrat no longer expected that TMV research alone would
provide the key to the "Rosetta stone" of life.

Conclusions

By the end of the 1950s, Stanley and his coworkers referred less often
to TMV as simply a nucleoprotein and more frequently to its functional
parts, the approximately 2,150 identical polypeptides of 158 amino acids

223. Tsugita et al., "Demonstration of the Messenger Role." According to Kay, these
experiments proved to be flawed and were subsequently retracted (*Who Wrote the Book of
Life?* 252, 373, n. 44).

224. Judson, *Eighth Day of Creation,* 464 ff. As Kay recounts (*Who Wrote the Book of
Life?* 254), Crick was particularly deflated by the breakthrough.

225. Kay, *Who Wrote the Book of Life?* 257.

226. Judson, *Eighth Day of Creation,* 468.

227. Stanley wrote, "Our own data on amino acid replacements in TMV following nitrous
acid treatment are proving quite useful as a means of checking the triplets that have been
proposed." Stanley to Thomas Jukes, 17 Apr 1962, Stanley papers, carton 9, folder Jukes,
Thomas Hughes.

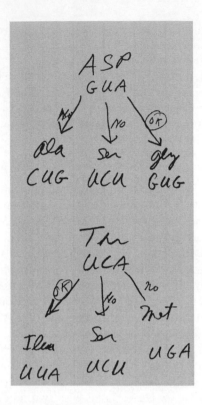

FIGURE 7.21 Stanley's hand-drawn illustration of coding possibilities. He suggests how specific amino acid replacements could result from single base replacements, according to the existing assignment of amino acids to base triplets. Some of his triplets are written backward according to current convention. (From letter to Thomas Jukes, 17 Apr 1962, Stanley papers, 78/18c, carton 9, folder Jukes, Thomas Hughes. Reprinted courtesy of the Bancroft Library, University of California, Berkeley.)

each and the approximately 6,000-nucleotide RNA (see fig. 7.22).[228] This conceptual shift coincided with the development of new biochemical techniques for separating and recombining the protein subunits and

228. On the number and length of virus coat protein subunits, see Ansevin and Lauffer, "Native Tobacco Mosaic Virus Protein." By the mid-1960s, the subunit molecular weight was revised down from 18,000 to 17,000, associated with a polypeptide of 158 amino acids. See Watson, *Molecular Biology of the Gene,* 134, 343. The current molecular weight listed for the coat protein is 17,500, obtained from the following website: http://biology.anu.edu.au/Groups/MES/vide/descr803.htm. Sequencing of the RNA showed its exact length to be 6,395 bases. Goelet et al., "Nucleotide Sequence."

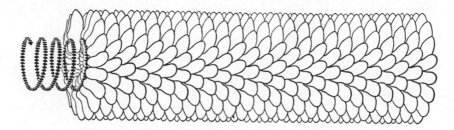

FIGURE 7.22 Schematic diagram of a portion of a TMV rod. The coat of protein subunits is depicted by the radially arranged white structures; these surround a strand of RNA, represented by the black helix. (From Fraenkel-Conrat, "Genetic Code of a Virus," [1964]: 47. Reprinted by permission of Irving Geis/The Geis Archives Trust. Rights owned by Howard Hughes Medical Institute—not to be used without permission.)

the nucleic acid, as well as with an increasing reliance on the electron microscope to visually "assay" for complete particles. The biological significance of the emphasis on functional parts was compelling:

> We have here what might be regarded as a beautifully perfected "mutual protection society." The protein units by themselves can aggregate, at about pH 5, to form the typical helical structure and rods of indeterminate length. But these aggregates are quite sensitive to pH and labile to heat and enzymes. The incorporation of RNA, however, produces a structure of finite length (3,000 Å) which is stable over the pH range of 3 to 9 and is remarkably resistant to proteolytic enzymes and heat. In addition, the RNA, which is normally very susceptible to attack by enzymes such as ribonuclease, is quite inaccessible to these agents when incorporated in the virus structure.[229]

The conceptualization of both protein and nucleic acid structure in terms of their sequences drew not only from the visibility of the "coding problem" but also from the rapid development of protein sequencing techniques in the 1950s. In the new emphasis on protein sequencing, TMV research followed the exemplar of Sanger's insulin, but the heightened interest in nucleic acids also opened new opportunities for TMV as a model system. As F. M. Burnet and Stanley asserted in 1959, "If insulin has been the protein par excellence for the unraveling of a structure which can serve as a model of other proteins of greater biological significance, so we can feel confident that the logical choice for nucleic acid is the RNA of tobacco mosaic virus or possibly that of one of the small spherical viruses,

229. Fraenkel-Conrat and Stanley, "Chemistry of Life. Part 2," 141.

such as tomato bushy stunt or turnip yellow mosaic."[230] Indeed, Fraenkel-Conrat and other protein chemists at the Virus Laboratory rapidly turned their attention to nucleic acids, particularly once techniques for mutagenizing the viral RNA had been developed by Mundry and Gierer. As the Virus Laboratory provided the nexus for the establishment of a new Department of Molecular Biology in Berkeley in 1964, virus structure and analysis of the role of nucleic acids were viewed as central and complementary aspects of the research program.[231]

More broadly, the shift to considering the functional pieces of the virus became part of the "informational turn" that was a distinctive feature of the emergence of molecular biology in the 1950s. Assessing the historical significance of information language and concepts in TMV research proves difficult, especially because, as Lily Kay has pointed out, this trend arrived relatively late to the Virus Laboratory, and did not disturb the prevailing laboratory practices.[232] It is instructive to compare the response of Stanley's laboratory to that of other prominent biochemical research centers of the late 1950s. In her study of the metaphors employed by the German virologists, Christina Brandt has argued that among the most important functions of the use of "code" and "information" by Schramm and then Wittmann was to connect their results to others in molecular biology, particularly with those of the phage group.[233] Specifically, an emphasis on the genetic information carried by TMV helped to foreground the similarities between their finding that TMV RNA was infective and the conclusion from the Hershey-Chase experiment that DNA was the genetic portion of bacteriophage. Looking at the British situation, Soraya de Chadarevian has analyzed the assimilation of notions of biological "information" at the MRC Unit in Cambridge University. She argues that an emphasis upon protein sequence provided important common ground central to defining the new Laboratory of Molecular Biology as an institutional marriage of structural biology (predominantly X-ray crystallography) and biochemistry, as represented by Sanger's investigation of insulin.[234] For Crick and Brenner, "sequence" referred to the material expression, in both nucleic acids and protein, of molecular genetic information. For Sanger, this emphasis on sequence became a new way to talk about long-standing interests in structure. Crick's "sequence hypothesis," articulated in 1957, thus served to provide an overarching

230. Burnet and Stanley, "Problems of Virology," 6.

231. For more on the institutional developments, see Creager, "Stanley's Dream," esp. 355–58.

232. Kay, *Who Wrote the Book of Life?* 186–87.

233. Brandt, "Tobacco Mosaic Virus."

234. de Chadarevian, "Sequences, Conformation, Information."

vision for the disparate tools and research practices folded into molecular biology.[235]

At the Virus Laboratory, the attention to sequence similarly drew on and redefined the emphasis on chemical structure characterizing Stanley's group since the 1930s. Moreover, the fact that TMV contained the nucleic acid that specified the virus polypeptide meant that all of the elements were present to approach the coding problem through chemical means. But the turn to information helped to consolidate TMV research and molecular biology in another way as well. For Fraenkel-Conrat and other protein chemists, the emphasis on "information" provided a way to reconceptualize the problem of protein structure. In their classic experiment published in 1959, F. H. White and Christian Anfinsen demonstrated that denatured RNAse would re-fold into its active form when placed into appropriate solution conditions (and this despite the absence of the sulfhydryl bond found in the native enzyme).[236] Not long afterward, TMV researchers recognized that many of the techniques they had used for separating virus protein from nucleic acid were denaturing the polypeptide. F. A. Anderer showed that following denaturation, a "structureless" TMV protein chain could reacquire its native structure.[237]

Thus biochemists began to speak of the "information" contained in the sequence of amino acids that directed the folding of the polypeptide into its active conformation. Crick, perhaps the most astute analyst of these developments, provided early theoretical justification for this trend:

> By information [in protein synthesis] I mean the specification of the amino acid sequence of the protein. It is conventional at the moment to consider separately the synthesis of the polypeptide chain and its folding. It is of course possible that there is a special mechanism for folding up the chain, but the more likely hypothesis is that the *folding is simply a function of the order of the amino acids,* provided it takes place as the newly formed chain comes off the template.[238]

This reframing of the problem of how proteins acquire their structure, while further confounding any attempts to systematize or mathematize the largely metaphorical uses of "information" in biology, helped to include proteins as well as nucleic acids in the new world of informational

235. Crick, "On Protein Synthesis."
236. White and Anfinsen, "Relationships of Structure to Function."
237. Anderer, "Reversible Denaturierung"; Tsugita and Fraenkel-Conrat, "Contributions from TMV Studies," 482.
238. Crick, "On Protein Synthesis," 144; emphasis in original.

agents.[239] The assimilation of "information" into research discourses and practices became part of the uneasy cohabitation of biochemists and self-described molecular biologists in the burgeoning world of biomedical research in the late 1950s and 1960s. TMV provided a ready experimental model that both biochemists and molecular geneticists could engage, as well as a public symbol of the scientific success of investigating life in molecular terms.

239. On the philosophically problematic uses of "information" in molecular biology, see Sarkar, "Biological Information."

TMV as an Experimental Model

To investigate the nature of a virus, the scientist must single out an appropriate one for study. Usually he speaks of his selection as a research tool, or model. A good many plant viruses had been discovered since the time of Iwanowski and Beijerinck, but the tobacco-mosaic virus—commonly known as TMV in the trade—was Stanley's choice for a number of reasons.

Greer Williams, *Virus Hunters*, 1959

When Stanley attempted the chemical isolation of a virus in the 1930s, TMV turned out to be a highly fortuitous choice. Because the virus is an abundant and very stable constituent of mosaic-diseased tobacco leaves, TMV could be isolated in pure form and characterized in physicochemical terms as a homogeneous, rod-shaped nucleoprotein. Yet the impact of Stanley's isolation of TMV and the productive research program it set in motion cannot be explained solely by the material properties of the virus. As we have seen, the direction of Stanley's research with TMV relied on the importation of specific tools and concepts, as well as on the analogies being made to other systems. Physicochemical instruments for biological research, many of which had been developed through the Rockefeller Foundation's sponsorship, were crucial for the investigation and representation of the virus as a macromolecule. The fact that TMV provided an exemplar for understanding other pathogens (such as poliovirus and influenza virus) and other biological entities (such as genes and cytoplasmic particles) affected both how TMV was investigated and how findings with TMV were generalized to other systems.

My argument that one can relate developments in different laboratories through the common use of exemplars such as TMV—without reducing research objects to their mere biological properties—speaks to an ongoing dialogue among historians about the process through which experimental practices and new concepts are disseminated. During the past two decades, historians and philosophers of science have turned their attention to scientific practice, that is, to what researchers actually do at a given place and time, rather than focusing on conceptual developments abstracted from laboratory life.[1] In part, this focus took inspiration from pioneering ethnographies of science, in which social scientists took the laboratory as a site for observation and analysis.[2] Historians of biology could not go "observe" their subjects in a similar way, but they could attend to the practical and material conditions of laboratory work, focusing on the local context in explaining historical developments. Studies of the dynamics of experimentation, the infrastructures necessary for conducting biomedical research, and the importance of scientific instruments took center stage.[3] Collections on both the "right tools" and the "right organisms" for the job epitomized the new historiography.[4] This chapter reflects on the story of TMV to probe the recent historical literature on biological experimentation, with an eye to the relevance of these scholarly discussions to the ascendance of the "model systems approach" in contemporary biomedical research.

From Practice to Experimental Systems

As Jan Golinski has observed, the increase in the size and scale of physics research in this century has been paralleled in biological experimentation by "an increasing complexity of its component elements."[5] The production of knowledge in contemporary experimental biology relies on systems that link together instruments, techniques, materials, specialized reagents, and skilled practitioners. In articles aimed at probing these experimental configurations, Robert Kohler and Hans-Jörg Rheinberger each looked to the biologists' idiomatic expression, "experimental system."[6] Other historians and philosophers of biology, myself included,

1. Here, of course, I am referring to Bruno Latour and Steve Woolgar's influential book *Laboratory Life.* Through both *Laboratory Life* and *Science in Action,* Latour inspired—and provoked—a generation of scholars to pay attention to what scientists actually do.

2. Knorr-Cetina, *Manufacture of Knowledge;* Lynch, *Art and Artifact;* Traweek, *Beamtimes and Lifetimes.*

3. For fine examples of studies of each of the three issues I mention, see Holmes, *Meselsohn, Stahl, and the Replication of DNA;* Clarke, "Research Materials"; and Rasmussen, *Picture Control.* Several essays in Gaudillière and Löwy, *Invisible Industrialist,* explore the last two topics as well. On experimentation and instruments beyond the life sciences, see Gooding, Pinch, and Schaffer, *Uses of Experiment;* Van Helden and Hankins, *Instruments;* Galison, *Image and Logic;* and Lenoir and Lécuyer, "Instrument Makers and Discipline Builders."

4. Clarke and Fujimura, *Right Tools;* Lederman and Burian, "Right Organism."

5. Golinski makes the point about increasing complexity in experimental science more generally, but his key examples are biological, and it is from this field that insights about "experimental systems" have emerged (*Making Natural Knowledge,* 139).

6. Kohler, "Systems of Production"; Rheinberger, "Experiment, Difference, and Writing."

have followed suit.[7] The term *system* draws attention to the heterogeneous ensembles of procedures, instruments, and objects involved in the process of research. The precise boundaries of the system remain difficult to specify. As Kohler comments, by *experimental system* biologists "mean something more than single instruments but less than the productive machinery of a laboratory or research field."[8] Rheinberger emphasizes their epistemological function: "Experimental systems are vehicles for materializing questions."[9]

This emphasis on experimental systems serves two important purposes for the historian. First, it provides a way to offer a more inclusive account of the social life of science than that available through conventional categories such as institutions, disciplines, and research schools. For example, throughout twentieth-century biology, scientific communities and networks of exchange (for both information and materials) have frequently centered on widely used organisms. The reliance on standardized organisms, which many researchers have used in constructing their experimental systems, is particularly visible in genetic research. At the beginning of the century, maize became the principal organism of choice for plant geneticists. Genetic standardization in the 1930s of *Mus musculus,* the laboratory mouse, enabled its increasingly dominant role in mammalian genetics.[10] Many of the achievements associated with the rise of bacterial genetics in the 1950s grew out of research on *Escherichia coli* (*E. coli*) and its associated viruses and episomes, making it "the species of choice for a wide range of biological problems."[11] The metabolic pathways and enzymes of *E. coli* were, in turn, objects of widespread biochemical investigation. Attention to these laboratory creatures, which were frequently standardized and sometimes commercialized, has been especially useful in accounting for how communities of researchers achieved consensus about biological knowledge in a world of diverse and complex organisms.

Second, the historiographical emphasis on research materials in experimentation—and in the resulting scientific knowledge—has been part of a response to (and, by some, a reaction against) the dominant

7. See for example, Rheinberger and Hagner, *Die Experimentalisierung des Lebens;* Hagner, Rheinberger, and Wahrig-Schmidt, *Objekte, Differenzen und Konjunkturen;* Creager and Gaudillière, "Meanings in Search of Experiments"; Fujimura, "Standardizing Practices"; Löwy, "Experimental Systems and Clinical Practices"; Stillwell, "Thymectomy as an Experimental System"; Silverstein, "Heuristic Value." See also Turnbull and Stokes, "Manipulable Systems."

8. Kohler, "Systems of Production," 88.

9. Rheinberger, *Toward a History of Epistemic Things,* 28.

10. On maize, see Comfort, *Tangled Field.* On the genetically standardized mouse, see Rader, *Making Mice;* and Löwy and Gaudillière, "Disciplining Cancer."

11. Lederberg, *"Escherichia coli,"* 230.

constructivist perspectives of the 1970s and 1980s.[12] Rather than accounting for science solely in terms of how its claims to truth have been *socially*
constructed, scholars have become interested in how knowledge-making
relates to the recalcitrance of "Nature" or, more agnostically, the material
world. Bruno Latour provided the most vivid example of this trend in his
insistence that Louis Pasteur "enrolled" the anthrax bacilli to partici-
pate in his successful demonstration of their role in disease causation.
Nonhuman actors, Latour provocatively asserted, deserve credit equal to
their human counterparts in historical accounts. However, Latour's semiotic interpretation of the roles of both the humans and nonhumans in his
story gave little regard to the actual practices of experimentation.[13]

More recently, Andrew Pickering has provided something of a manifesto for relating human and material agency in *The Mangle of Practice*.
In his view, the current preoccupation with the material world departs
from an older *representational* understanding of scientific activity that
remained essentially undisturbed by constructivism:

> The representational idiom casts science as, above all, an activity
> that seeks to represent nature, to produce knowledge that maps,
> mirrors, or corresponds to how the world really is. ... But there is
> quite another way of thinking about science. One can start from the
> idea that the world is filled not, in the first instance, with facts and
> observations, but with *agency*. The world, I want to say, is continually
> *doing things,* things that bear upon us not as observation statements
> upon disembodied intellects but as forces upon material beings.[14]

Whether one can entirely dispense with a representational view of science
remains to be seen. But, more important for our purposes, this perspective

12. Rheinberger, for instance, states, "Experimental scientists do not read the book of
nature, they do not depict reality. But they do not construct reality either. They are not engaged in platonistic exercises, in copy theory-guided asymptotic approximations to reality,
or in bluntly social constructivist endeavors" (*Toward a History of Epistemic Things*, 225). I
will not here attempt to summarize the development of constructivist approaches, particularly growing out of literature on the sociology of scientific knowledge; many other scholars
have already done so. For excellent overviews, see Golinski, *Making Natural Knowledge,*
esp. ch. 1; and Pickering, "From Science as Knowledge to Science as Practice." See also
Latour, *We Have Never Been Modern,* which I regard as an idiosyncratic but very insightful
overview.
13. Latour, "Give Me a Laboratory," and *Pasteurization of France*. In the latter, Latour
uses the semiotic term "actants" rather than actors. For further elaboration of Latour and
colleague Michel Callon's "actor network approach," see Callon, "Sociology of Translation";
and Callon and Latour, "Don't Throw the Baby Out." For criticisms of their ascription
of agency (and intentionality?) to nonhumans, see Schaffer, "Eighteenth Brumaire" and
Collins and Yearley, "Epistemological Chicken." For an account of Pasteur's work that
attends to experimental practice and detail, see Geison, *Private Science*.
14. Pickering, *Mangle of Practice,* 5–6; emphases in original.

insistently displaces agency in scientific knowledge-making from the researcher in favor of acknowledging the part played by the material world. Human and nonhuman agents are importantly different, in Pickering's view: nonhumans do not have intentionality, for instance.[15] Yet things *act* in the world (think of the weather, Pickering suggests), and in scientific settings, the activity of the world is often intertwined with the interventions of researchers.

Pickering's approach does not abandon the insights of constructivism altogether. Humans carefully construct and control the environments in which research takes place; scientific encounters with the agency of nature are thus modified by their occurrence in the simplified and standardized space of laboratories.[16] Neither is contingency in science downplayed in Pickering's view of the "mangle." Particularly important to his argument is the *temporality* of scientific practice—the fact that material agency emerges in "real time." Not only is there no way to predict at the outset of an experiment what the outcome will be, but goals are revised and strategies replaced in the process of human-material interaction.[17] As Pickering puts it, this means that "scientific practice is visibly path-dependent."[18] As Rheinberger notes, Pickering's insistence that scientific activity unfolds in real time agrees with Kuhn's argument against a teleological view of science. "Scientific development must be seen as a process driven from behind, not pulled from ahead—as evolution from, rather than evolution towards."[19]

Both of these points pertain directly to the analytic use of experimental systems. First, people assemble experimental systems; they are *material constructs* rather than natural objects.[20] In cases where experimental systems rely on genetically standardized organisms or biochemically purified

15. This asymmetry is important to the theory Pickering develops. As Theodore Schatzki notes, humans "remain the center of the action in Pickering's case studies. . . . 'Resistance,' for instance, is defined relative to human intentionality" (Schatzki, "To Mangle," 160).

16. As Latour and Woolgar asserted in 1979, "It is not simply that phenomena *depend on* certain material instrumentation; rather, the phenomena *are thoroughly constituted by* the material setting of the laboratory. The artificial reality, which participants describe in terms of an objective entity, has in fact been constructed by the use of inscription devices" (*Laboratory Life,* 64; emphases in original). Knorr-Cetina made the same point early on; see "Ethnographic Study of Scientific Work." For similar analysis of the field sciences, see Kohler, "Place and Practice."

17. Pickering, *Mangle of Practice,* 21–24. For analysis of this point, see Turner, "Practice in Real Time."

18. Pickering, "Explanation and the Mangle," 168.

19. Kuhn, "Road Since Structure," 7, as quoted in Rheinberger, "Reenacting History," 163.

20. I use the phrase *material construct* to point to both similarities and differences with the common statement that scientific facts are social constructs. Clearly, one cannot refer to a material construct as a fiction, but the role of human craftsmanship is equally undeniable.

fractions, the research subjects themselves do not exist (as "found" objects) outside the laboratory.[21] Second, experimental systems unfold in time, and in unpredictable ways. Both Kohler and Rheinberger emphasize the element of surprise inherent in biological experimental systems.[22] Because of these aspects of artificiality and contingency, an experimental systems approach does not reinstate a simple scientific realism.[23]

Kohler's 1994 book, *Lords of the Fly,* exemplified these insights through an account of how fruit flies came to play such a central role in experimental genetics.[24] Kohler draws on the approach of environmental historians in viewing the laboratory as yet another landscape of human habitation, one which was "colonized" by the fruit fly, *Drosophila melanogaster.*[25] Having "crossed the threshold" into Morgan's laboratory, *Drosophila* displaced other creatures inhabiting the workspace as it was domesticated and standardized as a scientific instrument for chromosomal mapping.[26] Kohler argues that by its material nature, *Drosophila* encouraged collaborative work in Morgan's laboratory and cooperative networks beyond it. In particular, the fecundity of the fly was such that each strain generated more mutants than any one scientist could study (Kohler calls the fly a "breeder reactor"); in order to stay on top of the game, researchers developed patterns of sharing materials and information and shipping fly strains to other interested researchers and teachers. In this respect, the social dynamics of experimental genetics as a field were derived in large part from the nature of the discipline's premier experimental system.[27] At the same time, the constraints of *Drosophila,* once

21. Laboratory organisms such as mice have been selected and inbred for generations, so that there are not genetic counterparts to them in the "wild." On "organisms as technology," see Kohler, *Lords of the Fly,* 6–8.

22. Kohler, "Systems of Production"; Rheinberger, "Experiment, Difference, and Writing." This issue is explored further below.

23. In addition, an interest in the agency of materials does not preclude an appreciation for constructivism's set of methodological tools; see Kohler, "Constructivists' Tool Kit," reviewing (approvingly) Golinski, *Making Natural Knowledge.*

24. Kohler, *Lords of the Fly.*

25. Indeed, as I have suggested elsewhere ("In the Fly Room"), the new relationship between scientist and fly depicted by Kohler might best be described as one of mutual subjugation.

26. Kohler's depiction of the fruit fly as both "natural" and "technological" draws inspiration from the work of environmental historians Donald Worster (*Dust Bowl;* "Transformations of the Earth"; "Seeing Beyond Culture") and William Cronon ("Modes of Prophecy"; *Nature's Metropolis;* "Place for Stories"). As Kohler puts it, "Landscapes are technologies; technological workplaces have natural histories. Laboratories of experimental biology have that same dual character" (*Lords of the Fly,* 11). Environmental historians, like historians of science, have sought to provide accounts that combine human and natural agency.

27. Drawing on E. P. Thompson ("Moral Economy") and, more recently, Shapin's "House of Experiment," Kohler refers to the norms prevailing in the Drosophilists'

reconstructed as a genetic instrument, led some researchers to move on to other organisms. For example, George Beadle, after struggling with Boris Ephrussi to train the fly for biochemical genetic experimentation, began working with *Neurospora crassa* (this time with collaborator Edward Tatum) because this bread mold proved more productive of biochemical knowledge.

This account of the emergence of a classic model organism largely shifts the cause of experimental developments (indeed, the motor of scientific change itself) from the ingenious researcher to the productive material— fly, mouse, mold. Rhetorically, Kohler's argument is carried in part by the fact that flies are organisms: they are mobile, they have a natural history, they have actual biological agency. Thus there is a congruence between the figural and the literal in speaking of the way in which *Drosophila* colonized the genetics research laboratory. Similarly, Karen Rader's account of the standardization of the laboratory mouse, Bonnie Clause's study of the Wistar rat, and Rachel Ankeny's history of developmental genetics based upon the nematode worm are all animated by actual reproducing organisms.[28] There are equally important differences emphasized by the historical accounts of these experimental organisms. For example, by the 1940s standardized laboratory mice were produced and sold by the Jackson Laboratory by the tens of thousands, yet, unlike in *Drosophila* research, scientists used them for a wide range of purposes—more as pure chemical reagents than as scientific instruments.[29] Even so, one can analyze major currents in the history of biology by focusing on the scientific stabilization of these creatures both within and beyond the laboratories that "domesticated" them for research.[30]

So historians have been interested in the degree to which organisms as experimental subjects direct the course of science. But there are also diverse experimental systems in biology that cannot be conflated with or signified by a living organism. Rheinberger's history of research on protein synthesis offers just such a case, recounting Paul Zamecnik's investigations of protein synthesis using cellular components of rat liver tissue. Rheinberger seeks to distinguish the experimental system, as a highly local configuration of instruments, assays, and materials, from the organism or material from which a system is constructed (be it rat liver, *E. coli,* or *Drosophila*). He refers to the ostensible object of study as the *epistemic thing,* which depends upon, but is not the same as, the system through

community in terms of a "moral economy" of science (*Lords of the Fly,* 11–13, and "Moral Economy, Material Culture").

28. Rader, *Making Mice;* Clause, "Wistar Rat"; Ankeny, *Conqueror Worm.*

29. Löwy and Gaudillière, "Disciplining Cancer."

30. Rader, "Mice, Medicine, and Genetics," esp. 321–23, and *Making Mice,* ch. 1.

which it is being investigated. Epistemic objects are not static, nor do they exist outside the experimental situation. In the case he examines, the epistemic thing was soluble RNA, eventually differentiated and reconceptualized into several molecular entities in the early 1960s, most notably transfer RNA. The experimental system in Zamecnik's laboratory encompassed the materials and practices through which the RNA was visualized: the rat liver tissue, ultracentrifuges, radioisotopes, scintillation counters, and other items and methods employed in the experimentation.

Rheinberger points out the complex epistemological relationship linking these concrete materials and operations to the elusive object (cytoplasmic RNA and its function) under investigation. As he puts it, "Paradoxically, epistemic things embody what one does not yet know. Scientific objects have the precarious status of being absent in their experimental presence; they are not simply hidden things to be brought to light through sophisticated manipulations."[31] But if they stand for unknown entities, epistemic objects are not merely scientists' conceptual constructs; rather Rheinberger insists that "the reality of epistemic things lies in their resistance, their capacity to turn around the (im)precisions of our foresight and understanding."[32]

Particularly important to Rheinberger's scientific epistemology is the way in which the experimental system leads the researcher in unforeseen directions, toward conceptual transformations that the scientist could not have anticipated. Such shifts and unexpected turns often appear in the laboratory's space of representation through inscriptions, or *graphemes*.[33] The process of connecting these emerging results with the shifting conception of the object is a scientific "tracing game," in analogy with Ludwig's Wittgenstein's memorable description of writing as a "language-game."[34] This aspect of Rheinberger's depiction of research draws from his serious engagement with postpositivist and poststructuralist philosophy.[35] At the simplest level, the lack of fixed external referents in his vision of scientific experimentation has its counterpart in the slippery nature of language, in which meanings cannot be fixed, and unanticipated readings subvert the author's intentions. For Rheinberger, it is the inherent *instability* of an experimental system, the constant displacements, which make it useful for research. Indeed, as soon as "nature" is stabilized sufficiently

31. Rheinberger, *Toward a History of Epistemic Things,* 28.
32. Ibid., 23.
33. The emphasis on inscriptions is indebted in part to Latour and Woolgar, *Laboratory Life*.
34. Rheinberger, *Toward a History of Epistemic Things,* 21; Wittgenstein, *Philosophical Investigations*.
35. In addition to the relevance of Rheinberger's training and work as a philosopher, he was the translator of Derrida's *De la grammatologie* into German.

for an epistemic object to be understood, it has ceased to be scientifically interesting. At this point, the epistemic object may become a technical object, used in the experimental setup to search out new epistemic objects. In this sense, Rheinberger suggests that standardization itself provides little explanatory power for understanding how new scientific knowledge emerges from biological experimentation.[36]

From Systems as Machines to Systems as Models

Kohler and Rheinberger offer two distinct conceptions of the experimental system as a historical category, and their methodological approaches draw on very different literatures—Kohler using environmental history and work-oriented sociology of science, and Rheinberger invoking deconstruction. Yet both present the experimental system as a kind of *machine* for research: once established, the system exhibits a self-generating nature. Kohler elaborates the analogy between experimental creatures and scientific instruments, arguing that genetically standardized organisms in the laboratory "most resemble spectrophotometers, bubble chambers, ultracentrifuges, and other physical instruments."[37] Rheinberger is warier of technological language, for he maintains that scientific and technical objects are epistemologically different, yet he espouses François Jacob's memorable line that experimental systems are "machines for making the future."[38]

In agreement with this depiction of the experimental system as sitting in the driver's seat of biological research, we have seen several instances in which viruses as experimental subjects can be said to have taken researchers to unanticipated destinations. Most notably, Max Delbrück's experimental system of synchronized bacteriophage growth, designed to enable direct access to the mechanism of virus reproduction, unexpectedly developed into an extraordinary tool for understanding bacterial genetics. Delbrück and Luria's experiments did not resolve the "riddle of life" as quickly as they had planned, but they did open up an extensive and highly productive field of research for a new generation of geneticists. Another unanticipated opening in virus research occurred when Ralph Wyckoff recognized that the crystalline precipitate pulled down by sedimenting tobacco-mosaic-infected sap in the air-driven ultracentrifuge was purified virus. This unexpected result led to the use of the ultracentrifuge to isolate TMV and a wide range of other viruses.

36. For the historiographical context of this issue, see Laubichler and Creager, "How Constructive Is Deconstruction?" 134–35.
37. Kohler, *Lords of the Fly,* 6. The genetically standardized creatures he has in mind here are "*Drosophila,* white mice and rats, maize, *E. coli,* or *Neurospora.*"
38. Rheinberger, *Toward a History of Epistemic Things,* 28; Jacob, *Statue Within,* 9.

However, the second example also points to the problem with ceding all of the interesting moves to the experimental system itself. Once Wyckoff saw the potential utility of the ultracentrifuge for isolating viruses, he and many others adapted this machine to lines of research far removed from Stanley's TMV laboratory. Scientists do not work with their experimental systems in isolation, either from the institutions that support them or, even more important, from their colleagues and competitors. Rarely is the "next step" in an experimental trajectory dictated by the experimental system. Many innovations in experimentation arise when the researcher adopts strategies, methods, or concepts from someone else's experimental system.

We have seen numerous examples of the strategic reconfiguring of one experimental system in an attempt to benefit from the successes achieved with another. By treating his virus as an enzyme, after the example of digestive enzymes isolated as crystalline proteins by his colleague John Northrop, Stanley obtained TMV as a pure protein. (Granted, in doing so he missed the nucleic acid, an oversight whose significance grew over time.) Rather than perceiving this kind of emulation as "derivative" science, we should recognize the scientific ingenuity in the everyday practice of finding and identifying workable models for innovative experimentation. (Note that I use the term *model* in the sense of *example* or exemplar, not a three-dimensional structural model, mechanical model, symbolic generalization, or mathematical model.)

A related point is that adopting an experimental model nearly always requires adaptation—model systems are not strict templates but rather are resources for creative borrowing and the elaboration of previously unseen analogies. The comparisons being made in the 1940s between viruses and microsomal particles reinforced shared practices for isolating cytoplasmic entities as macromolecular particles (particularly through the use of the ultracentrifuge and electron microscope), but the material model of viruses did not "pathologize" microsomes. In fact, the analogy in the end reshaped the model: perceived similarities between animal tumor viruses and normal cell constituents tended to undermine the presumption that viruses were exogenous agents.

Of course, it should also be noted that having a good model or experimental referent does not guarantee that similar results and successes can be achieved in another system. When Northrop attempted to utilize the air-driven ultracentrifuge to isolate phage, in the wake of Stanley and Wyckoff's success with TMV, the results were disappointing. To take a second example, clear-cut chemical differences between the virus strains that Knight was able to identify using TMV and its variants could not be easily detected among strains of T3 bacteriophage and poliovirus. The material properties of experimental subjects matter as much as instruments and analogies do. The point of my emphasis on the activity of modeling in experimentation is not to deny the agency of materials, but rather to argue that it is intertwined with human choices in the management of experimental systems.[39]

By drawing attention to these kinds of modeling activities in research, my motivation has been to link up the focus on experimental systems as units of research with larger scientific changes and transformations. When following a single research trajectory, one can subsume even the scientists' actions into the "experimental system." However, when looking to experimental systems as interacting elements in a broader scientific story, one must be able to speak to the decisions of scientists, who are the principal agents operating on and between different experimental systems.[40] Rheinberger refers to the mutual reshaping of experimental systems in terms of their "conjunctures," and speaks of the way systems are "defined with respect to, and in distinction from, other neighboring experimental systems."[41] I view the kind of modeling we have seen with TMV as one form of the conceptual analogies and material adaptations that scientists normally engage in, enabling such conjunctures to occur. One way that agreement between laboratories is achieved is that experimental systems are modeled upon each other, contributing to the observations of similar biological phenomena in many different species, cells, pathways, or genes.[42] Thus, results "travel" and are generalized through these kinds of material convergences.

At another level, the adoption of a particular experimental system as a model for investigating another object or process often provides the link between "basic" research and "real-world" problems. We saw several examples in which TMV research provided an example for laboratory work on human pathogens. As head of the Committee on Scientific Research for the National Foundation for Infantile Paralysis in 1938,

39. Pickering refers to this back-and-forth between human intervention and material agency as a "dialectic of resistance and accommodation" (*Mangle of Practice*, 22–23). Pickering also refers to this process as a kind of "modeling," but my emphasis here is not so much on the interactions between scientist and material as on the resources the researcher brings to bear on his or her system from other examples and experimental setups—as models.

40. My interest here in the day-to-day choices and activities of science as work is deeply indebted to the writings of those who have extended symbolic interactionism to understand science (e.g., Fujimura, "Constructing Doable Problems," and "Molecular Biology Bandwagon"; Clarke and Gerson, "Symbolic Interactionism in Science Studies"; Star and Griesemer, "Institutional Ecology"). Kohler similarly acknowledges these sociologists of science in providing the production-oriented concepts important in his history of fly genetics; see *Lords of the Fly*, 5.

41. Rheinberger, *Toward a History of Epistemic Things*, 135.

42. For a detailed example of this point, see Creager and Gaudillière, "Meanings in Search of Experiments."

Thomas Rivers sought to fund research on poliovirus in biophysical and biochemical terms, as with TMV. Not long after, as American scientists were mobilized for World War II, Stanley began investigating influenza along the lines of his work on TMV, extending the ultracentrifuge's success in isolating pure viruses to the task of preparing a cleaner influenza vaccine. The continuities were carried only in part by his instrument, the air-driven ultracentrifuge; this machine was not standardized, and Stanley chose to adapt the procedure to a less-specialized Sharples centrifuge. The analogy of influenza to TMV as a macromolecule crucially sustained the shift to flu vaccine development. Stanley's postwar research program then used TMV as both a substitute and a guide for investigating poliovirus, although the approach from TMV research had to be modified (ultracentrifugation failed to concentrate poliovirus). Pointing to the role of TMV as a model is not to suggest that clinical research is a derivative of "basic" laboratory research, but to emphasize key points of continuity and overlap in the back-and-forth of experimental practice between these different problems and domains.

To sum up, one of the unacknowledged virtues of focusing on experimental systems has to do with their frequent roles in producing exemplars of theory and material models for research on other biological and clinical objects.[43] Analyzing life science in terms of particular experimental systems not only brings into focus the material realities and constraints of biological research but also illuminates the interchanges, both instrumental and conceptual, between different lines of work, and between different research spaces—university laboratory, hospital, and agricultural experiment station.

Model Systems

Throughout the book, I have used the terms *model system* and *model object* to highlight the incidents in which TMV was taken as a model for research on other experimental systems. This terminology is admittedly imprecise. Just as the boundaries of an experimental system may be hard to specify (e.g., Are human actions inclusive or exclusive? Do experimental systems stop at the laboratory door?), the term *model system* may too easily conflate the research object (fly, virus, gene) with the research apparatus required to produce or visualize it.[44] Indeed, this kind

43. For a consonant vision of mathematical models as representing both theory and the world at the same time, see Morgan and Morrison, *Models as Mediators*. Their emphasis on the partial autonomy of models also mirrors the experimental systems approach.

44. Rheinberger's differentiation of the "epistemic object" from the experimental system provides one way to tame this chronic conflation of object and system, although "epistemic object" is similarly both material and representational; see *Toward a History of Epistemic Things*.

of slippage is ubiquitous in how biologists actually speak of experimental systems and model systems.[45] This ambiguity, however, points to an important issue in the use of such systems as referents. While model systems remain deeply rooted in their material instantiations, they are, in part, idealizations. To refer to one's system as *Drosophila melanogaster* is not to call up one specific fly. Neither is the referent entirely abstract. Standing behind the "fly" are many actual flies, and likely many different strains of flies, in conjunction with the relevant practices of genetic experimentation. As Kohler puts it, "Genetic instruments [like *Drosophila*] are extended systems, consisting of large and fecund families of mutants designed for particular purposes and freighted with ever-growing bodies of craft knowledge and skills."[46] Thus, when researchers refer to their organism as a model system, the experimental system lies just in the background; a set of practices is implicit.[47]

The accumulation of practice and knowledge that becomes embedded in the use of particular organisms gives model systems their self-reinforcing quality. As Joshua Lederberg has observed in accounting for the postwar popularity of *E. coli*, "Soon, the very accumulation of knowledge, mostly concentrated on a single strain, 'K-12,' made it more likely that it would be a prototype for still further studies."[48] Two important issues should be noted with reference to the model system as "prototype." The more a model system is investigated, and the more numerous the perspectives from which it is understood, the more useful it becomes as a tool for asking detailed and complex questions.[49] Second, model systems are often treated as "typical," in terms of the vital process they exemplify or the disease they manifest.[50]

If the power of model systems derives in part from the practices and know-how that have accrued with their use, they are also supported by the analogies drawn to humans and their health problems. In the world

45. See Bowker, "Model Systems"; Coleman, Goff, and Stein-Taylor, *Algae as Experimental Systems;* Hanken, "Model Systems versus Outgroups"; Kellogg and Shaffer, "Model Organisms"; Koshland, "Biological Systems"; and White and Caskey, "Human as an Experimental System." I am indebted to Rachel Ankeny, Emily Brock, and Gail Schmitt for bringing some of these references to my attention.

46. Kohler, "Systems of Production," 127.

47. One might include even low-level description as part of a model system. Rachel Ankeny argues that the wiring diagram specifying all the connections between neurons in the nematode (*Caenorhabditis elegans*) is itself part of the model system. This wiring diagram is not based on one particular worm; it is a canonical nervous system based on the neurological connections observed in many standard specimens (Ankeny, "Model Organisms").

48. Lederberg, *"Escherichia coli,"* 231.

49. I am indebted to discussions with Norton Wise on this point.

50. For scientific criticisms of this assumption in the field of developmental biology, see Bowker, "Model Systems."

of postwar biomedicine, the way in which a researcher's system has been made into a *model* for a particular disease or vital process is key to its place in biomedical research; it represents the concrete justification linking the government-sponsored mission of improving health and the day-to-day activities of the laboratory scientists fulfilling this purpose through basic research. As we have seen, voluntary health agencies played a historic role in shaping the emergence of this kind of ecology of knowledge for virus research.[51] Prior to the pronounced growth of federal government support for biomedical research in the 1950s, which employed a disease, or "categorical," framework for funding basic research, public philanthropies such as the National Foundation for Infantile Paralysis and the American Cancer Society began funding research on selected "basic" systems that could be viewed as models for pathogens or disease processes. Such a funding strategy encouraged the investigation of actual disease agents along the lines of the basic models while setting a precedent for a funding policy that recognized the value of model systems. Historians are only beginning to relate specific developments in postwar life science to the flowering of federal support for biomedical research; analyzing specific model systems is one approach to examining the connections forged between laboratory practice and the postwar political economy of health-related research.

The manner in which the relevance of a model system is established may not be straightforward. Interestingly, Stanley himself used the term *model system* most explicitly when justifying the use of laboratory viruses to understand human cancer. In this case, the analogy between the laboratory situation and the disease being modeled was particularly contested.[52] Human cancer, after all, fails to exhibit the patterns of contagion or epidemics typical of other viral diseases. Moreover, cancer is not a single disease, but a cluster of diseases having to do with unregulated cell growth. Thus one etiology would be unlikely to explain all incidents of human cancer, a fact not lost on medical practitioners and clinical researchers who questioned viral explanations of cancer. Nonetheless, Stanley strongly advocated the importance to cancer research of laboratory models based on tumor viruses, testifying before Congress in 1959 to urge their support of tumor virus research through the National Cancer Institute.[53] As he explained a few years later in a lecture for the American Cancer Society,

51. On "ecology of knowledge," see Rosenberg, "Toward an Ecology of Knowledge."
52. Jean-Paul Gaudillière has analyzed the historical complexities of developing animal models for human cancer, particularly an animal model for the viral transmission of cancer, in "Circulating Mice and Viruses."
53. See Gaudillière, "Molecularization of Cancer Etiology," esp. 155.

> I continue to be especially interested in model systems in which the conversion from the normal cell to a cancer cell by means of a virus can be studied in the laboratory in great detail under exactly controlled conditions. Two systems which are especially favorable are concerned with the polyoma virus and with the Rous sarcoma virus.... I believe that work on model systems such as those of polyoma and the Rous sarcoma virus may provide ideas useful in studies on human cancer-virus relationships.[54]

What these model systems offered, in Stanley's view, were experimental tools and concepts that might be used to understand human cancer, even where tumor viruses would not provide a comprehensive model for the disease.

The basic study of tumor viruses did in fact become an intensive area of research in the 1960s and 1970s, although its significance as a source for cancer prevention and therapy remained contested politically and medically.[55] Moreover, the model systems devised to study tumor viruses, including SV40 and adenovirus, became principal tools in molecular biology for understanding eukaryotic gene regulation. Tumor viruses were also at the center of debates in the 1970s over the possible public health dangers associated with genetic engineering, resulting in calls for a voluntary moratorium on cloning with cancer-related animal viruses.[56] Thus a more detailed story of tumor viruses as laboratory materials and model systems would illuminate the connections between Nixon's "War against Cancer," the changing practices of molecular genetics (particularly associated with the emergence of cloning), and the newly recognized significance of introns, exons, and transcriptional regulators in the expression of genes. In my examples of the use of TMV as a model system for research on human pathogens, the analogies between the plant virus and the human viruses have been quite direct. However, even in the many situations in which the relationship of a model system to health or illness is indirect—or mediated by additional animal or statistical models—historical analysis of the materials of research will illuminate previously invisible connections between laboratory practices, changing disease etiologies, and science policy.[57]

54. Stanley, "Viruses and Cancer," lecture presented at the 1962 Scientific Session of the American Cancer Society, New York City, 22 Oct 1962, Stanley papers, carton 20, folder "Manuscripts for talks."

55. See Patterson, *Dread Disease*.

56. See Wright, *Molecular Politics*. SV40 stands for simian virus 40.

57. I am indebted to Jean-Paul Gaudillière on this point. In addition to the articles cited above, see his insightful essay, "Oncogenes as Metaphors."

TWO SYSTEMATIC BIOLOGISTS, commenting upon the reliance of most of their laboratory colleagues on just a few species, opened an article with the following observations:

> Mendel used peas. Morgan used *Drosophila*. Delbrück used T4 phages. The idea was to learn about the general by studying the particular, to use a specific organism as a model for all others. The approach has been unquestionably powerful and successful. Laboratory-based studies of molecular and cellular biology have come to dominate the field of biology, using a handful of familiar laboratory models: *Escherichia coli,* yeast, maize, *Arabidopsis* (a cabbage relative), *Drosophila, Caenorhabditis* (nematode), *Xenopus* [a frog], and mouse. These are commonly thought of as *the* model organisms. The result has been a very large community of scientists using a very small group of organisms to answer questions fundamental to much of biology.[58]

The preferential use of a few special organisms to understand life in general has increasingly characterized twentieth-century biology. Researchers and science journalists frequently enumerate the current class of featured biological organisms as if their selection as exemplars was foreordained. The Human Genome Project, which has sequenced the genomes of a few laboratory organisms alongside that of representative *Homo sapiens,* has further reinforced the privilege accorded a few model creatures.[59] Hagiographic and teleological accounts of their centrality to biology and medicine are commonplace.[60]

The story of TMV might be told similarly, beginning with its priority as the first virus discovered. But, as we have seen, the circumstances that made TMV a valuable model system were diverse and unpredictable. Technological developments such as the appearance of the ultracentrifuge and electron microscope; conceptual developments such as the perceived similarity to genes that gave viruses a singular biological importance; and political developments—particularly the efforts by voluntary health agencies to raise public support for virus research—all contributed to the value of TMV. A closer view reveals that these more general factors played out through a series of specific analogies between TMV and other pathogens, objects, and systems. The result was that TMV was not a

58. Kellogg and Shaffer, "Model Organisms," 409.

59. The nonhuman organisms whose genomes were selected for complete sequencing through the Human Genome Project are *E. coli,* yeast (*Sacchomyces cerevisiae*), *Caenorhabditis elegans* (nematode worm), *Drosophila melanogaster,* and *Mus musculus* (mouse).

60. For one example of the public presentation of the key model organisms, see W. Maxwell Cowan, "Flies, Worms, and Human Genes" on the website of the Howard Hughes Medical Institute: http://www.hhmi.org/annual/research/fwhg.htm.

model for one thing, but for many, including primordial life forms, biological macromolecules, genes, plasmagenes, human pathogens, the genetic code, and molecular self-assembly. Today the list of its uses is different, but no less significant—TMV is a model of choice for studying gene expression in transgenic plants, cell-to-cell movement in plant cells, and virus resistance. It has also been developed into a commercial expression vector for the production of valuable therapeutic proteins in tobacco.[61] In the age of molecular genetics, TMV's usefulness as both a probe of plant physiology and a genetic instrument for biotechnology has taken this long-time resident of biomedical laboratories back out to the field. The diverse career of TMV demonstrates that long-lived model systems are continually reinvented in response to unexpected results, new strategies, and unseen opportunities. In other words, the experimental life of this virus has been illustrious, but never predestined.

61. See Atabekov et al., "Identification and Study of Tobacco Mosaic Virus Movement Function"; Citovsky, "Tobacco Mosaic Virus"; Erickson et al., "Interactions between Tobacco Mosaic Virus Protein and the Tobacco *N* Gene"; Beachy, "Coat-Protein-Mediated Resistance"; and Turpen, "Tobacco Mosaic Virus and the Virescence of Biotechnology."

Bibliography

Abir-Am, Pnina G. "The Discourse of Physical Power and Biological Knowledge in the 1930s: A Reappraisal of the Rockefeller Foundation's 'Policy' in Molecular Biology." *Social Studies of Science* 12 (1982): 341–82.

————. "Themes, Genres, and Orders of Legitimation in the Consolidation of New Scientific Disciplines: Deconstructing the Historiography of Molecular Biology." *History of Science* 23 (1985): 73–117.

————. "Synergy or Clash: Disciplinary and Marital Strategies in the Career of Mathematical Biologist Dorothy Wrinch." In *Uneasy Careers and Intimate Lives: Women in Science, 1789–1979*, ed. Pnina G. Abir-Am and Dorinda Outram. New Brunswick: Rutgers University Press, 1987, 239–80.

————. "The Politics of Macromolecules: Molecular Biologists, Biochemists, and Rhetoric." *Osiris* (2d ser.) 7 (1992): 164–91.

————. "From Multidisciplinary Collaboration to Transnational Objectivity: International Spaces as Constitutive of Molecular Biology, 1930–1970." In *Denationalizing Science: The Contexts of International Scientific Practice,* ed. Elisabeth Crawford, Terry Shinn, and Sverker Sörlin. Sociology of the Sciences Yearbook. Vol. 16. Dordrecht, the Netherlands: Kluwer Academic Publishers, 1993, 153–86.

————. " 'New' Trends in the History of Molecular Biology." *Historical Studies in the Physical and Biological Sciences* 26 (1995): 167–96.

Abraham, Tara. "From Plasmagenes to Steady States: Cybernetic Metaphors and the Emergence of Genetic Regulation, 1945–1961." Unpublished paper presented at the Joint Atlantic Seminar for the History of Biology, University of Pennsylvania, 10 Apr 1999.

Adams, Mark H. "Cold Spring Harbor Symposium on the Biological Nature of Viruses." *Science* 118 (1953): 66–67.

Ainsworth, Geoffrey C. *Introduction to the History of Plant Pathology.* Cambridge: Cambridge University Press, 1981.

Alexander, Jerome. *Colloid Chemistry: Theoretical and Applied.* 7 vols. New York: The Chemical Catalog, 1926–1950.

Alexander, Jerome, and Calvin B. Bridges. "Some Physicochemical Aspects of Life, Mutation, and Evolution." *Science* 70 (1929): 508–10.

Allard, H. A. "Effect of Dilution upon the Infectivity of the Virus of the Mosaic Disease of Tobacco." *Journal of Agricultural Research* 3 (1915): 295–99.

———. "Some Properties of the Virus of the Mosaic Disease of Tobacco." *Journal of Agricultural Research* 6 (1916): 649–74.

———. "A Specific Mosaic Disease in *Nicotiana viscosum* Distinct from the Mosaic Disease of Tobacco." *Journal of Agricultural Research* 7 (1916): 481–86.

———. "The Mosaic Disease of Tomatoes and Petunias." *Phytopathology* 6 (1916): 328–35.

———. "Effects of Various Salts, Acids, Germicides, etc., upon the Infectivity of the Virus Causing the Mosaic Disease of Tobacco." *Journal of Agricultural Research* 13 (1918): 619–37.

Allen, W. M. "The Isolation of Crystalline Progestin." *Science* 82 (1935): 89–93.

Altenburg, Edgar. "The Viroid Theory in Relation to Plasmagenes, Viruses, Cancer, and Plastids." *American Naturalist* 80 (1946): 559–67.

———. "The Symbiont Theory in Explanation of the Apparent Cytoplasmic Inheritance in Paramecium." *American Naturalist* 80 (1946): 661–62.

American Cancer Society. "Foreword." In *The Research Attack on Cancer, 1946.* A Report on the American Cancer Society Research Program by the Committee on Growth of the National Research Council. Washington, DC: National Research Council, 1946.

American Foundation. *Medical Research: A Midcentury Survey.* 2 vols. Boston: Little, Brown, 1955.

"The American Foundation Proposals for Medical Care." *Journal of the American Medical Association* 109 (1937): 1280–81.

Amsterdamska, Olga. "Beneficent Microbes: The Delft School of Microbiology and Its Industrial Connections." In Bos and Theunissen, *Beijerinck and the Delft School of Microbiology,* 193–213.

Anderer, F. A. "Reversible Denaturierung des Proteins aus Tabakmosaikvirus." *Zeitschrift für Naturforschung* 14B (1959): 642–47.

Anderer, F. A.; H. Uhlig; E. Weber; and G. Schramm. "Primary Structure of the Protein of Tobacco Mosaic Virus." *Nature* 186 (1960): 922–25.

Anderson, Philip W. "More Is Different." *Science* 177 (1972): 393–96.

Anderson, Thomas F. "Morphological and Chemical Relations in Viruses and Bacteriophages." *Cold Spring Harbor Symposia on Quantitative Biology* 11 (1946): 1–13.

———. "Electron Microscopy of Phages." In Cairns, Stent, and Watson, *Phage and the Origins,* 63–78.

Anderson, Thomas F., and Wendell M. Stanley. "A Study by Means of the Electron Microscope of the Reaction between Tobacco Mosaic Virus and Its Antiserum." *Journal of Biological Chemistry* 139 (1941): 339–44.

Andriewsky, P. "L'ultrafiltration et les microbes invisibles. Part 1. Communication: La peste des poules." *Zentralblatt für Bakteriologie, Parasitenkunde und Infektionskrankheiten,* Abt. I, Orig., 75 (1914): 90–93.

Anfinsen, Christian B.; Robert R. Redfield; Warren L. Choate; Juanita Page; and William R. Carroll. "Studies on the Gross Structure, Cross-Linkages, and Terminal Sequences in Ribonuclease." *Journal of Biological Chemistry* 207 (1954): 201–10.

Ankeny, Rachel. *The Conqueror Worm: An Historical and Philosophical Examination of the Use of the Nematode C. elegans as a Model Organism.* Ph.D. diss., University of Pittsburgh, 1997.

———. "Model Organisms as Case-Based Reasoning: Worms in Contemporary Biomedical Science." Unpublished paper presented at workshop entitled "Model Systems, Cases, and Exemplary Narratives," Program in History of Science, Princeton University, 2 Oct 1999.

Ansevin, Allan T., and Max A. Lauffer. "Native Tobacco Mosaic Virus Protein of Molecular Weight 18,000." *Nature* 183 (1959): 1601–2.

Appel, Toby A. *Shaping Biology: The National Science Foundation and American Biological Research, 1945–1975.* Baltimore: Johns Hopkins University Press, 2000.

Atabekov, J. G.; S. I. Malyshenko; S. Yu. Morozov; M. E. Taliansky; A. G. Solovyev; A. A. Agranovsky; and N. A. Shapka. "Identification and Study of Tobacco Mosaic Virus Movement Function by Complementation Tests." *Philosophical Transactions of the Royal Society of London* 354B (1999): 629–35.

Avery, Oswald T.; Colin M. MacLeod; and Maclyn McCarty. "Studies on the Chemical Nature of the Substance Inducing Transformation of Pneumococcal Types." *Journal of Experimental Medicine* 79 (1944): 137–58.

Ayres, Arthur U. "80,000 Revs per Minute? Modern Design Does It!" *Machine Design* 10 (Apr 1938): 17–19.

Bachrach, Howard L., and Carlton Schwerdt. "Purification Studies on Lansing Poliomyelitis Virus: pH Stability, CNS Extraction and Butanol Purification Experiments." *Journal of Immunology* 69 (1952): 551–61.

Barton-Wright, E., and Alan M. McBain. "Possible Chemical Nature of Tobacco Mosaic Virus." *Nature* 132 (1933): 1003–4.

———. "Possible Chemical Nature of Tobacco Mosaic Virus (response to Caldwell)." *Nature* 133 (1933): 260.

Bauer, J. H., and E. G. Pickels. "A High Speed Vacuum Centrifuge Suitable for the Study of Filterable Viruses." *Journal of Experimental Medicine* 64 (1936): 503–28.

Bawden, F. C. *Plant Viruses and Virus Diseases.* Leiden, Holland: Chronica Botanica Company, 1939.

———. *Plant Viruses and Virus Diseases.* 2d ed. Leiden, Holland: Chronica Botanica Company, 1943.

———. "Virus Diseases of Plants." *Journal of the Royal Society of Arts* 94 (1946): 136–68.

———. "Discussion: Mechanisms of Virus and Rickettsial Infections." In *International Symposium: The Dynamics of Virus and Rickettsial Infections,* ed. Frank W. Hartman, Frank L. Horsfall Jr., and John G. Kidd. New York: Blakiston, 1954, 52.

———. "Effect of Nitrous Acid on Tobacco Mosaic Virus: Mutation or Selection?" *Nature* 184 (1959): 27–29.

———. "Musings of an Erstwhile Plant Pathologist." *Annual Review of Phytopathology* 8 (1970): 1–12.

Bawden, F. C., and N. W. Pirie. "Experiments on the Chemical Behavior of Potato Virus 'X'." *British Journal of Experimental Pathology* 17 (1936): 64–74.

————. "The Isolation and Some Properties of Liquid Crystalline Substances from Solanaceous Plants Infected with Three Strains of Tobacco Mosaic Virus." *Proceedings of the Royal Society of London* 123B (1937): 274–320.

————. "Crystalline Preparations of Tomato Bushy Stunt Virus." *British Journal of Experimental Pathology* 19 (1938): 251–63.

————. "Contribution to Aggregation of Purified Tobacco Mosaic Virus." *Nature* 142 (1938): 842–43.

————. "Virus Multiplication Considered as a Form of Protein Synthesis." In *The Nature of Virus Multiplication,* Proceedings of the Second Symposium of the Society for General Microbiology. Cambridge: Cambridge University Press, 1953, 21–45.

Bawden, F. C.; N. W. Pirie; J. D. Bernal; and I. Fankuchen. "Liquid Crystalline Substances from Virus-Infected Plants." *Nature* 138 (1936): 1051–52.

Beachy, Roger N. "Coat-Protein-Mediated Resistance to Tobacco Mosaic Virus: Discovery Mechanisms and Exploitation." *Philosophical Transactions of the Royal Society of London* 354B (1999): 659–64.

Beadle, George W. "Biochemical Genetics." *Chemical Reviews* 37 (1945): 15–96.

Beale, Helen Purdy. *See* Purdy Beale, Helen.

Beams, Jesse W. "A Simple Ultracentrifuge." *Physical Review* 39 (1932): 858.

————. "High Speed Centrifuging." *Review of Modern Physics* 10 (1938): 245–63.

Beams, Jesse W., and Edward G. Pickels. "The Production of High Rotational Speeds." *Review of Scientific Instruments* 6 (1935): 299–308.

Beams, Jesse W., and A. J. Weed. "A Simple Ultra-Centrifuge." *Science* 74 (1931): 44–46.

Beams, J. W.; A. J. Weed; and E. G. Pickels. "The Ultracentrifuge." *Science* 78 (1933): 338–40.

Beard, Joseph W. "Review: Purified Animal Viruses." *Journal of Immunology* 58 (1948): 49–108.

Beard, Joseph W.; W. Ray Bryan; and Ralph W. G. Wyckoff. "The Isolation of the Rabbit Papilloma Virus." *Journal of Infectious Diseases* 65 (1939): 43–52.

Beard, Joseph W., and Ralph W. G. Wyckoff. "The Isolation of a Homogeneous Heavy Protein from Virus-Induced Rabbit Papillomas." *Science* 85 (1937): 201–2.

Beatty, John. "Opportunities for Genetics in the Atomic Age." Paper presented at the Fourth Mellon Workshop on Comparative Perspectives on the History and Social Studies of Modern Life Science, 10 Apr 1994.

Beijerinck, M. W. "Über ein Contagium vivum fluidum als Ursache der Fleckenkrankheit der Tabaksblätter." *Verhandelingen der Koninklijke Akademie van Wetenschappen te Amsterdam, Afdeeling Natuurkunde* 6 (1898): 3–21. Trans. James Johnson and reprinted as "Concerning a Contagium vivum fluidum as Cause of the Spot Disease of Tobacco Leaves." *Phytopathological Classics* 7 (1942): 33–52.

————. "Nachweis der Violaceusbakterien." *Folia Microbiologica* 4 (1916): 207–10.

Benison, Saul. *Tom Rivers: Reflections on a Life in Medicine and Science: An Oral History Memoir.* Cambridge, MA: MIT Press, 1967.

———. "The History of Polio Research in the United States: Appraisal and Lessons." In Holton, *Twentieth-Century Sciences,* 308–43.

Benzer, Seymour. "The Elementary Units of Heredity." In *A Symposium on the Chemical Basis of Heredity,* ed. William D. McElroy and Bentley Glass. Baltimore: Johns Hopkins University Press, 1957, 70–93.

Bernal, J. D., and I. Fankuchen. "X-Ray and Crystallographic Studies of Plant Virus Preparations, Part 1. Introduction and Preparation of Specimens, Part 2. Modes of Aggregation of the Virus Particles, Part 3." *Journal of General Physiology* 25 (1944): 111–65.

Berol, David N. *Living Materials and the Structural Ideal: The Development of the Protein Crystallography Community in the Twentieth Century.* Ph.D. diss., Princeton University, 2001.

Beurton, Peter J.; Raphael Falk; and Hans-Jörg Rheinberger, eds. *The Concept of the Gene in Development and Evolution: Historical and Epistemological Perspectives.* Cambridge: Cambridge University Press, 2000.

Beutner, R. *Life's Beginnings on the Earth.* Baltimore: Williams & Wilkins, 1938.

Biscoe, J.; Edward G. Pickels; and Ralph W. G. Wyckoff. "An Air-Driven Ultracentrifuge for Molecular Sedimentation." *Journal of Experimental Medicine* 64 (1936): 39–45.

———. "Light Metal Rotors for the Molecular Ultracentrifuge." *Review of Scientific Instruments* 7 (1936): 246–50.

Bos, Lute. "The Embryonic Beginning of Virology: Unbiased Thinking and Dogmatic Stagnation." *Archives of Virology* 140 (1995): 613–19.

———. "100 Years of Virology: From Vitalism via Molecular Biology to Genetic Engineering." *Trends in Microbiology* 8 (2000): 82–87.

Bos, Pieter, and Bert Theunissen. *Beijerinck and the Delft School of Microbiology.* Delft, the Netherlands: Delft University Press, 1995.

Bowker, Jessica. "Model Systems in Developmental Biology." *Bioessays* 17 (1995): 451–55.

Brandt, Christina. "Tobacco Mosaic Virus and the Genetic Code: A Case Study in the Epistemology of Metaphors." Unpublished paper presented at the meeting of the International Society for the History, Philosophy, and Social Studies of Biology in Oaxaca, Mexico, 7–10 Jul 1999.

Brenner, Sydney. "On the Impossibility of All Overlapping Triplet Codes in Information Transfer from Nucleic Acid to Proteins." *Proceedings of the National Academy of Sciences, USA* 43 (1957): 687–94.

Brock, Thomas D. *The Emergence of Bacterial Genetics.* Cold Spring Harbor, NY: Cold Spring Harbor Laboratory Press, 1990.

Bulloch, William. *The History of Bacteriology.* London and New York: Oxford University Press, 1938.

Burian, Richard M. "How the Choice of Experimental Organism Matters: Epistemological Reflections on an Aspect of Biological Practice." *Journal of the History of Biology* 26 (1993): 351–68.

———. "'The Tools of the Discipline: Biochemists and Molecular Biologists': A Comment." *Journal of the History of Biology* 29 (1996): 451–62.

Burian, Richard M.; Jean Gayon; and Doris Zallen. "The Singular Fate of Genetics in the History of French Biology, 1900–1940." *Journal of the History of Biology* 21 (1988): 357–402.

Burnet, F. M. "Influenza Virus on the Developing Egg. Part 1. Changes Associated with the Development of an Egg-Passage Strain of Virus." *British Journal of Experimental Pathology* 17 (1936): 282–93.

Burnet, F. M., and Wendell M. Stanley. "The Problems of Virology." In *The Viruses: Biochemical, Biological, and Biophysical Properties,* ed. F. M. Burnet and W. M. Stanley, vol. 1, *General Virology.* New York: Academic Press, 1959, 1–14.

Bush, Vannevar. *Science—The Endless Frontier . . . July 1945.* Washington, DC: National Science Foundation, 1960.

Butenandt, Adolf. "The Historical Development of Modern Virus Research in Germany, Especially in the Kaiser-Wilhelm-/Max Planck-Society, 1936–1954." *Medical Microbiology and Immunology* 164 (1977): 3–14.

Butenandt, Adolf, and Gerhard Schramm. "Über die Bromierung des Δ^5-cholestenon-dibromids." *Berichte der deutschen chemischen Gesellschaft* 69B (1936): 2289–99, and *Chemical Abstracts* 31 (1937): 697.

Butenandt, Adolf; Gerhard Schramm; Alexander Wolff; and Helmut Kudszus. "Einige Bemerkungen zur Kenntnis gebromter Sterinketon." *Berichte der deutschen chemischen Gesellschaft* 69B (1936): 2779–83, and *Chemical Abstracts* 31 (1937): 1425.

Butler, J. A. V. "Rockefeller Institute for Medical Research, Princeton." *Nature* 162 (1948): 479–80.

Cairns, John; Gunther S. Stent; and James D. Watson, eds. *Phage and the Origins of Molecular Biology.* Cold Spring Harbor, NY: Cold Spring Harbor Laboratory Press, [1966] 1992.

Caldwell, John. "Studies in the Physiology of Virus Diseases in Plants. Part 4. The Nature of the Virus Agent of Aucuba or Yellow Mosaic of Tomato." *Annals of Applied Biology* 20 (1933): 100–116.

———. "Possible Chemical Nature of Tobacco Mosaic Virus." *Nature* 133 (1934): 177.

———. "The Agent of Virus Disease in Plants." *Nature* 138 (1936): 1065–66.

Callon, Michel. "Some Elements of a Sociology of Translation: Domestication of the Scallops and the Fishermen of St. Brieuc Bay." In *Power, Action, and Belief: A New Sociology of Knowledge?* ed. John Law. Boston: Routledge, 1986, 196–233.

Callon, Michel, and Bruno Latour. "Don't Throw the Baby Out with the Bath School! A Reply to Collins and Yearley." In Pickering, *Science as Practice and Culture,* 343–68.

Campbell, C. Lee; Paul D. Peterson; and Clay S. Griffith. *The Formative Years of Plant Pathology in the United States.* St. Paul, MN: American Phytopathological Society Press, 1999.

Carlson, Elof Axel. "An Unacknowledged Founding of Molecular Biology: H. J. Muller's Contributions to Gene Theory, 1910–1936." *Journal of the History of Biology* 4 (1971): 149–70.

———. *Genes, Radiation, and Society: The Life and Work of H. J. Muller.* Ithaca, NY: Cornell University Press, 1981.

Carlson, W. Bernard. "Innovation and the Modern Corporation." In *Science in the Twentieth Century,* ed. John Krige and Dominique Pestre. Amsterdam: Harwood Academic Publishers, 1997, 203–26.

Carsner, Eubanks. "Attenuation of the Virus of Sugar Beet Curly-Top." *Phytopathology* 15 (1925): 745–57.

Carson, John. "Minding Matter/Mattering Mind: Knowledge and the Subject in Nineteenth-Century Psychology." *Studies in History and Philosophy of Biological and Biomedical Sciences* 30C (1999): 345–76.

Caspar, Donald L. D. "Radial Density Distribution in the Tobacco Mosaic Virus Particle." *Nature* 177 (1956): 928.

Caspersson, Torbjörn, and Jack Schultz. "Nucleic Acid Metabolism of the Chromosomes in Relation to Gene Reproduction." *Nature* 142 (1938): 294–95.

Cedarbaum, Daniel Goldman. "Paradigms." *Studies in History and Philosophy of Science* 14 (1983): 173–213.

Centanni, Eugenio. "Die Vogelpest: Beitrag zu dem durch Kerzen filtrierbaren Virus." *Zentralblatt für Bakteriologie, Parasitenkunde und Infektionskrankheiten,* Abt. I, Orig., 31 (1902): 145–52, 182–201.

Chantrenne, H. "Problems of Protein Synthesis." In *The Nature of Virus Multiplication: Second Symposium of the Society for General Microbiology Held at Oxford University, April 1952,* ed. Paul Fildes and W. E. van Heyningen. Cambridge: Cambridge University Press, 1953, 1–20.

Chargaff, Erwin. "On the Chemistry and Function of Nucleoproteins and Nucleic Acids." *Rendiconti dell' Istituto Lombardo Accademia di Scienze e Lettere. Parte generale e atti ufficiali* 89 (1955): 101–15.

———. "Amphisbaena." In *Essays on Nucleic Acids.* New York: Elsevier, 1963, 174–99.

Citovsky, Vitaly. "Tobacco Mosaic Virus: A Pioneer of Cell-to-Cell Movement." *Philosophical Transactions of the Royal Society of London* 354B (1999): 637–43.

Clarke, Adele E. "Research Materials and Reproductive Science in the United States, 1910–1940." In *Physiology in the American Context, 1850–1940,* ed. Gerald L. Geison. Bethesda, MD: American Physiological Society, 1987, 323–50. Reprinted in *Ecologies of Knowledge: Work and Politics in Science and Technology,* ed. Susan Leigh Star. Albany: SUNY Press, 1995, 183–225.

Clarke, Adele E., and Joan H. Fujimura, eds. *The Right Tools for the Job: At Work in Twentieth-Century Life Sciences.* Princeton, NJ: Princeton University Press, 1992.

Clarke, Adele E., and Elihu Gerson. "Symbolic Interactionism in Science Studies." In *Symbolic Interactionism and Cultural Studies,* ed. Howard S. Becker and Michael McCall. Chicago: University of Chicago Press, 1990, 170–214.

Claude, Albert. "Properties of the Causative Agent of a Chicken Tumor. Part 13. Sedimentation of the Tumor Agent and Separation from the Associated Inhibitor." *Journal of Experimental Medicine* 66 (1937): 59–72.

———. "Particulate Components of Cytoplasm." *Cold Spring Harbor Symposia on Quantitative Biology* 9 (1941): 263–71.

———. "Distribution of Nucleic Acids in the Cell and the Morphological Constitution of Cytoplasm." *Biological Symposia* 10 (1943): 111–29.

———. "Studies on Cells: Morphology, Chemical Constitution, and Distribution of Biochemical Functions." *Harvey Lectures* 43 (1948): 121–64.

Clause, Bonnie Toucher. "The Wistar Rat as a Right Choice: Establishing Mammalian Standards and the Ideal of a Standardized Mammal." *Journal of the History of Biology* 26 (1993): 329–49.

Cochran, W.; Francis H. C. Crick; and V. Vand. "The Structure of Synthetic Polypeptides. Part 1. The Transform of Atoms on a Helix." *Acta Crystallographica* 5 (1952): 581–86.

Cohen, Seymour S. "Growth Requirements of Bacterial Viruses." *Bacteriological Reviews* 13 (1949): 1–24.

――――. *Virus-Induced Enzymes.* New York: Columbia University Press, 1968.

――――. "Stanley, Wendell Meredith." In *Dictionary of Scientific Biography, Supplement,* ed. Charles C. Gillispie. New York: Scribner, 1990, 841–48.

Cohen, Seymour S., and Thomas F. Anderson. "Chemical Studies on Host-Virus Interactions. Part 1. The Effect of Bacteriophage Adsorption on the Multiplication of Its Host, Escherichia coli B." *Journal of Experimental Medicine* 84 (1940): 511–23.

Cohen, Seymour S., and W. M. Stanley. "The Action of Intestinal Nucleophosphatase on Tobacco Mosaic Virus." *Journal of Biological Chemistry* 142 (1942): 863–70.

――――. "The Molecular Size and Shape of the Nucleic Acid of Tobacco Mosaic Virus." *Journal of Biological Chemistry* 144 (1942): 589–98.

Coleman, Annette W.; Lynda J. Goff; and Janet R. Stein-Taylor, eds. *Algae as Experimental Systems.* New York: Alan R. Liss, 1989.

Collins, H. M., and Steven Yearley. "Epistemological Chicken." In Pickering, *Science as Practice and Culture,* 301–26.

Comfort, Nathaniel C. *The Tangled Field: Barbara McClintock's Search for the Patterns of Genetic Control.* Cambridge, MA: Harvard University Press, 2001.

Commoner, Barry. "The Biochemical Basis of Tobacco Mosaic Virus Infectivity." In *Biochemistry of Viruses,* Proceedings of the Fourth International Congress of Biochemistry, vol. 7, ed. E. Broda and W. Frisch-Niggemeyer. New York: Pergamon, 1959, 17–30.

Commoner, Barry; J. A. Lippincott; G. B. Shearer; E. E. Richman; and J.-H. Wu. "Reconstitution of Tobacco Mosaic Virus Components." *Nature* 178 (1956): 767–71.

Commoner, Barry; Masashi Yamada; Sidney D. Rodenberg; Tung-Yue Wang; and Eddie Basler Jr. "The Proteins Synthesized in Tissue Infected with Tobacco Mosaic Virus." *Science* 118 (1953): 529–34.

Corner, George W. *A History of the Rockefeller Institute, 1901–1953: Origins and Growth.* New York: Rockefeller Institute Press, 1964.

Cranefield, Paul F. "The Glass Electrode, the pH Meter, and Ion-Selective Electrodes." In *The Beckman Symposium on Biomedical Instrumentation,* ed. Carol L. Moberg. New York: The Rockefeller University in association with Beckman Instruments, Inc., 1986, 19–26.

Creager, Angela N. H. "In the Fly Room (Essay Review of *Lords of the Fly* by Robert E. Kohler)." *Historical Studies in the Physical and Biological Sciences* 25 (1995): 357–60.

――――. "Wendell Stanley's Dream of a Free-Standing Biochemistry Department at the University of California, Berkeley." *Journal of the History of Biology* 29 (1996): 331–60.

———. "Producing Molecular Therapeutics from Human Blood: Edwin Cohn's Wartime Enterprise." In de Chadarevian and Kamminga, *Molecularizing Biology and Medicine,* 107–38.

Creager, Angela N. H., and Jean-Paul Gaudillière. "Meanings in Search of Experiments and Vice-Versa: The Invention of Allosteric Regulation in Paris and Berkeley, 1959–1968." *Historical Studies in the Physical and Biological Sciences* 27 (1996): 1–89.

———. "Experimental Arrangements and Technologies of Visualization: Cancer as a Viral Epidemic, 1930–1960." In Löwy and Gaudillière, *Heredity and Infection,* 203–41.

Creager, Angela N. H., and Judith A. Swan. "Fashioning the Virus as a Chemical Object: Stanley, Authorship, and TMV." Unpublished paper presented at "What Is a Scientific Author?" Harvard University, 8 Mar 1997.

Crick, Francis H. C. "Nucleic Acids." *Scientific American* 197, no. 3 (Sep 1957): 188–200.

———. "On Protein Synthesis." In *The Biological Replication of Macromolecules.* Symposia of the Society for Experimental Biology. Cambridge: Cambridge University Press, 1958, 138–63.

———. "The Present Position of the Coding Problem." In *Structure and Function of Genetic Elements.* Proceedings of the June 1–3, 1959, Symposium. Upton, NY: Brookhaven National Laboratory Biology Department, 1959, 35–39.

Crick, Francis H. C.; John S. Griffith; and Leslie E. Orgel. "Codes without Commas." *Proceedings of the National Academy of Sciences, USA* 43 (1957): 416–21.

Crick, Francis H. C., and James D. Watson. "Structure of Small Viruses." *Nature* 177 (1956): 473–75.

Cronon, William. "Modes of Prophecy and Production: Placing Nature in History." *Journal of American History* 76 (1990): 1122–31.

———. *Nature's Metropolis: Chicago and the Great West.* New York: Norton, 1991.

———. "A Place for Stories: Nature, History, and Narrative." *Journal of American History* 78 (1992): 1347–76.

Dale, H. H. "The Biological Nature of the Viruses." *Nature* 128 (1931): 599–602.

Darlington, C. D. "Heredity, Development, and Infection." *Nature* 154 (1944): 164–69.

de Chadarevian, Soraya. "Sequences, Conformation, Information: Biochemists and Molecular Biologists in the 1950s." *Journal of the History of Biology* 29 (1996): 361–86.

———. "Following Molecules: Hemogloblin between the Clinic and the Laboratory." In de Chadarevian and Kamminga, *Molecularizing Biology and Medicine,* 171–201.

———. *Designs for Life: Molecular Biology after World War II.* Cambridge: Cambridge University Press, 2001.

de Chadarevian, Soraya, and Jean-Paul Gaudillière, "The Tools of the Discipline: Biochemists and Molecular Biologists." Introduction to a special issue under the same name. *Journal of the History of Biology* 29 (1996): 327–30.

de Chadarevian, Soraya, and Harmke Kamminga, eds. *Molecularizing Biology and Medicine: New Practices and Alliances, 1910s–1970s.* Amsterdam: Harwood Academic Publishers, 1998.

Deichman, Ute. *Biologists under Hitler.* Trans. Thomas Dunlap. Cambridge, MA: Harvard University Press, 1996.

De Kruif, Paul. *Microbe Hunters.* San Diego: Harcourt, Brace, Jovanovich, [1926] 1996.

———. *The Sweeping Wind: A Memoir.* New York: Harcourt, Brace & World, 1962.

Delbrück, Max. "Radiation and the Hereditary Mechanism." *American Naturalist* 74 (1940): 350–62.

———. "The Growth of Bacteriophage and Lysis of the Host." *Journal of General Physiology* 23 (1940): 643–60.

———. "A Theory of Autocatalytic Synthesis of Polypeptides and Its Application to the Problem of Chromosome Reproduction." *Cold Spring Harbor Symposia for Quantitative Biology* 9 (1941): 122–24.

———. "Bacterial Viruses (Bacteriophages)." *Advances in Enzymology and Related Subjects* 2 (1942): 1–32.

———. "Experiments with Bacterial Viruses (Bacteriophages)." *Harvey Lectures* 41 (1946): 161–87.

———. "Biochemical Mutants of Bacterial Viruses." *Journal of Bacteriology* 56 (1948): 1–16.

———. "Discussion." Colloques Internationaux du Centre National de la Recherche Scientifique, VIII, *Unités biologiques douées de continuité génétique.* Paris: Publications du Centre National de la Recherche Scientifique, 1949, 33–34.

———, ed. *Viruses 1950.* Pasadena, CA: California Institute of Technology, 1950.

———. "Introductory Remarks about the Program." *Cold Spring Harbor Symposia on Quantitative Biology* 18 (1953): 1–2.

———. "A Physicist's Renewed Look at Biology: Twenty Years Later [Nobel address delivered 10 Dec 1969]." *Science* 168 (1970): 1312–15.

Delbrück, Max, and W. T. Bailey. "Induced Mutations in Bacterial Viruses." *Cold Spring Harbor Symposia for Quantitative Biology* 11 (1946): 33–37.

Delbrück, Max, and Salvador E. Luria. "Interference between Bacterial Viruses. Part 1. Interference between Two Bacterial Viruses Acting upon the Same Host, and the Mechanism of Virus Growth." *Archives of Biochemistry* 1 (1942): 111–41.

Demerec, Milislav. "Foreword." *Cold Spring Harbor Symposia on Quantitative Biology* 18 (1953): v.

Dennis, Michael Aaron. "Accounting for Research: New Histories of Corporate Laboratories and the Social History of American Science." *Social Studies of Science* 17 (1987): 479–518.

Derrida, Jacques. *De la grammatologie.* Paris: Editions de Minuit, 1967.

d'Herelle, Félix. "Sur un microbe invisible antagoniste des bacilles dysentériques." *Comptes rendus hebdomadaires des Séances de l'Académie des Sciences* 165 (1917): 373–75.

———. *Le bactériophage: Son rôle dans l'immunité.* Paris: Masson et cie, 1921. Issued in English as *The Bacteriophage: Its Role in Immunity,* trans. George H. Smith. Baltimore: Williams & Wilkins, 1922.

———. *Le bactériophage et son comportement.* Paris: Masson et cie, 1926. Issued in English as *The Bacteriophage and Its Behavior,* trans. George H. Smith. Baltimore: Williams & Wilkins, 1926.

d'Herelle, Félix, and E. Peyre. "Contribution à l'étude des tumeurs expérimentales." *Comptes rendus hebdomadaires des Séances de l'Académie des Sciences* 185 (1927): 227–30.

———. "Contribution à l'étude des tumeurs spontanées." *Comptes rendus hebdomadaires des Séances de l'Académie des Sciences* 185 (1927): 513–15.

Doermann, A. H. "The Vegetative State in the Life Cycle of Bacteriophage: Evidence for Its Occurrence and Its Genetic Characterization." *Cold Spring Harbor Symposia on Quantitative Biology* 18 (1953): 3–11.

Drew, Elizabeth Brenner. "The Health Syndicate: Washington's Noble Conspirators." *Atlantic Monthly* 220 (Dec 1967): 75–82.

Dubuy, H. G., and M. W. Woods. "Evidence for the Evolution of Phytopathogenic Viruses from Mitochondria and Their Derivatives. Part 2. Chemical Evidence." *Phytopathology* 33 (1943): 766–77.

Duclaux, Emile. *Traité de microbiologie.* Paris: Masson, 1898.

Duggar, B. M., and Joanne L. Karrer. "The Sizes of the Infective Particles in the Mosaic Disease of Tobacco." *Annals of the Missouri Botanical Garden* 8 (1921): 343–56.

Duggar, B. M., and Joanne Karrer Armstrong, "Indications Respecting the Nature of the Infective Particles in the Mosaic Disease of Tobacco." *Annals of the Missouri Botanical Garden* 10 (1923): 191–212.

Dulbecco, Renato. "From Lysogeny to Animal Viruses." In Monod and Borek, *Of Microbes and Life,* 110–13.

Dulbecco, Renato, and Marguerite Vogt. "Plaque Formation and Isolation of Pure Lines with Poliomyelitis Viruses." *Journal of Experimental Medicine* 99 (1954): 167–82.

Dunn, L. C., ed. *Genetics in the Twentieth Century: Essays on the Progress of Genetics During Its First Fifty Years.* New York: Macmillan, 1951.

Dupree, A. Hunter. "The Great Instauration of 1940: The Organization of Scientific Research for War." In Holton, *Twentieth-Century Sciences,* 443–67.

Dvorak, Mayme. "The Effect of Mosaic on the Globulin of Potato." *Journal of Infectious Diseases* 41 (1927): 215–21.

Eckert, E. A.; Dorothy Beard; and Joseph W. Beard. "Dose-Response Relations in Experimental Transmission of Avian Erythromyeloblastic Leukosis. Part 3. Titration of the Virus." *Journal of the National Cancer Institute* 14 (1954): 1055–66.

Edsall, John T. "Proteins as Macromolecules: An Essay on the Development of the Macromolecular Concept and Some of Its Vicissitudes." *Archives of Biochemistry and Biophysics, Supplement* 1 (1962): 12–20.

Elford, W. J., and C. H. Andrewes. "The Sizes of Different Bacteriophages." *British Journal of Experimental Pathology* 13 (1932): 446–56.

Ellis, Emory L. "Bacteriophage: One-Step Growth." In Cairns, Stent, and Watson, *Phage and the Origins,* 53–62.

Ellis, Emory L., and Max Delbrück. "The Growth of Bacteriophage." *Journal of General Physiology* 22 (1939): 365–84.

Elzen, Boelie. "Two Ultracentrifuges: A Comparative Study of the Social Construction of Artefacts." *Social Studies of Science* 16 (1986): 621–62.

———. *Scientists and Rotors: The Development of Biochemical Ultracentrifuges,* Ph.D. diss., University of Twente, the Netherlands, 1988.

Endicott, Kenneth M., and Ernest M. Allen. "The Growth of Medical Research 1941–1953 and the Role of Public Health Service Research Grants." *Science* 118 (1953): 337–43.

Ephrussi, Boris. "Remarks on Cell Heredity." In Dunn, *Genetics in the Twentieth Century,* 241–62.

Erickson, F. L.; S. P. Dinesh-Kumar; S. Holzberg; C. V. Ustach; M. Dutton; V. Handley; C. Corr; and B. J. Baker. "Interactions between Tobacco Mosaic Virus and the Tobacco *N* Gene." *Philosophical Transactions of the Royal Society of London* 354B (1999): 653–58.

Eriksson-Quensel, Inga-Britta, and Theodor Svedberg. "Sedimentation and Electrophoresis of the Tobacco-Mosaic Virus Protein." *Journal of the American Chemical Society* 58 (1936): 1863–67.

Falk, Raphael. "What Is a Gene?" *Studies in History and Philosophy of Science* 17 (1986): 133–73.

Farley, John. *The Spontaneous Generation Controversy from Descartes to Oparin.* Baltimore: Johns Hopkins University Press, 1977.

———. "The International Health Division of the Rockefeller Foundation: The Russell Years, 1920–1934." In *International Health Organisations and Movements, 1918–1939,* ed. Paul Weindling. Cambridge: Cambridge University Press, 1995, 203–21.

Feffer, Stuart M. "Atoms, Cancer, and Politics: Supporting Atomic Science at the University of Chicago, 1944–1950." *Historical Studies in the Physical and Biological Sciences* 22 (1992): 233–61.

Fischer, Ernst Peter, and Carol Lipson. *Thinking about Science: Max Delbrück and the Origins of Molecular Biology.* New York: Norton, 1988. Based on an earlier German version published in 1985 by Universitätsverlag Konstanz.

Fleming, Donald. "Emigré Physicists and the Biological Revolution." *Perspectives in American History* 2 (1968): 152–89.

Flexner, Simon, and P. A. Lewis. "The Nature of the Virus of Epidemic Poliomyelitis." *Journal of the American Medical Association* 53 (1909): 2095.

Forman, Paul. "Behind Quantum Electronics: National Security as Basis for Physical Research in the United States, 1940–1960." *Historical Studies in the Physical and Biological Sciences* 18 (1987): 149–229.

Fox, Daniel M. "The Politics of the NIH Extramural Program, 1937–1950." *Journal of the History of Medicine and Allied Sciences* 42 (1987): 447–66.

Fraenkel-Conrat, Heinz. "The Reaction of Proteins with ^{14}C-Labelled N-Carboxyleucine Anhydride." *Biochimica et Biophysica Acta* 10 (1953): 180–82.

———. "Reaction of Nucleic Acid with Formaldehyde." *Biochimica et Biophysica Acta* 15 (1954): 307–9.

———. "The Reaction of Tobacco Mosaic Virus with Iodine." *Journal of Biological Chemistry* 217 (1955): 373–81.

———. "The Role of the Nucleic Acid in the Reconstitution of Active Tobacco Mosaic Virus." *Journal of the American Chemical Society* 78 (1956):
882–83.

———. "Rebuilding a Virus." *Scientific American* 194, no. 6 (Jun 1956): 42–47.

———. "Degradation and Structure of Tobacco Mosaic Virus." *Federation
Proceedings* 16 (1957): 810–15.

———. "The Infectivity of Tobacco Mosaic Virus Nucleic Acid." In *Cellular
Biology, Nucleic Acids, and Viruses: A Symposium in Honor of Basil O'Connor.*
Special Publications. Vol. 5. New York: New York Academy of Sciences, 1957,
217–25.

———. "Tobacco Mosaic, a Molecular Infection." In *Immunity and Virus Infection,* Symposium held at Vanderbilt University School of Medicine, 1–2 May
1958, ed. Victor A. Najjar. New York: John Wiley & Sons, 1959, 206–10.

———. "The Genetic Code of a Virus." *Scientific American* 211, no. 4 (Oct 1964):
47–54.

———. "The Impact of Wendell M. Stanley on the Biochemical Sciences." *Proceedings of the Robert A. Welch Foundation Conferences on Chemical Research*
20 (1977): 253–61.

———. "Protein Chemists Encounter Viruses." *Annals of the New York Academy
of Sciences* (special issue: *The Origins of Modern Biochemistry: A Retrospect
on Proteins,* ed. P. R. Srinivasan, Joseph S. Fruton, and John T. Edsall) 325
(1979): 309–15.

———. "The History of Tobacco Mosaic Virus and the Evolution of Molecular
Biology." In *The Plant Viruses,* ed. M. H. V. van Regenmortel and Heinz
Fraenkel-Conrat, vol. 2, *The Rod-Shaped Plant Viruses.* New York: Plenum
Press, 1986, 5–17.

Fraenkel-Conrat, Heinz; J. Ieuan Harris; and A. L. Levy. "Recent Developments
in Techniques for Terminal and Sequence Studies in Peptides and Proteins."
In *Methods of Biochemical Analysis,* vol. 2, ed. David Glick. New York: Interscience, 1955, 359–425.

Fraenkel-Conrat, Heinz, and Beatrice A. Singer. "The Peptide Chains of Tobacco
Mosaic Virus." *Journal of the American Chemical Society* 76 (1954): 180–83.

———. "Virus Reconstitution. Part 2. Combination of Protein and Nucleic Acid
from Different Strains." *Biochimica et Biophysica Acta* 24 (1957): 540–48.

———. "The Structural Basis of the Activity of Tobacco Mosaic Virus." In
Biochemistry of Viruses, Proceedings of the Fourth International Congress
of Biochemistry, vol. 7, ed. E. Broda and W. Frisch-Niggemeyer. New York:
Pergamon, 1959, 9–16.

Fraenkel-Conrat, Heinz; Beatrice A. Singer; and Robley C. Williams. "The
Nature of the Progeny of Virus Reconstituted from Protein and Nucleic Acid
of Different Strains of Tobacco Mosaic Virus." In *The Chemical Basis of
Heredity,* Symposium of the McCollum-Pratt Institute, ed. William D.
McElroy and Bentley Glass. Baltimore: Johns Hopkins University Press, 1957,
501–12.

Fraenkel-Conrat, Heinz, and Wendell M. Stanley. "The Chemistry of Life. Part 2.
Implications of Recent Studies of a Simple Virus." *Chemical and Engineering
News* 39 (15 May 1961): 136–44.

Fraenkel-Conrat, Heinz, and Robley C. Williams. "Reconstitution of Active Tobacco Mosaic Virus from Its Inactive Protein and Nucleic Acid Components." *Proceedings of the National Academy of Sciences, USA* 41 (1955): 690–98.

Fraile, Aurora; Fernando Escrui; Miguel A. Aranda; José M. Malpica; Adrian J. Gibbs; and Fernando García-Arenal. "A Century of Tobamovirus Evolution in an Australian Population of *Nicotiana glauca*." *Journal of Virology* 71 (1997): 8316–20.

Frampton, Vernon L. "On the Molecular Weight of the Tobacco Mosaic Virus Protein." *Science* 90 (1939): 305–6.

Frampton, Vernon L., and A. M. Saum. "An Estimate of the Maximum Value for the Molecular Weight of the Tobacco Mosaic Virus Protein." *Science* 89 (1939): 84–85.

Francis, Thomas, Jr. "Transmission of Influenza by Filtrable Virus." *Science* 80 (1934): 457–59.

Francis, Thomas, Jr.; Jonas Salk; and J. J. Quilligan Jr. "Experience with Vaccination Against Influenza in the Spring of 1947: A Preliminary Report." *American Journal of Public Health* 37 (1947): 1013–16.

Franklin, Rosalind E. "Structure of Tobacco Mosaic Virus." *Nature* 175 (1955): 379–81.

————. "Location of the Ribonucleic Acid in the Tobacco Mosaic Virus Particle." *Nature* 177 (1956): 928–30.

Franklin, Rosalind E.; Aaron Klug; and Kenneth C. Holmes. "X-ray Diffraction Studies of the Structure and Morphology of Tobacco Mosaic Virus." In *Ciba Foundation Symposium on the Nature of Viruses*, ed. G. E. W. Wolstenholme and E. C. P. Millar. London: Churchill Press, 1956, 39–55.

Fraser, Dean. *Viruses and Molecular Biology*. New York: Macmillan, 1967.

Fraser, Dean, and Robley C. Williams. "Details of Frozen-Dried T3 and T7 Bacteriophages as Shown by Electron Microscopy." *Journal of Bacteriology* 65 (1953): 167–70.

————. "Electron Microscopy of the Nucleic Acid Released from Individual Bacteriophage Particles." *Proceedings of the National Academy of Sciences, USA* 39 (1953): 750–53.

Friedewald, W. F., and Edward G. Pickels. "Centrifugation and Ultrafiltration Studies on Allantoic Fluid Preparations of Influenza Virus." *Journal of Experimental Medicine* 79 (1944): 301–17.

Friedrich-Freksa, Hans; Georg Melchers; and Gerhard Schramm. "Biologischer, chemischer und serologischer Vergleich zweier Parallelmutanten phytopathogener Viren mit ihren Ausgangsformen." *Biologisches Zentralblatt* 65 (1946): 187–222.

Fruton, Joseph S. *Molecules and Life: Historical Essays on the Interplay of Chemistry and Biology*. New York: Wiley-Interscience, 1972.

————. *A Skeptical Biochemist*. Cambridge, MA: Harvard University Press, 1992.

Fujimura, Joan H. "Constructing Doable Problems in Cancer Research: Articulating Alignment." *Social Studies of Science* 17 (1987): 257–93.

————. "The Molecular Biology Bandwagon in Cancer Research: Where Social Worlds Meet." *Social Problems* 35 (1988): 261–83.

———. "Standardizing Practices: A Socio-History of Experimental Systems in Classical Genetic and Virological Cancer Research, ca. 1920–1978." *History and Philosophy of the Life Sciences* 18 (1996): 3–54.

———. *Crafting Science: A Sociohistory of the Quest for the Genetics of Cancer.* Cambridge, MA: Harvard University Press, 1996.

Galison, Peter. *Image and Logic: A Material Culture of Microphysics.* Chicago: University of Chicago Press, 1997.

Galison, Peter, and Bruce Hevly, eds. *Big Science: The Growth of Large-Scale Research.* Stanford, CA: Stanford University Press, 1992.

Galperin, Charles. "Le bactériophage, la lysogénie et son déterminisme génétique." *History and Philosophy of the Life Sciences* 9 (1987): 175–224.

Gamow, George. "Possible Relation between Deoxyribonucleic Acid and Protein Structures." *Nature* 173 (1954): 318.

———. "Possible Mathematical Relations between Deoxyribonucleic Acid and Proteins." *Det Kongelige Danske Videnskabernes Selkab, Biologiske Meddelelsker,* bind 22 (1954): 1–13.

Gamow, George; Alexander Rich; and Martynas Yčas. "The Problem of Information Transfer from the Nucleic Acids to Proteins." *Advances in Biological and Medical Physics* 4 (1956): 23–68.

Gardner, David P. *The California Oath Controversy.* Berkeley: University of California Press, 1967.

Gaudillière, Jean-Paul. *Biologie moléculaire et biologistes dans les années soixante: La naissance d'une discipline. Le cas français.* Ph.D. diss., University of Paris, 1991.

———. "J. Monod, S. Spiegelman et l'adaptation enzymatique. Programmes de recherche, cultures locales et traditions disciplinaires." *History and Philosophy of the Life Sciences* 14 (1992): 23–71.

———. "Oncogenes as Metaphors for Human Cancer: Articulating Laboratory Practices and Medical Demands." In *Medicine and Social Change: Historical and Sociological Studies of Medical Innovation,* ed. Ilana Löwy. Montrouge, France, and London: John Libbey Eurotext, 1993, 213–47.

———. "Molecular Biology in the French Tradition? Redefining Local Traditions and Disciplinary Patterns." *Journal of the History of Biology* 26 (1993): 473–98.

———. "Wie man Labormodelle für Krebsentstehung konstruiert: Viren und Transfektion am (US) National Cancer Institute." In Hagner, Rheinberger, and Wahrig-Schmidt, *Objekte, Differenzen und Konjunkturen,* 233–57.

———. "Molecular Biologists, Biochemists, and Messenger RNA: The Birth of a Scientific Network." *Journal of the History of Biology* 29 (1996): 417–45.

———. "The Molecularization of Cancer Etiology in the Postwar United States: Instruments, Politics and Management." In de Chadarevian and Kamminga, *Molecularizing Biology and Medicine,* 139–70.

———. "Circulating Mice and Viruses: The Jackson Memorial Laboratory, the National Cancer Institute, and the Genetics of Breast Cancer, 1930–1965." In *The Practices of Human Genetics,* ed. Michael Fortun and Everett Mendelsohn. Sociology of the Sciences Yearbook. Vol. 21. Dordrecht, the Netherlands: Kluwer Academic Publishers, 1999, 89–124.

Gaudillière, Jean-Paul, and Ilana Löwy, eds. *The Invisible Industrialist: Manufactures and the Production of Scientific Knowledge.* London: Macmillan, 1998.

Geiger, Roger L. *Research and Relevant Knowledge: American Research Universities Since World War II.* New York and Oxford: Oxford University Press, 1993.

Geison, Gerald L. "The Protoplasmic Theory of Life and the Vitalist-Mechanist Debate." *Isis* 60 (1969): 273–92.

———. *The Private Science of Louis Pasteur.* Princeton, NJ: Princeton University Press, 1995.

Geison, Gerald L., and Angela N. H. Creager. "Introduction: Research Materials and Model Organisms in the Biological and Biomedical Sciences." *Studies in History and Philosophy of Biological and Biomedical Sciences* 30C (1999): 315–18.

Geison, Gerald L., and Manfred D. Laubichler. "The Varied Lives of Organisms: Variation in the Historiography of the Biological Sciences." *Studies in History and Philosophy of Biological and Biomedical Sciences* 32C (2001): 1–29.

Gibbs, Adrian. "Evolution and Origins of Tobamoviruses." *Philosophical Transactions of the Royal Society of London* 354B (1999): 593–602.

Gierer, Alfred. "Structure and Biological Function of Ribonucleic Acid from Tobacco Mosaic Virus." *Nature* 179 (1957): 1297–99.

Gierer, Alfred, and Karl-Wolfgang Mundry. "Production of Mutants of Tobacco Mosaic Virus by Chemical Alteration of its Ribonucleic Acid *in vitro.*" *Nature* 182 (1958): 1457–58.

Gierer, Alfred, and Gerhard Schramm. "Infectivity of Ribonucleic Acid from Tobacco Mosaic Virus." *Nature* 177 (1956): 702–3.

———. "Die Infektiosität der Nucleinsäure aus Tabakmosaikvirus." *Zeitschrift für Naturforschung* 11B (1956): 138–42.

Goelet, P.; G. P. Lomonossoff; P. J. G. Butler; M. E. Akam; M. J. Gait; and J. Karn. "Nucleotide Sequence of Tobacco Mosaic Virus RNA." *Proceedings of the National Academy of Sciences, USA* 79 (1982): 5818–22.

Goldsworthy, M. C. "Attempts to Cultivate the Tobacco Mosaic Virus." *Phytopathology* 16 (1926): 873–75.

Golinski, Jan. *Making Natural Knowledge: Constructivism and the History of Science.* Cambridge: Cambridge University Press, 1998.

Golomb, Solomon W.; Basil Gordon; and Lloyd R. Welch. "Comma-Free Codes." *Canadian Journal of Mathematics* 10 (1958): 202–9.

Gooding, David; Trevor Pinch; and Simon Schaffer, eds. *The Uses of Experiment: Studies in the Natural Sciences.* Cambridge: Cambridge University Press, [1989] 1993.

Gowen, John W. "Mutation in *Drosophila,* Bacteria, and Viruses." *Cold Spring Harbor Symposia on Quantitative Biology* 9 (1941): 189–93.

Gowen, John W., and W. C. Price. "Inactivation of Tobacco-Mosaic Virus by X-rays." *Science* 84 (1936): 536–37.

Grande, Francisco, and Carlos Asensio. "Severo Ochoa and the Development of Biochemistry." In *Reflections on Biochemistry: In Honour of Severo Ochoa,* ed. Arthur Kornberg, B. L. Horecker, L. Cornudella, and J. Oro. New York: Pergamon Press, 1976, 1–14.

Granick, S. "Plastid Structure, Development, and Inheritance." In *Handbuch der Pflanzenphysiologie,* ed. W. Ruhland. Berlin: Springer-Verlag, 1955, 507–64.

Gray, George W. "The Ultracentrifuge." *Scientific American* 184, no. 6 (Jun 1951): 42–51.

Green, R. H.; Thomas F. Anderson; and Joseph E. Smadel. "Morphological Structure of the Virus of Vaccinia." *Journal of Experimental Medicine* 75 (1942): 651–56.

Green, Robert G. "On the Nature of Filterable Viruses." *Science* 82 (1935): 443–45.

Greenberg, Daniel S. *The Politics of Pure Science.* Chicago: University of Chicago Press, [1968] 1999.

Grunberg-Manago, Marianne, and Severo Ochoa. "Enzymatic Synthesis and Breakdown of Polynucleotides; Polynucleotide Phosphorylase." *Journal of the American Chemical Society* 77 (1955): 3165–66.

Haddow, Alexander. "Transformation of Cells and Viruses." *Nature* 154 (1944): 194–99.

Hagner, Michael; Hans-Jörg Rheinberger; and Bettina Wahrig-Schmidt, eds. *Objekte, Differenzen und Konjunkturen: Experimentalsysteme im historischen Kontext.* Berlin: Akademie-Verlag, 1994.

Haldane, J. B. S. "The Origin of Life." In *Origin of Life* by J. D. Bernal. London: Weidenfeld and Nicolson, 1967. Originally published as "The Origin of Life," in *Rationalist Annual* (1929): 3–10.

———. *The Causes of Evolution.* London: Harper, 1932.

———. *Keeping Cool and Other Essays.* London: Chatto & Windus, 1940.

Hanken, James. "Model Systems versus Outgroups: Alternative Approaches to the Study of Head Development and Evolution." *American Zoologist* 33 (1993): 448–56.

Haraway, Donna Jeanne. *Crystals, Fabrics, and Fields: Metaphors of Organicism in Twentieth-Century Developmental Biology.* New Haven, CT: Yale University Press, 1976.

Harden, Victoria A. *Inventing the NIH: Federal Biomedical Research Policy, 1887–1937.* Baltimore: Johns Hopkins University Press, 1986.

Harding, T. Swann. "What Is Life?" *Scientific American* 156, no. 4 (Apr 1937): 234–36.

Harrington, William F., and Howard K. Schachman. "Studies on the Alkaline Degradation of Tobacco Mosaic Virus. Part 1. Ultracentrifugal Analysis." *Archives of Biochemistry and Biophysics* 65 (1956): 278–95.

Harris, J. Ieuan, and C. Arthur Knight. "Action of Carboxypeptidase on Tobacco Mosaic Virus." *Nature* 170 (1952): 613–14.

Harris, J. Ieuan, and Choh Hao Li. "The Biological Activity of Enzymatic Digests of Insulin." *Journal of the American Chemical Society* 74 (1952): 2945–46.

Harrison, Bryan D. "Frederick Charles Bawden: Plant Pathologist and Pioneer in Plant Virus Research." *Annual Review of Phytopathology* 32 (1994): 39–47.

———. "The Chemical Nature of Tobacco Mosaic Virus Particles." In Scholthof, Shaw, and Zaitlin, *Tobacco Mosaic Virus,* 71–73.

Hart, Roger G. "Electron-Microscopic Evidence for the Localization of Ribonucleic Acid in the Particles of Tobacco Mosaic Virus." *Proceedings of the
National Academy of Sciences, USA* 41 (1955): 261–64.

Hayes, William. "The Mechanism of Genetic Recombination in Escherichia coli."
Cold Spring Harbor Symposia on Quantitative Biology 18 (1953): 75–93.

Helvoort, Ton van. "What Is a Virus? The Case of Tobacco Mosaic Disease."
Studies in History and Philosophy of Science 22 (1991): 557–88.

――――. "The Controversy between John H. Northrop and Max Delbrück on the
Formation of Bacteriophage: Bacterial Synthesis or Autonomous Multiplication?" *Annals of Science* 49 (1992): 545–75.

――――. "A Bacteriological Paradigm in Influenza Research in the First Half of
the Twentieth Century." *History and Philosophy of the Life Sciences* 15 (1993):
3–21.

――――. "The Construction of Bacteriophage as Bacterial Virus: Linking Endogenous and Exogenous Thought Styles." *Journal of the History of Biology* 27
(1994): 91–139.

――――. "A Comment on the Early Influenza Virus Vaccines: The Role of the Con-
cept of Virus." In *Vaccinia, Vaccination and Vaccinology: Jenner, Pasteur and
Their Successors,* ed. Stanley A. Plotkin and Bernardino Fantini. Amsterdam:
Elsevier, 1996, 193–97.

――――. "A Century of Research into the Cause of Cancer." *History and Philosophy of the Life Sciences* 21 (1999): 293–330.

Herriott, Roger M. "Nucleic Acid-Free T2 Virus 'Ghosts' with Specific Biological
Action." *Journal of Bacteriology* 61 (1951): 752–54.

Hershey, Alfred D. "Spontaneous Mutations in Bacterial Viruses." *Cold Spring
Harbor Symposia for Quantitative Biology* 11 (1946): 67–77.

――――. "The Injection of DNA into Cells by Phage." In Cairns, Stent, and Watson,
Phage and the Origins, 100–108.

Hershey, Alfred D., and Martha Chase. "Independent Functions of Viral Protein
and Nucleic Acid in Growth of Bacteriophage." *Journal of General Physiology*
36 (1952): 39–56.

Hershey, Alfred D., and Raquel Rotman. "Linkage among Genes Controlling
Inhibition of Lysis in a Bacterial Virus." *Proceedings of the National Academy
of Sciences, USA* 34 (1948): 89–96.

――――. "Genetic Recombination between Host-Range and Plaque-Type Mutants
of Bacteriophage in Single Bacterial Cells." *Genetics* 34 (1949): 44–71.

Hesse, Mary B. *Models and Analogies in Science.* London: Sheed and Ward,
1963.

Hollinger, David A. "Free Enterprise and Free Inquiry: The Emergence of
Laissez-Faire Communitarianism in the Ideology of Science in the United
States." *New Literary History* 21 (1990): 897–919.

Holmes, Francis O. "Accuracy in Quantitative Work with Tobacco Mosaic Virus."
Botanical Gazette 86 (1928): 66–81.

――――. "Local Lesions in Tobacco Mosaic." *Botanical Gazette* 87 (1929): 39–55.

Holmes, Frederic L. "Seymour Benzer and the Definition of the Gene." In
Beurton, Falk, and Rheinberger, *Concept of the Gene,* 115–55.

———. *Meselsohn, Stahl, and the Replication of DNA: A History of "The Most Beautiful Experiment in Biology."* New Haven, CT: Yale University Press, 2001.

Holton, Gerald J., ed. *The Twentieth-Century Sciences: Studies in the Biography of Ideas.* New York: Norton, 1972.

Horgan, John. *The End of Science: Facing the Limits of Knowledge in the Twilight of the Scientific Age.* Reading, MA: Addison-Wesley, 1996.

Hotchkiss, Rollin D. "Discussion [of Fraenkel-Conrat's Paper]." In *Cellular Biology, Nucleic Acids, and Viruses, A Symposium in Honor of Basil O'Connor.* New York: New York Academy of Sciences, 1957, 226–27.

Hughes, Sally Smith. *The Origins and Development of the Concept of the Virus in the Late Nineteenth Century.* Ph.D. diss., University of London, 1972.

———. *The Virus: A History of the Concept.* New York: Science History Publications, 1977.

Hunger, F. W. T. "Untersuchungen und Betrachtungen über die Mosaikkrankheit der Tabakspflanze." *Zeitschrift für Pflanzenkrankheiten* 15 (1905): 257–311.

———. "Neue Theorie zur Ätiologie der Mosaikkrankheit des Tabaks." *Berichte der deutschen botanischen Gesellschaft* 23 (1905): 415–18.

Huxley, Thomas H. "On the Physical Basis of Life." *Fortnightly Review* n.s. 5 (1869): 129–45.

"Influenza Vaccines." *Journal of the American Medical Association* 134 (2 Aug 1947): 1177.

Iterson, G. van, Jr.; L. E. den Dooren de Jong; and A. J. Kluyver. *Martinus Willem Beijerinck: His Life and His Work.* Madison, WI: Science Tech., [1940] 1983.

Ivanovskii, Dmitri. "O dvukh bolezniakh tabaka. Tabachnaia pepelitsa. Mozaichnaia bolezn' tabaka." *Sel'skoe khoziiastvo i lesovodstvo* 189 (1892): 104–21.

———. "Über die Mosaikkrankheit der Tabakspflanze."*Bulletin de l'Académie Impériale des Sciences de St. Pétersbourg,* Nouv. Sér. 3, 35 (1892): 67–70. Translated and reprinted as "Concerning the Mosaic Disease of the Tobacco Plant." *Phytopathological Classics* 7 (1942): 27–30.

———. "Über die Mosaikkrankheit der Tabakspflanze." *Zentralblatt für Bakteriologie, Parasitenkunde und Infektionskrankheiten,* Abt. II 5 (1899): 250–54.

Jacob, François. *The Statue Within: An Autobiography.* Trans. Franklin Phillip. New York: Basic Books, 1988.

Jacob, François, and Elie L. Wollman. *Sexuality and the Genetics of Bacteria.* New York: Academic Press, 1961.

James, Laylin K., ed. *Nobel Laureates in Chemistry, 1901–1992.* Washington, DC: American Chemical Society and the Chemical Heritage Foundation, 1993.

Jeener, R.; P. Lemoine; and C. Lavand'Homme. "Détection et propriétés de formes du virus de la mosaïque du tabac dépourvues d'acide ribonucléique et non infectieuses." *Acta Biochimica et Biophysica* 14 (1954): 321–34.

Jenner, Edward. *An Inquiry into the Causes and Effects of the Variolae Vaccinae.* London: Sampson Low, 1798.

Jensen, J. H. "Isolation of Yellow-Mosaic Viruses from Plants Infected with Tobacco Mosaic." *Phytopathology* 23 (1933): 964–74.

Johannsen, Wilhelm. *Elemente der exakten Erblichkeitslehre.* Jena: Gustav Fischer, 1909.

Johnson, James, and Theodore J. Grant. "The Properties of Plant Viruses from Different Host Species." *Phytopathology* 22 (1932): 741–57.

Judson, Horace Freeland. *The Eighth Day of Creation: Makers of the Revolution in Biology.* Cold Spring Harbor, NY: Cold Spring Harbor Laboratory Press, [1979] 1996.

Kamminga, Harmke. "Studies in the History of Ideas on the Origin of Life." Ph.D. diss., University of London, 1980.

———. "The Protoplasm and the Gene." In *Clay Minerals and the Origins of Life,* ed. G. Cairns-Smith and H. Hartman. Cambridge: Cambridge University Press, 1986, 1–10.

Karl, Barry D., and Stanley M. Katz. "The American Private Philanthropic Foundation and the Public Sphere, 1890–1930." *Minerva* 14 (1981): 236–70.

Kausche, G. A.; E. Pfankuch; and H. Ruska. "Die Sichtbarmachung von pflanzlichem Virus im Übermikroskop." *Naturwissenschaften* 27 (1939): 292–99.

Kay, Lily E. "Conceptual Models and Analytical Tools: The Biology of Physicist Max Delbrück." *Journal of the History of Biology* 18 (1985): 207–46.

———. "The Secret of Life: Niels Bohr's Influence on the Biology Program of Max Delbrück." *Rivista di Storia della Scienza* 2 (1985): 487–510.

———. *Cooperative Individualism and the Growth of Molecular Biology at the California Institute of Technology.* Ph.D. diss., John Hopkins University, 1986.

———. "W. M. Stanley's Crystallization of the Tobacco Mosaic Virus, 1930–1940." *Isis* 77 (1986): 450–72.

———. "Laboratory Technology and Biological Knowledge: The Tiselius Electrophoresis Apparatus, 1930–1945." *History and Philosophy of the Life Sciences* 10 (1988): 51–72.

———. "Selling Pure Science in Wartime: The Biochemical Genetics of G. W. Beadle." *Journal of the History of Biology* 22 (1989): 73–101.

———. "Virus, enzyme ou gène? Le problème du bactériophage (1917–1947)." In *L'Institut Pasteur: Contributions à son histoire,* ed. Michel Morange. Paris: La Découverte, 1991, 187–97.

———. *The Molecular Vision of Life: Caltech, the Rockefeller Foundation, and the Rise of the New Biology.* New York and Oxford: Oxford University Press, 1993.

———. " 'Biochemists and Molecular Biologists: Laboratories, Networks, Disciplines': Comments." *Journal of the History of Biology* 29 (1996): 447–50.

———. *Who Wrote the Book of Life? A History of the Genetic Code.* Stanford, CA: Stanford University Press, 2000.

Keller, Evelyn Fox. "Physics and the Emergence of Molecular Biology: A History of Cognitive and Political Synergy." *Journal of the History of Biology* 23 (1990): 389–409.

———. "From Secrets of Life to Secrets of Death." In *Secrets of Life, Secrets of Death: Essays on Language, Gender, and Science.* New York: Routledge, 1992.

———. *Refiguring Life: Metaphors of Twentieth-Century Biology.* New York: Columbia University Press, 1995.

———. *A Century of the Gene.* Cambridge, MA: Harvard University Press, 2001.

Kellogg, Elizabeth A., and H. Bradley Shaffer. "Model Organisms in Evolutionary Studies." *Systematic Biology* 42 (1993): 409–14.

Kendrew, John C. "How Molecular Biology Started." *Scientific American* 216, no. 3 (Mar 1967): 141–44.

Kennedy, Donald. "What Animal Research Says about Cancer." *Human Nature* 1, no. 5 (1978): 84–89.

Kerker, Milton. "The Svedberg and Molecular Reality." *Isis* 77 (1986): 278–82.

Kevles, Daniel J. *The Physicists: The History of a Scientific Community in Modern America.* Cambridge, MA: Harvard University Press, [1971] 1995.

———. "Scientists, the Military, and the Control of Postwar Defense Research: The Case of the Research Board for National Security, 1944–46." *Technology and Culture* 16 (1975): 20–47.

———. "The National Science Foundation and the Debate over Postwar Research Policy, 1942–45: A Political Interpretation of *Science—The Endless Frontier.*" *Isis* 68 (1977): 5–26.

———. "Foundations, Universities, and Trends in Support for the Physical and Biological Sciences, 1900–1992." *Daedalus* 121 (1992): 195–235.

———. "Renato Dulbecco and the New Animal Virology: Medicine, Methods, and Molecules." *Journal of the History of Biology* 26 (1993): 409–42.

———. "Pursuing the Unpopular: A History of Courage, Viruses, and Cancer." In *Hidden Histories of Science,* ed. Robert B. Silvers. New York: New York Review, 1995, 69–112.

Kleinman, Daniel Lee. "Layers of Interests, Layers of Influence: Business and the Genesis of the National Science Foundation." *Science, Technology, and Human Values* 19 (1994): 259–82.

Klopsteg, Paul E. "Role of Government in Basic Research." *Science* 121 (1955): 781–84.

Klug, Aaron. "The Tobacco Mosaic Virus Particle: Structure and Assembly." *Philosophical Transactions of the Royal Society of London* 354B (1999): 531–35.

Knight, C. Arthur. "The Nature of Some of the Chemical Differences among Strains of Tobacco Mosaic Virus." *Journal of Biological Chemistry* 171 (1947): 297–308.

———. "Amino Acid Composition of Highly Purified Viral Particles of Influenza A and B." *Journal of Experimental Medicine* 86 (1947): 125–29.

———. "The Nucleic Acids of Some Strains of Tobacco Mosaic Virus." *Journal of Biological Chemistry* 197 (1952): 241–49.

———. "The Chemical Constitution of Viruses." *Advances in Virus Research* 2 (1954): 153–82.

Knight, C. Arthur, and Dean Fraser. "The Mutation of Viruses." *Scientific American* 193, no. 1 (July 1955): 74–78.

Knorr-Cetina, Karin D. *The Manufacture of Knowledge: An Essay on the Constructivist and Contextual Nature of Science.* Oxford: Pergamon Press, 1981.

———. "The Ethnographic Study of Scientific Work: Towards a Constructivist Interpretation of Science." In Knorr-Cetina and Mulkay, *Science Observed,* 115–40.

Knorr-Cetina, Karin D., and Michael Mulkay, eds. *Science Observed: Perspectives on the Social Study of Science.* Beverly Hills, CA: Sage, 1983.

Kohler, Robert E. "The Reception of Eduard Buchner's Discovery of Cell-Free Fermentation." *Journal of the History of Biology* 5 (1972): 327–53.

———. "The Enzyme Theory and the Origin of Biochemistry." *Isis* 64 (1973): 181–96.

———. "The Management of Science: The Experience of Warren Weaver and the Rockefeller Foundation Programme in Molecular Biology." *Minerva* 14 (1976): 279–306.

———. *From Medical Chemistry to Biochemistry: The Making of a Biomedical Discipline.* Cambridge and New York: Cambridge University Press, 1982.

———. "Systems of Production: *Drosophila, Neurospora,* and Biochemical Genetics." *Historical Studies in the Physical and Biological Sciences* 22 (1991): 87–130.

———. *Partners in Science: Foundations and Natural Scientists, 1900–1945.* Chicago: University of Chicago Press, 1991.

———. "*Drosophila:* A Life in the Laboratory." *Journal of the History of Biology* 26 (1993): 281–310.

———. *Lords of the Fly:* Drosophila *Genetics and the Experimental Life.* Chicago: University of Chicago Press, 1994.

———. "Place and Practice in American Field Biology." Unpublished paper prepared for "The Life Sciences in the United States and Russia: Biology, Ecology, Agriculture, Medicine," a meeting at the American Philosophical Society, 10–12 Aug 1998.

———. "Moral Economy, Material Culture, and Community in *Drosophila* Genetics." In *The Science Studies Reader,* ed. Mario Biagioli. New York: Routledge, 1999, 243–57.

———. "The Constructivists' Tool Kit." *Isis* 90 (1999): 329–31.

Koprowski, Hilary. "Latent or Dormant Viral Infections." *Annals of the New York Academy of Sciences* 54 (1952): 963–76.

Koshland, Daniel E. "Biological Systems." *Science* 240 (1988): 1385.

Kozloff, L.; F. W. Putnam; and E. A. Evans Jr. "Precursors of Bacteriophage Nitrogen and Carbon." In Delbrück, *Viruses 1950,* 55–63.

Kraemer, Elmer. "Discussion [of "The Ultracentrifuge and Its Field of Research" by The Svedberg]." *Industrial and Engineering Chemistry, Analytical Edition* 10 (1938): 128.

Krueger, Albert P., and John H. Northrop. "The Kinetics of the Bacterium-Bacteriophage Reaction." *Journal of General Physiology* 14 (1931): 223–54.

Kuhn, Thomas S. *The Structure of Scientific Revolutions.* 2d ed. Chicago: University of Chicago Press, [1962] 1970.

———. "The Road Since Structure." In *Proceedings of the 1990 Biennial Meeting of the Philosophy of Science Association,* vol. 2, ed. A. Fine, M. Forbes, and L. Wessels. East Lansing, MI: Philosophy of Science Association, 1991, 3–13.

Kunitz, Moses, and John H. Northrop. "Crystalline Chymo-Trypsin and Chymo-Trypsinogen. Part 1. Isolation, Crystallization, and General Properties of a New Proteolytic Enzyme and Its Precursor." *Journal of General Physiology* 18 (1935): 433–58.

Kunkel, Louis Otto. "Studies on Acquired Immunity with Tobacco and Aucuba Mosaics." *Phytopathology* 24 (1934): 437–66.

_____. "Virus Diseases of Plants: Twenty-Five Years of Progress, 1910–1935." *Memoirs of the Brooklyn Botanical Gardens* 4 (1936): 51–55.

Laidlaw, Sir Patrick. *Virus Diseases and Viruses.* Cambridge: Cambridge University Press, 1939.

Landsteiner, Karl, and E. Popper. "I. Übertragung der Poliomyelitis acuta auf Affen." *Zeitschrift für Immunitätsforschung* 2 (1909): 377–90.

Landsteiner, Karl, and H. Raubitschek. "Demonstriert mikroskopische Präparate von einem menschlichen und zwei Affenrückenmarken." *Wiener klinische Wochenschrift* 21 (1908): 1830.

Langer, James. "Non-Equilibrium Physics and the Origins of Complexity in Nature." In *Critical Problems in Physics,* ed. Val L. Fitch, Daniel R. Marlow, and Margit A. E. Dementi. Princeton, NJ: Princeton University Press, 11–27.

Latour, Bruno. "Give Me a Laboratory and I Will Raise the World." In Knorr-Cetina and Mulkay, *Science Observed,* 141–70.

_____. *Science in Action: How to Follow Scientists and Engineers Through Society.* Cambridge, MA: Harvard University Press, 1987.

_____. *The Pasteurization of France.* Trans. Alan Sheridan and John Law. Cambridge, MA: Harvard University Press, 1988.

_____. *We Have Never Been Modern.* Trans. Catherine Porter. Cambridge, MA: Harvard University Press, 1993.

Latour, Bruno, and Steve Woolgar. *Laboratory Life: The Construction of Scientific Facts.* 2d ed. Princeton, NJ: Princeton University Press, [1979] 1986.

Laubichler, Manfred D., and Angela N. H. Creager. "How Constructive Is Deconstruction? A Review Essay of *Toward a History of Epistemic Things,* by Hans-Jörg Rheinberger." *Studies in History and Philosophy of Biological and Biomedical Sciences* 30C (1999): 129–42.

Lauffer, Max A. "The Viscosity of Tobacco Mosaic Virus Protein Solutions." *Journal of Biological Chemistry* 126 (1938): 443–53.

Lauffer, Max A., and Wendell M. Stanley. "Stream Double Refraction of Virus Proteins." *Journal of Biological Chemistry* 123 (1938): 507–25.

_____. "The Denaturation of Tobacco Mosaic Virus by Urea. Part 1. Biochemical Aspects." *Archives of Biochemistry* 2 (1943): 413–24.

_____. "Biophysical Properties of Preparations of PR8 Influenza Virus." *Journal of Experimental Medicine* 80 (1944): 531–48.

Lavin, G. I.; Hubert S. Loring; and Wendell M. Stanley. "Ultraviolet Absorption Spectra of Latent Mosaic and Ring Spot Viruses and of their Nucleic Acid and Protein Components." *Journal of Biological Chemistry* 130 (1939): 259–68.

Lavin, G. I., and Wendell M. Stanley. "The Ultraviolet Absorption Spectrum of Crystalline Tobacco Mosaic Virus Protein." *Journal of Biological Chemistry* 118 (1937): 269–74.

Lazarow, Arnold. "The Chemical Structure of Cytoplasm as Investigated in Professor Bensley's Laboratory During the Past Ten Years." *Biological Symposia* 10 (1943): 9–26.

Lechevalier, Hubert. "Dmitri Iosifovich Ivanovski." *Bacteriological Reviews* 36 (1972): 135–45.

Lederberg, Esther M. "Lysogenicity in *E. coli* K-12." *Genetics* 36 (1951): 560.

Lederberg, Esther M., and Joshua Lederberg. "Genetic Studies of Lysogenicity in *Escherichia coli*." *Genetics* 38 (1953): 51–64.

Lederberg, Joshua. "Gene Recombination and Linked Segregations in *Escherichia coli*." *Genetics* 32 (1947): 505–25.

———. "Genetic Studies with Bacteria." In Dunn, *Genetics in the Twentieth Century*, 263–89.

———. "Cell Genetics and Hereditary Symbiosis." *Physiological Reviews* 32 (1952): 403–30.

———. "Forty Years of Genetic Recombination in Bacteria: A Fortieth Anniversary Reminiscence." *Nature* 324 (1986): 627–28.

———. "The Transformation of Genetics by DNA: An Anniversary Celebration of Avery, MacLeod, and McCarty (1944)." *Genetics* 136 (1994): 423–26.

———. "Greetings." *Annals of the New York Academy of Sciences* (special issue: *DNA: The Double Helix. Perspective and Prospective at Forty Years,* ed. D. A. Chambers) 758 (1995): 176–79.

———. "*Escherichia coli.*" In *Instruments of Science: An Historical Encyclopedia,* ed. Robert Bud and Deborah Jean Warner. New York: Garland Publishing, 1998, 230–32.

———. "Plasmid (1952–1997)." *Plasmid* 39 (1998): 1–9.

Lederman, Muriel, and Richard M. Burian, eds. "The Right Organism for the Job." *Journal of the History of Biology* (special section) 26 (1993): 233–367.

Lederman, Muriel, and Sue A. Tolin. "OVATOOMB: Other Viruses and the Origins of Molecular Biology." *Journal of the History of Biology* 26 (1993): 239–54.

Lenoir, Timothy. "Models and Instruments in the Development of Electrophysiology, 1845–1912." *Historical Studies in the Physical and Biological Sciences* 17 (1986): 1–54.

Lenoir, Timothy, and Christophe Lécuyer. "Instrument Makers and Discipline Builders: The Case of Nuclear Magnetic Resonance." *Perspectives on Science* 3 (1995): 276–345.

Levine, Arnold J. *Viruses.* New York: Scientific American Library, 1992.

———. "The Origins of the Small DNA Tumor Viruses." *Advances in Cancer Research* 65 (1994): 141–68.

L'Héritier, Philippe. "Sensitivity to CO_2 in *Drosophila*—A Review." *Heredity* 2 (1948): 325–48.

Lindegren, Carl C. "A New Gene Theory and an Explanation of the Phenomenon of Dominance to Mendelian Segregation of the Cytogene." *Proceedings of the National Academy of Sciences, USA* 32 (1946): 68–70.

———. *The Cold War in Biology.* Ann Arbor, MI: Planarian Press, 1966.

Lippincott, J. A., and Barry Commoner. "Reactivation of Tobacco Mosaic Virus Infectivity in Mixtures of Virus Protein and Nucleic Acid." *Biochimica et Biophysica Acta* 19 (1956): 198–99.

Lode, A., and J. Gruber. "Bakteriologische Studien über die Ätiologie einer epidemischen Erkrankung der Hühner in Tirol (1901)." *Zentralblatt für Bakteriologie, Parasitenkunde und Infektionskrankheiten,* Abt. 1, 30 (1901): 593–604.

Loeffler, Friedrich, and Robert Doerr. "Über filtrierbares Virus." *Zentralblatt für Bakteriologie, Parasitenkunde und Infektionskrankheiten,* Abt. 1, Referate 50 (1911): 1–12 (Loeffler), 12–23 (Doerr).

Loeffler, Friedrich, and Paul Frosch. "Berichte der Kommission zur Erforschung der Maul- und Klauenseuche bei dem Institut für Infektionskrankheiten in Berlin." *Zentralblatt für Bakteriologie, Parasitenkunde und Infektionskrankheiten,* Abt. 1, 23 (1898): 371–91.

Lojkin, Mary, and Carl G. Vinson. "Effect of Enzymes upon the Infectivity of the Virus of Tobacco Mosaic." *Contributions from the Boyce Thompson Institute* 3 (1931): 147–62.

Lorch, J. "The Charisma of Crystals in Biology." In *The Interaction between Science and Philosophy,* ed. Y. Elkana. Atlantic Highlands, NJ: Humanities Press, 1974, 445–61.

Loring, Hubert S.; Max A. Lauffer; and Wendell M. Stanley. "Aggregation of Purified Tobacco Mosaic Virus." *Nature* 142 (1938): 841–42.

Löwy, Ilana. "The Strength of Loose Concepts—Boundary Concepts, Federative Experimental Strategies, and Disciplinary Growth: The Case of Immunology." *History of Science* 30 (1992): 371–96.

———. "Experimental Systems and Clinical Practices: Tumor Immunology and Cancer Immunotherapy, 1895–1980." *Journal of the History of Biology* 27 (1994): 403–35.

Löwy, Ilana, and Jean-Paul Gaudillière. "Disciplining Cancer: Mice and the Practice of Genetic Purity." In Gaudillière and Löwy, *Invisible Industrialist,* 209–49.

———, eds. *Heredity and Infection: A History of Disease Transmission.* Amsterdam: Harwood Academic Publishers, 2001.

Luria, Salvador E. "Genetics of Bacterium-Bacterial Virus Relationship." *Annals of the Missouri Botanical Garden* 32 (1945): 235–42.

———. "Bacteriophage: An Essay on Virus Reproduction." In Delbrück, *Viruses 1950,* 7–15.

———. "Mutations of Bacteria and of Bacteriophage." In Cairns, Stent, and Watson, *Phage and the Origins,* 173–79.

———. *A Slot Machine, a Broken Test Tube: An Autobiography.* New York: Harper & Row, 1984.

Luria, Salvador E., and Thomas F. Anderson. "The Identification and Characterization of Bacteriophages with the Electron Microscope." *Proceedings of the National Academy of Sciences, USA* 28 (1942): 127–30.

Luria, Salvador E., and Max Delbrück. "Mutations of Bacteria: From Virus Sensitivity to Virus Resistance." *Genetics* 28 (1943): 491–511.

Luria, Salvador E.; Max Delbrück; and Thomas F. Anderson. "Electron Microscope Studies of Bacterial Viruses." *Journal of Bacteriology* 46 (1943): 57–77.

Lustig, Alice, and Arnold J. Levine. "One Hundred Years of Virology." *Journal of Virology* 66 (1992): 4629–31.

Lwoff, André. "The Concept of Virus. The Third Marjory Stephenson Memorial Lecture." *Journal of General Microbiology* 17 (1957): 239–53.

———. "From Protozoa to Bacteria and Viruses: Fifty Years with Microbes." *Annual Review of Microbiology* 25 (1971): 1–26.

Lwoff, André, and A. Gutmann. "Recherches sur un *Bacillus megatherium* lysogène." *Annales de l'Institut Pasteur* 78 (1950): 711–39.

Lwoff, André; L. Siminovitch; and N. Kjeldgaard. "Induction de la production de bactériophages chez une bactérie lysogène." *Annales de l'Institut Pasteur* 79 (1950): 815–59.

Lwoff, André, and Agnes Ullman. *Origins of Molecular Biology: A Tribute to Jacques Monod.* New York: Academic Press, 1979.

Lynch, Michael. *Art and Artifact in Laboratory Science: A Study of Shop Work and Shop Talk in a Research Laboratory.* Boston: Routledge, 1985.

Macallum, Archibald B. "On the Origin of Life on the Globe." *Transactions of the Canadian Institute* 8 (1908): 435–36.

Macrakis, Kristie. *Surviving the Swastika: Scientific Research in Nazi Germany.* Oxford: Oxford University Press, 1993.

———. "Adolf Butenandt." In James, *Nobel Laureates in Chemistry,* 253–58.

Maggioria, A., and G. L. Valenti. "Über eine Seuche von exsuditivem Typhus bei Hühnern." *Zeitschrift für Hygiene und Infektionskrankheiten* 42 (1903): 185–243.

Marks, Harry M. "Cortisone, 1949: A Year in the Political Life of a Drug." *Bulletin of the History of Medicine* 66 (1992): 419–39.

———. "Leviathan and the Clinic." Unpublished paper prepared for the History of Science Society meeting, 27–30 Dec 1992.

Marton, L. *Early History of the Electron Microscope.* San Francisco: San Francisco Press, 1968.

Mason, Max, and Warren Weaver. "The Settling of Small Particles in a Fluid." *Physical Review* 23 (1924): 412–26.

Mayer, Adolf. "Über die Mosaikkrankheit des Tabaks." *Landwirtschaftlichen Versuchs-Stationen* 32 (1886): 451–67. Trans. James Johnson and reprinted as "Concerning the Mosaic Disease of Tobacco." *Phytopathological Classics* 7 (1942): 11–24.

McBain, James W. "Some Uses of the Air-Driven Spinning Top." *Nature* 135 (1935): 831.

McIntosh, J., and R. F. Selbie. "The Application of the Sharples Centrifuge to the Study of Viruses." *Journal of Experimental Pathology* 21 (1940): 153–60.

McKinney, H. H. "Virus Mixtures That May Not Be Detected in Young Tobacco Plants." *Phytopathology* 16 (1926): 893.

———. "Quantitative and Purification Methods in Virus Studies." *Journal of Agricultural Research* 35 (1927): 13–38.

———. "Mosaic Diseases in the Canary Islands, West Africa, and Gibraltar." *Journal of Agricultural Research* 39 (1929): 557–78.

———. "The Inhibiting Influence of a Virus on One of Its Mutants." *Science* 82 (1935): 463–64.

Medawar, Peter B. "Cellular Inheritance and Transformation." *Biological Reviews* 22 (1947): 360–89.

Mendelsohn, John Andrew. *Cultures of Bacteriology: Formation and Transformation of a Science in France and Germany, 1870–1914.* Ph.D. diss., Princeton University, 1996.

————. "Medicine and the Making of Bodily Inequality in Twentieth-Century Europe." In Löwy and Gaudillière, *Heredity and Infection,* 21–79.

Meselsohn, Matthew, and Franklin W. Stahl. "The Replication of DNA in *Escherichia coli.*" *Proceedings of the National Academy of Sciences, USA* 44 (1958): 671–82.

M'Fadyean, John. "African Horse-Sickness." *Journal of Comparative Pathology and Therapeutics* 13 (1900): 1–20.

————. "The Ultravisible Viruses." *Journal of Comparative Pathology and Therapeutics* 21 (1908): 58–68, 168–75, 232–42.

Mider, G. Burroughs. "The Federal Impact on Biomedical Research." In *Advances in American Medicine: Essays at the Bicentennial,* vol. 2, ed. John Z. Bowers and Elizabeth F. Purcell. New York: Josiah Macy Jr. Foundation and the National Library of Medicine, 1976, 806–71.

Mills, Clarence A. "Distribution of American Research Funds." *Science* 107 (1948): 127–30.

Monod, Jacques. "The Phenomenon of Enzymatic Adaptation and Its Bearings on Problems of Genetics and Cellular Differentiation." *Growth Symposium* 11 (1947): 223–89.

————. "Du microbe à l'homme." In Monod and Borek, *Of Microbes and Life,* 1–9.

Monod, Jacques, and Ernest Borek, eds. *Of Microbes and Life.* New York: Columbia University Press, 1971.

Morange, Michel. *A History of Molecular Biology.* Trans. Matthew Cobb. Cambridge, MA: Harvard University Press, 1998.

Morgan, Mary S. "The Technology of Analogical Models: Irving Fisher's Monetary Worlds." *Philosophy of Science* 64 (1997): S304–14.

Morgan, Mary S., and Margaret Morrison, eds. *Models as Mediators: Perspectives on Natural and Social Science.* Cambridge: Cambridge University Press, 1999.

Morgan, Neil. "The Strategy of Biological Research Programmes: Reassessing the 'Dark Age' of Biochemistry, 1910–1930." *Annals of Science* 47 (1990): 139–50.

Morgan, T. H. "The Relation of Genetics to Physiology and Medicine." In *Nobel Lectures in Molecular Biology.* New York: Elsevier North-Holland, 1977, 3–18.

Mrowka. "Das Virus der Hühnerpest ein Globulin." *Zentralblatt für Bakteriologie, Parasitenkunde und Infektionskrankheiten,* Abt. I, Orig. 67 (1912): 249–68.

Muller, H. J. "Variation Due to Change in the Individual Gene. Part 1. The Relation between the Genes and the Characters of the Organisms." *American Naturalist* 56 (1922): 32–50.

————. "Artificial Transmutations of the Gene." *Science* 66 (1927): 84–87.

————. "The Gene as the Basis of Life." *Proceedings of the International Congress of Plant Sciences.* Vol. 1. Menasha, WI: George Banta, 1929, 897–921.

————. "Physics in the Attack on the Fundamental Problems of Genetics." *Scientific Monthly* 44 (1936): 210–14.

————. "Résumé and Perspectives of the Symposium on Genes and Chromosomes." *Cold Spring Harbor Symposia on Quantitative Biology* 9 (1941): 290–308.

———. "The Development of the Gene Theory." In Dunn, *Genetics in the Twentieth Century*, 77–99.

Mullins, Nicholas C. "The Development of a Scientific Specialty: The Phage Group and the Origins of Molecular Biology." *Minerva* 10 (1972): 51–82.

Mulvania, Maurice. "Cultivation of the Virus of Tobacco Mosaic by the Method of Olitsky." *Science* 62 (1925): 37.

———. "Studies on the Nature of the Virus of Tobacco Mosaic." *Phytopathology* 16 (1926): 853–71.

Mundry, Karl-Wolfgang. "The Effect of Nitrous Acid on Tobacco Mosaic Virus: Mutation, Not Selection." *Virology* 9 (1959): 722–26.

———. "TMV in Tübingen (1945–1965)." In Scholthof, Shaw, and Zaitlin, *Tobacco Mosaic Virus*, 155–60.

Mundry, Karl-Wolfgang, and Alfred Gierer. "Erzeugung von Mutanten des Tabakmosaikvirus durch chemische Veränderung seiner Nukleinsäure *in vitro*." In *Biochemistry of Viruses*, Proceedings of the Fourth International Congress of Biochemistry, vol. 7, ed. E. Broda and W. Frisch-Niggemeyer. New York: Pergamon, 1959, 62–65.

"Mutations in Tobacco Mosaic Virus." *Nature* 137 (1936): 540.

Nanney, David L. "The Role of the Cytoplasm in Heredity." In *The Chemical Basis of Heredity*, ed. William D. McElroy and Bentley Glass. Baltimore: Johns Hopkins Press, 1957, 134–64.

———. "Epigenetic Control Systems." *Proceedings of the National Academy of Sciences, USA* 44 (1958): 712–17.

Narita, Kozo. "Isolation of Acetylpeptide from Enzymic Digests of TMV-Protein." *Biochimica et Biophysica Acta* 28 (1958): 184–91.

Neurath, Hans; Gerald R. Cooper; D. G. Sharp; A. R. Taylor; Dorothy Beard; and J. W. Beard. "Molecular Size, Shape, and Homogeneity of the Rabbit Papilloma Protein." *Journal of Biological Chemistry* 140 (1941): 293–306.

Neushul, Peter. "Science, Government, and the Mass Production of Penicillin." *Journal of the History of Medicine and Allied Sciences* 48 (1993): 371–95.

Newman, Barclay Moon. "The Smooth Slide Up to Life." *Scientific American* 156, no. 5 (May 1937): 304–6.

Niu, Ching-I., and Heinz Fraenkel-Conrat. "C-Terminal Amino-Acid Sequence of Tobacco Mosaic Virus Protein." *Biochimica et Biophysica Acta* 16 (1955): 597–98.

———. "Determination of C-Terminal Amino Acids and Peptides by Hydrazinolysis." *Journal of the American Chemical Society* 77 (1955): 5882–85.

Nocard, E.; E. Roux; A. Borrel; A. T. Salimbeni; and E. Dujardin-Beaumetz. "Le microbe de la péripneumonie." *Annales de l'Institut Pasteur* 12 (1898): 240–62.

Nordenskiöld, Erik. *The History of Biology: A Survey*. Trans. Leonard Buckhall Eyre. New York: Alfred A. Knopf, 1928.

Northrop, John H. "Crystalline Pepsin. Part 1. Isolation and Tests of Purity." *Journal of General Physiology* 13 (1930): 730–66.

———. "Isolation and Properties of Pepsin and Trypsin." *The Harvey Lectures* (1934–35). Series 30. Baltimore: Williams & Wilkins, 1936, 229–70.

———. "Concentration and Partial Purification of Bacteriophage." *Science* 84 (1936): 90–91.

———. "Chemical Nature and Mode of Formation of Pepsin, Trypsin, and Bacteriophage." *Science* 86 (1937): 479–83.

Northrop, John H., and Moses Kunitz. "Crystalline Trypsin. Part 1. Isolation and Tests of Purity." *Journal of General Physiology* 16 (1932): 267–94.

Nye, Mary Jo. *Molecular Reality: A Perspective on the Scientific Work of Jean Perrin.* London: Macdonald; New York: American Elsevier, 1972.

Ochoa, Severo. "Biosynthesis of Ribonucleic Acid." In *Cellular Biology, Nucleic Acids, and Viruses: A Symposium in Honor of Basil O'Connor.* New York: New York Academy of Sciences, 1957, 191–200.

O'Connor, Basil. "Research and Physical Energy." In *Poliomyelitis: Papers and Discussions Presented at the Third International Poliomyelitis Conference, 1954.* Philadelphia: J. B. Lippincott, 1955.

Olby, Robert. "The Macromolecular Concept and the Origins of Molecular Biology." *Journal of Chemical Education* 47 (1970): 168–74.

———. "Schrödinger's Problem: What Is Life?" *Journal of the History of Biology* 4 (1971): 119–48.

———. *The Path to the Double Helix: The Discovery of DNA.* New York: Dover, [1974] 1994.

Olitsky, Peter K. "Experiments on the Cultivation of the Active Agent of Mosaic Disease of Tobacco and Tomato." *Science* 60 (1924): 593–94.

———. "Experiments on the Cultivation of the Active Agent of Mosaic Disease in Tobacco and Tomato Plants." *Journal of Experimental Medicine* 41 (1925): 129–36.

"1,200,000 R.P.M." *Scientific American* 150, no. 4 (Apr 1934): 208.

Oparin, Alexander. *The Origin of Life.* Trans. Sergius Morgulis. New York: Macmillan, 1938.

Oreskes, Naomi, and Ronald Rainger. "Science and Security before the Atomic Bomb: The Loyalty Case of Harald U. Sverdrup." *Studies in History and Philosophy of Modern Physics* 31B (2000): 309–69.

Oster, Gerald; C. Arthur Knight; and Wendell M. Stanley. "Electron Microscope Studies of Dahlem, Rothamsted, and Princeton Samples of Tobacco Mosaic Virus." *Archives of Biochemistry* 15 (1947): 279–88.

Osterhout, W. J. V., and Wendell M. Stanley. "The Accumulation of Electrolytes. Part 5. Models Showing Accumulation and a Steady State." *Journal of General Physiology* 15 (1932): 667–89.

Palade, George E. "An Electron Microscope Study of the Mitochondrial Structure." *Journal of Histochemistry and Cytochemistry* 1 (1953): 188–211.

Pasteur, Louis. "La rage" (1890). In *Oeuvres de Pasteur réunies par Pasteur Vallery-Radot,* vol. 6, 1922–39. Paris: Masson et cie, 672–88.

Patterson, James T. *The Dread Disease: Cancer and Modern American Culture.* Cambridge, MA: Harvard University Press, 1987.

Paul, Diane B., and Barbara A. Kimmelman. "Mendel in America: Theory and Practice, 1900–1919." In *The American Development of Biology,* ed. Ronald Rainger, Keith R. Benson, and Jane Maienschein. New Brunswick, NJ: Rutgers University Press, [1988] 1991, 281–310.

Paul, John R. *A History of Poliomyelitis.* New Haven, CT: Yale University Press, 1971.

Pedersen, Kai O. "The Development of Svedberg's Ultracentrifuge." *Biophysical Chemistry* 5 (1976): 3–18.

―――. "The Svedberg and Arne Tiselius: The Early Development of Modern Protein Chemistry at Uppsala." In *Comprehensive Biochemistry,* vol. 35, ed. Marcel Florkin and E. H. Stotz. Amsterdam: Elsevier Science Publishers, 1983, 235–82.

Penick, J. L., Jr.; C. W. Pursell Jr.; M. B. Sherwood; and D. C. Swain, eds. *The Politics of American Science: 1939 to the Present.* Chicago: Rand McNally, 1965.

Pennazio, Sergio. "A Short History of Plant Virology. Part 4. Twenty Years of Change (1940–1960)." *Rivista di Biologia/Biology Forum* 92 (1999): 119–42.

Perutz, Max. "Erwin Schrödinger's *What Is Life?* and Molecular Biology." In *Schrödinger: Centenary Celebration of a Polymath,* ed. C. W. Kilmister. Cambridge: Cambridge University Press, 1987, 234–51.

Pfankuch, Edgar. "Über die Spaltung von Virusproteinen der Tabakmosaik-Gruppe." *Biochemische Zeitschrift* 306 (1940): 125–29.

Pfankuch, Edgar, and G. A. Kausche. "Isolierung und übermikroskopische Abbildung eines Bakteriophagen." *Naturwissenschaften* 28 (1940): 46.

Pfankuch, Edgar; G. A. Kausche; and H. Stubbe. "Über die Entstehung, die biologische und physikalisch-chemische Charakterisierung von Röntgen- und γ-Strahlen induzierten 'Mutationen' des Tabakmosaikvirusproteins." *Biochemische Zeitschrift* 304 (1940): 238–58.

Pfankuch, Edgar, and F. Piedenbrock. "Zur Spaltung von Virusproteinen der Tabakmosaik-Gruppe." *Naturwissenschaften* 31 (1943): 94.

Pickels, Edward G. *The Air-Driven Ultra-centrifuge.* Ph.D. diss., University of Virginia, Charlottesville, 1935.

―――. "Practical Speed-Measuring Devices for High Speed Centrifuges." *Review of Scientific Instruments* 9 (1938): 354–64.

―――. "An Improved Type of Electrically Driven High Speed Laboratory Centrifuge." *Review of Scientific Instruments* 13 (1942): 93–100.

Pickels, Edward G., and Johannes H. Bauer. "Ultracentrifugation Studies of Yellow Fever Virus." *Journal of Experimental Medicine* 71 (1940): 703–17.

Pickels, Edward G.; William F. Harrington; and Howard K. Schachman, "An Ultracentrifuge Cell for Producing Boundaries Synthetically by a Layering Technique." *Proceedings of the National Academy of Sciences, USA* 38 (1952): 943–48.

Pickels, Edward G., and Joseph E. Smadel. "Ultracentrifugation Studies on the Elementary Bodies of Vaccine Virus." *Journal of Experimental Medicine* 68 (1938): 583–606.

Pickering, Andrew. "From Science as Knowledge to Science as Practice." In Pickering, *Science as Practice,* 1–26.

―――, ed. *Science as Practice and Culture.* Chicago: University of Chicago Press, 1992.

―――. *The Mangle of Practice: Time, Agency, & Science.* Chicago: University of Chicago Press, 1995.

_____. "Explanation and the Mangle: A Response to My Critics." *Studies in History and Philosophy of Science* 30A (1999): 167–71.

Pierpont, W. S. "Norman Wingate Pirie, 1907–1997." *Biographical Memoirs of Fellows of the Royal Society* 45 (1999): 397–415.

Pinkerton, Henry. "The Pathogenesis and Pathology of Virus Infections." *Annals of the New York Academy of Sciences* 54 (1952): 874–81.

Piper, Anne. "Light on a Dark Lady." *Trends in Biochemical Sciences* 23 (1998): 151–54.

Pirie, N. W. "The Meaninglessness of the Terms Life and Living." In *Perspectives in Biochemistry: Thirty-One Essays Presented to Sir Frederick Gowland Hopkins.* Cambridge: Cambridge University Press, 1938.

_____. "The Viruses." *Annual Review of Biochemistry* 15 (1946): 573–92.

_____. "Discussion [of Papers by Commoner and by Schramm and Anderer]." In *Cellular Biology, Nucleic Acids and Viruses: A Symposium in Honor of Basil O'Connor.* New York: New York Academy of Sciences, 1957, 247–48.

_____. "The Viruses." In *Scientific Thought, 1900–1960: A Selected Survey,* ed. Rom Harré. Oxford: Clarendon Press, 1969, 227–37.

_____. "Frederick Charles Bawden, Thirty-eight Years of Collaboration: A Personal Appreciation." *Report of Rothamsted Experimental Station for 1971.* 1972.

_____. "Frederick Charles Bawden, 1908–1972." *Biographical Memoirs of Fellows of the Royal Society* 19 (1973): 19–63.

_____. "Recurrent Luck in Research." In *Selected Topics in the History of Biochemistry: Personal Recollections,* ed. G. Semenza, vol. 36 of *Comprehensive Biochemistry,* ed. Marcel Florkin and Elmer Stotz. Amsterdam: Elsevier, 1986, 491–522.

Podolsky, Scott. "The Role of the Virus in Origin-of-Life Theorizing." *Journal of the History of Biology* 29 (1996): 79–126.

Pollock, Martin R. "The Changing Concept of Organism in Microbiology." *Progress in Biophysics and Molecular Biology* 19 (1969): 273–305.

Pontecorvo, G. "Microbiology, Biochemistry, and the Genetics of Micro-Organisms." *Nature* 157 (1946): 95–96.

Porter, Keith R., and Frances L. Kallman. "Significance of Cell Particulates as Seen by Electron Microscopy." *Annals of the New York Academy of Sciences* 54 (1952): 882–91.

Portugal, Franklin H., and Jack S. Cohen. *A Century of DNA: A History of the Discovery of the Structure and Function of the Genetic Substance.* Cambridge, MA: MIT Press, 1977.

Preer, John R. "The Killer Cytoplasmic Factor Kappa: Its Rate of Reproduction, the Number of Particles per Cell, and Its Size." *American Naturalist* 82 (1948): 35–42.

Price, W. C. "Acquired Immunity from Plant Virus Diseases." *Quarterly Review of Biology* 15 (1940) 338–61.

Price, W. C., and Ralph W. G. Wyckoff. "The Ultracentrifugation of the Proteins of Cucumber Viruses 3 and 4." *Nature* 141 (1937): 685–86.

_____. "Ultracentrifugation of Juices from Plants Affected by Tobacco Necrosis." *Phytopathology* 29 (1939): 83–94.

"Prof. Beijerinck, on a contagium vivum fluidum, causing the spot-disease of tobacco leaves." *Nature* 59 (1898): 216.

Punnett, R. C., ed. *Proceedings of the Seventh International Genetical Congress, Edinburgh, Scotland, 23–30 August 1939.* Cambridge: Cambridge University Press, 1941.

Purdy, Helen A. "Attempt to Cultivate an Organism from Tomato Mosaic." *Botanical Gazette* 81 (1926): 210–17.

———. "Immunologic Reactions with Tobacco Mosaic Virus." *Proceedings of the Society for Experimental Biology and Medicine* 25 (1928): 702–3.

Purdy Beale, Helen. "Specificity of the Precipitin Reaction in Tobacco Mosaic Disease." *Contributions from the Boyce Thompson Institute* 3 (1931): 529–39.

Rader, Karen A. *Making Mice: C. C. Little, the Jackson Laboratory, and the Standardization of* Mus musculus *for Research.* Ph.D. diss., Indiana University, 1995.

———. "Of Mice, Medicine, and Genetics: C. C. Little's Creation of the Inbred Laboratory Mouse, 1909–1918." *Studies in History and Philosophy of Biological and Biomedical Sciences* 30C (1999): 319–43.

Rasmussen, Nicolas. "Facts, Artifacts, and Mesosomes: Practicing Epistemology with the Electron Microscope." *Studies in History and Philosophy of Science* 24 (1993): 227–65.

———. "Freund's Adjuvant and the Realization of Questions in Postwar Immunology." *Historical Studies in the Physical and Biological Sciences* 23 (1993): 337–66.

———. "Mitochondrial Structure and the Practice of Cell Biology in the 1950s." *Journal of the History of Biology* 28 (1995): 381–429.

———. "Making a Machine Instrumental: RCA and the Wartime Origins of Biological Electron Microscopy in America, 1940–1945." *Studies in History and Philosophy of Science* 27 (1996): 311–49.

———. "The Midcentury Biophysics Bubble: Hiroshima and the Biological Revolution in America, Revisited." *History of Science* 35 (1997): 245–93.

———. *Picture Control: The Electron Microscope and the Transformation of Biology in America, 1940-1960.* Stanford, CA: Stanford University Press, 1997.

———. "Instruments, Scientists, Industrialists, and the Specificity of 'Influence': The Case of RCA and Biological Electron Microscopy." In Gaudillière and Löwy, *Invisible Industrialist,* 173–208.

Ravin, Arnold W. "The Gene as Catalyst; the Gene as Organism." *Studies in the History of Biology* 1 (1977): 1–45.

Reed, Walter, and James Carroll. "The Etiology of Yellow Fever (A Supplemental Note)." *American Medicine* 3 (1902): 980.

Reich, Leonard S. *The Making of American Industrial Research: Science and Business at Bell and GE, 1876–1926.* Cambridge: Cambridge University Press, 1985.

Reingold, Nathan. "Vannevar Bush's New Deal for Research: Or, the Triumph of the Old Order." *Historical Studies in the Physical and Biological Sciences* 17 (1987): 299–344.

_____. "Science and Government in the United States since 1945." *History of Science* 32 (1994): 361–86.

Remlinger, Paul. "Le passage du Virus rabique à travers les filtres." *Annales de l'Institut Pasteur* 17 (1903): 834–49.

_____. "Les microbes filtrants." *Bulletin de l'Institut Pasteur* 3 (1906): 337–45, 385–92.

Renner, Otto. "Zur Kenntnis der nichtmendelnden Buntheit der Laubblätter." *Flora* 30 (1936): 218–90.

Rheinberger, Hans-Jörg. "Experiment, Difference, and Writing. Part 1. Tracing Protein Synthesis. Part 2. The Laboratory Production of Transfer RNA." *Studies in History and Philosophy of Science* 23 (1992): 305–31, 389–422.

_____. "From Microsomes to Ribosomes: 'Strategies' of 'Representation.' " *Journal of the History of Biology* 28 (1995): 49–89.

_____. "Comparing Experimental Systems: Protein Synthesis in Microbes and in Animal Tissue at Cambridge (Ernest F. Gale) and at Massachusetts General Hospital (Paul C. Zamecnik), 1945–1960." *Journal of the History of Biology* 29 (1996): 387–416.

_____. *Toward a History of Epistemic Things: Synthesizing Proteins in the Test Tube.* Stanford, CA: Stanford University Press, 1997.

_____. "Reenacting History." *Studies in History and Philosophy of Science* 30A (1999): 163–66.

_____. "Virus Research at the Kaiser Wilhelm Institutes for Biochemistry and for Biology, 1937–1945." Unpublished paper presented at the workshop "Geschichte der Kaiser-Wilhelm-Gesellschaft im Nationalsozialismus Bestandsaufnahme und Perspektiven der Forschung." Berlin, 10–12 Mar 1999.

_____. "Gene Concepts: Fragments from the Perspective of Molecular Biology." In Beurton, Falk, and Rheinberger, *Concept of the Gene,* 219–39.

_____. "Putting Isotopes to Work: Liquid Scintillation Counters, 1950–1970." In *Instrumentation: Between Science, State, and Industry,* ed. Bernward Joerges and Terry Shinn. Sociology of the Sciences Yearbook. Vol. 22. Dordrecht, the Netherlands: Kluwer Academic Publishers, 2000, 143–74.

Rheinberger, Hans-Jörg, and Michael Hagner, eds. *Die Experimentalisierung des Lebens: Experimentalsysteme in den biologischen Wissenschaften 1850/1950.* Berlin: Akademie Verlag, 1993.

Richards, A. N. "The Impact of the War on Medicine." *Science* 103 (1946): 575–78.

Richter, Curt P. "Free Research versus Design Research." *Science* 118 (1953): 91–93.

Rivers, Thomas M. "Filterable Viruses: A Critical Review." *Journal of Bacteriology* 14 (1927): 217–57.

_____. "Some General Aspects of Filterable Viruses." In *Filterable Viruses,* ed. Thomas M. Rivers. Baltimore: Williams & Wilkins, 1928, 3–52.

_____. "The Nature of Viruses." *Physiological Reviews* 12 (1932): 423–52.

_____. "Viruses and Koch's Postulates (Presidential address delivered before the Society of American Bacteriologists at its Thirty-eighth Annual Meeting, Indianapolis, Dec. 29, 1936)." *Journal of Bacteriology* 33 (1937): 1–12.

————. *Lane Medical Lectures: Viruses and Virus Diseases.* Stanford, CA: Stanford University Press, 1939.

————. "The Infinitely Small in Biology." *Science* 93 (1941): 143–45.

Rodgers, Andrew Denny. *Erwin Frink Smith: A Story of North American Plant Pathology.* Philadelphia: American Philosophical Society, 1952.

Rolle, Andrew. *California: A History.* 5th ed. Wheeling, IL: Harlan Davidson, 1998.

Roll-Hansen, Nils. "The Genotype Theory of Wilhelm Johannsen and Its Relation to Plant Breeding and the Study of Evolution." *Centaurus* 22 (1978): 201–35.

————. "The Application of Complementarity to Biology: From Niels Bohr to Max Delbrück." *Historical Studies in the Physical and Biological Sciences* 30 (2000): 417–42.

Roosevelt, J. "A Tribute to Basil O'Connor." In *Memorials to Basil O'Connor on His Seventy-fifth Birthday.* New York: National Foundation, 1967.

Rosenberg, Charles E. "Toward an Ecology of Knowledge: On Discipline, Context, and History." In *The Organization of Knowledge in Modern America, 1860–1920,* ed. Alexandra Oleson and John Voss. Baltimore: Johns Hopkins University Press, 1979, 440–55. Reprinted in *No Other Gods: On Science and American Social Thought,* rev. ed. Baltimore: Johns Hopkins University Press, 1997, 225–39.

Rosenfeld, Léon. "Niels Bohr in the Thirties: Consolidation and Extension of the Conception of Complementarity." In *Niels Bohr: His Life and Work as Seen by Friends and Colleagues,* ed. S. Rozental. Amsterdam: North-Holland; New York: Wiley, 1967, 114–36.

Ross, A. Frank, and Wendell M. Stanley. "Partial Reactivation of Formalized Tobacco Mosaic Virus Protein." *Proceedings of the Society for Experimental Biology and Medicine* 38 (1938): 260–63.

Rous, Francis Peyton. "Transmission of a Malignant New Growth by Means of a Cell-Free Filtrate." *Journal of the American Medical Association* 56 (1911): 198.

————. "The Challenge to Man of the Neoplastic Cell." *Science* 157 (1967): 24–28.

Roux, Emile. "Sur les microbes dit 'invisibles.'" *Bulletin de l'Institut Pasteur* 1 (1903): 7–12, 49–56.

Roux, Emile, and A. Yersin. "Contribution à l'étude de la diphthérie." *Annales de l'Institut Pasteur* 2 (1888): 629–61.

Ruska, H. "Die Sichtbarmachung der bakteriophagen Lyse im Übermikroskop." *Naturwissenschaften* 28 (1940): 45–46.

————. "Über ein neues bei der bakteriophagen Lyse auftretendes Formelement." *Naturwissenschaften* 29 (1941): 367–68.

Samuel, Geoffrey, and J. G. Bald. "On the Use of the Primary Lesions in Quantitative Work with Two Plant Viruses." *Annals of Applied Biology* 20 (1933): 70–99.

Sanarelli, Giuseppe. "Das myxomatogene Virus: Beitrag zum Studium der Krankheitserreger ausserhalb des Sichtbaren." *Zentralblatt für Bakteriologie, Parasitenkunde und Infectionskrankheiten,* Abt. I, 23 (1898): 865–73.

Sanger, Frederick. "The Free Amino Groups of Insulin." *Biochemical Journal* 39 (1945): 507–15.

Santesmases, María Jesús. "Severo Ochoa and the Biomedical Sciences in Spain under Franco, 1959–1975." *Isis* 91 (2000): 706–34.

Santesmases, María Jesús, and Emilio Muñoz. "Scientific Organizations in Spain (1950–1970): Social Isolation and International Legitimation of Biochemists and Molecular Biologists on the Periphery." *Social Studies of Science* 27 (1997): 187–219.

Sapp, Jan. *Beyond the Gene: Cytoplasmic Inheritance and the Struggle for Authority in Genetics.* New York: Oxford University Press, 1987.

————. *Evolution by Association: A History of Symbiosis.* New York: Oxford University Press, 1994.

Sarkar, Sahotra. "Biological Information: A Skeptical Look at Some Central Dogmas of Molecular Biology." In *The Philosophy and History of Molecular Biology: New Perspectives,* ed. Sahotra Sarkar. Dordrecht, the Netherlands: Kluwer Academic Publishers, 1996, 187–231.

Schachman, Howard K. "Physical Chemical Studies on Rabbit Papilloma Virus." *Journal of the American Chemical Society* 73 (1951): 4453–55.

————. *Ultracentrifugation in Biochemistry.* New York: Academic Press, 1959.

————. "Development of the Ultracentrifuge and Its Application to Biological Systems." *The Beckman Symposium on Biomedical Instrumentation,* ed. Carol L. Moberg. Fullerton, CA: The Rockefeller University in association with Beckman Instruments, 1986, 37–58.

————. "Is There a Future for the Ultracentrifuge?" In *Analytical Ultracentrifugation in Biochemistry and Polymer Science,* ed. S. E. Harding, A. J. Rowe, and J. C. Horton. Cambridge, U.K.: The Royal Society of Chemistry, 1992, 3–15.

————. "Still Looking for the Ivory Tower." *Annual Review of Biochemistry* 69 (2000): 1–29.

Schachman, Howard K., and William F. Harrington. "On Viscosity Measurement in the Ultracentrifuge." *Journal of the American Chemical Society* 74 (1952): 3965–66.

————. "Ultracentrifuge Studies with a Synthetic Boundary Cell. Part 1. General Applications." *Journal of Polymer Science* 12 (1954): 379–90.

Schäfer, Werner. "Structure of Some Animal Viruses and Significance of Their Components." *Bacteriological Reviews* 27 (1963): 1–17.

Schaffer, Frederick L. "The Purification and Physicochemical Properties of Plant and Animal Viruses." In *Immunity and Virus Infection,* Symposium held at Vanderbilt University School of Medicine, 1–2 May 1958, ed. Victor A. Najjar. New York: John Wiley & Sons, 1959, 176–87.

Schaffer, Frederick L., and Carl F. T. Mattern. "Infectivity and Physicochemical Studies on RNA Preparations from Highly Purified Poliomyelitis and Coxsackie Viruses." *Federation Proceedings* 18 (1959): 317.

Schaffer, Frederick L., and Carlton E. Schwerdt. "Crystallization of Purified MEF-1 Poliomyelitis Virus Particles." *Proceedings of the National Academy of Sciences, USA* 41 (1955): 1020–23.

————. "Nucleic Acid Composition of Purified Preparations of Poliomyelitis Virus." *Federation Proceedings* 14 (1955): 275.

Schaffer, Simon. "The Eighteenth Brumaire of Bruno Latour." *Studies in History and Philosophy of Science* 22 (1991): 174–92.

Schaffner, Kenneth F. *Discovery and Explanation in Biology and Medicine.* Chicago: University of Chicago Press, 1993.

Schatzki, Theodore R. "To Mangle: Emergent, Unconstrained, Posthumanist?" *Studies in History and Philosophy of Science* 30A (1999): 157–61.

Schlesinger, Max. "Reindarstellung eines Bakteriophagen in mit freiem Auge sichtbaren Mengen." *Biochemische Zeitschrift* 264 (1933): 6–12.

Schlesinger, R. Walter. "Virus Research at the Princeton Rockefeller Institute, 1930–1945: Recalling a Wellspring of Discoveries." *Infectious Agents and Diseases* 1 (1992): 325–37.

Schloegel, Judy Johns. "From Anomaly to Unification: Tracy Sonneborn and the Species Problem in Protozoa, 1954–1957." *Journal of the History of Biology* 32 (1999): 93–132.

Schmiedebach, H.-P. "The Prussian State and Microbiological Research— Friedrich Loeffler and His Approach to the 'Invisible' Virus." *Archives of Virology* 15 (supp.): 9–23 (1999).

Scholthof, Karen-Beth G.; John G. Shaw; and Milton Zaitlin, eds. *Tobacco Mosaic Virus: One Hundred Years of Contributions to Virology.* St. Paul, MN: American Phytopathological Society Press, 1999.

Schramm, Gerhard. "Die luftgetriebene Ultrazentrifuge." *Kolloid-Zeitschrift* 97 (1941): 106–15.

————. "Über die enzymatische Abspaltung der Nucleinsäure aus dem Tabakmosaikvirus." *Berichte der deutschen chemischen Gesellschaft* 74B (1941) 532–36; *Chemical Abstracts* 35 (1941): 5536.

————. "Über die Spaltung des Tabakmosaikvirus in niedermolekulare Proteine und die Rückbildung hochmolekularen Proteins aus den Spaltstücken." *Naturwissenschaften* 31 (1943): 94–96.

————. "Über die Konstitution des Tabakmosaikvirus." *Angewandte Chemie* 57 (1944): 109–13.

————. "Über periodische Fällungen in nativen und regenerierten Zellulosefasern." *Kolloid-Zeitschrift* 108 (1944): 166–69.

————. "Über die Spaltung des Tabakmosaikvirus und die Wiedervereinigung der Spaltstücke zu höhermolekularen Proteinen. [Part 1.] Die Spaltungsreaktion." *Zeitschrift für Naturforschung* 2B (1947): 112–21.

————. "Über die Spaltung des Tabakmosaikvirus und die Wiedervereinigung der Spaltstücke zu höhermolekularen Proteinen. [Part 2.] Versuche zur Wiedervereinigung der Spaltstücke." *Zeitschrift für Naturforschung* 2B (1947): 249–57.

————. "Die Struktur des Tabakmosaikvirus und seiner Mutanten." *Advances in Enzymology* 15 (1954): 449–84.

————. "Neue Untersuchungen über die Struktur des Tabakmosaikvirus und ihre biologische Bedeutung." *Zentralblatt für Bakteriologie, Parasitenkunde und Infektionskrankheiten,* Abt. 2, 109 (1956): 322–24.

———. "Aufbau und Vermehrung phytopathogener Viren." *Nova Acta Leopold-ina* 19 (1957): 29–37.

———. "Biosynthese des Tabakmosaikvirus." In *Biochemistry of Viruses,* Proceedings of the Fourth International Congress of Biochemistry, vol. 7, ed. E. Broda and W. Frisch-Niggemeyer. New York: Pergamon, 1959, 1–8.

Schramm, Gerhard; H. J. Born; and A. Lang. "Versuch über den Phosphoraustausch zwischen radiophosphorhaltigem Tabakmosaikvirus und Natriumphosphat." *Naturwissenschaften* 30 (1942): 170–71.

Schramm, Gerhard, and Gerhard Braunitzer. "Prolin als Endgruppe des Tabakmosaikvirus." *Zeitschrift für Naturforschung* 8B (1953): 61–66.

Schramm, Gerhard; Gerhard Braunitzer; and Juan W. Schneider. "Zur Bestimmung der Amino-Endgruppe im Tabakmosaikvirus." *Zeitschrift für Naturforschung* 9B (1954): 298–301.

Schramm, Gerhard, and Hans Friedrich-Freksa. "Die Präcipitinreaktion des Tabakmosaikvirus mit Kaninchen- und Schweineantiserum." *Zeitschrift für physiologische Chemie* 270 (1941): 233–46.

Schramm, Gerhard, and Hans Müller. "Über die Konfiguration der im Tabakmosaikvirus enthaltenen Aminosäuren." *Naturwissenschaften* 28 (1940): 223–24.

Schramm, Gerhard, and Malene Wiedemann. "Größenverteilung des Tabakmosaikvirus in der Ultrazentrifuge und im Elektronenmikroskop." *Zeitschrift für Naturforschung* 6B (1951): 379–83.

Schrödinger, Erwin. *What Is Life? The Physical Aspect of the Living Cell.* Cambridge: Cambridge University Press, [1944] 1967.

Schultz, Jack. "The Question of Plasmagenes." *Science* 111 (1950): 403–7.

Schuster, Heinz, and Gerhard Schramm. "Bestimmung der biologisch wirksamen Einheit in der Ribosenucleinsäure des Tabakmosaikvirus auf chemischem Wege." *Zeitschrift für Naturforschung* 13B (1958): 697–704.

Schweber, Silvan S. "The Mutual Embrace of Science and the Military: ONR and the Growth of Physics in the United States after World War II." In *Science, Technology, and the Military,* ed. Everett Mendelsohn, Merritt Roe Smith, and Peter Weingart. Sociology of the Sciences Yearbook. Vol. 12. Dordrecht, the Netherlands: Kluwer Academic Publishers, 1988, 3–45.

Schwerdt, Carlton E., and Arthur B. Pardee. "The Intracellular Distribution of Lansing Poliomyelitis Virus in the Central Nervous System of Infected Cotton Rats." *Journal of Experimental Medicine* 96 (1952): 121–35.

Schwerdt, Carlton E.; Robley C. Williams; Wendell M. Stanley; Frederick L. Schaffer; and Mary E. McClain. "Morphology of Type II Poliomyelitis Virus (MEF1) as Determined by Electron Microscopy." *Proceedings of the Society for Experimental Biology and Medicine* 86 (1954): 310–12.

Sédillot, C. "De l'influence des découvertes de M. Pasteur sur le progrès de la chirurgie." *Comptes rendus hebdomadaires des Séances de l'Académie des Sciences* 86 (1878): 634–40.

Servos, John W. *Physical Chemistry from Ostwald to Pauling: The Making of a Science in America.* Princeton, NJ: Princeton University Press, 1990.

Shapin, Steven. "The House of Experiment in Seventeenth-Century England." *Isis* 79 (1988): 373–404.

———. *A Social History of Truth: Civility and Science in Seventeenth-Century England.* Chicago: University of Chicago Press, 1994.

Sharp, D. G.; A. R. Taylor; Dorothy Beard; and Joseph W. Beard. "Study of the Papillomatosis Virus Protein with the Electron Microscope." *Proceedings of the Society for Experimental Biology and Medicine* 50 (1942): 205–7.

———. "Electron Micrography of the Western Strain Equine Encephalomyelitis Virus." *Proceedings of the Society for Experimental Biology and Medicine* 51 (1942): 206–7.

Sharp, D. G.; A. R. Taylor; and Joseph W. Beard. "The Density and Size of the Rabbit Papilloma Virus." *Journal of Biological Chemistry* 163 (1946): 289–99.

Sharp, D. G.; A. R. Taylor; A. R. Hook; and Joseph W. Beard. "Rabbit Papilloma and Vaccinia Viruses and T2 Bacteriophage of *E. coli* in 'Shadow' Electron Micrographs." *Proceedings of the Society for Experimental Biology and Medicine* 61 (1946): 259–65.

Shaughnessy, Donald F. *The Story of the American Cancer Society.* Ph.D. diss., Columbia University, 1957.

Shope, Richard E. "Immunization of Rabbits to Infectious Papillomatosis." *Journal of Experimental Medicine* 65 (1937): 219–31.

———. " 'Masking,' Transformation, and Interepidemic Survival of Animal Viruses." In Delbrück, *Viruses 1950,* 79–92.

Shriner, Ralph L. "The William H. Nichols Medalist for 1946: Wendell M. Stanley, His Career." *Chemical and Engineering News* 24 (25 Mar 1946): 750–52.

Shryock, Richard Harrison. *American Medical Research Past and Present.* New York: Commonwealth Fund, 1947.

———. *National Tuberculosis Association 1904–1954: A Study of the Voluntary Health Movement in the United States.* New York: Arno, 1977; originally published in 1957 by the National Tuberculosis Association.

Siedentopf, H., and R. Zsigmondy. "Über die Sichtbarmachung und Grössenbestimmung ultramikroskopisches Teilchen, mit besonderer Anwendung auf Goldrubingläser." *Annalen der Physik* 10 (1903): 1–39.

Sigel, M. M.; F. W. Shaffer; M. Wiener Kirber; A. B. Light; and W. Henle. "Influenza A in a Vaccinated Population." *Journal of the American Medical Association* 136 (1948): 437–41.

Silverman, Milton. "Now—Man-Made Virus—First Step in Controlling Heredity?" *Collier's Magazine,* 31 Aug 1956, 74–77.

Silverstein, Arthur M. "The Heuristic Value of Experimental Systems: The Case of Immune Hemolysis." *Journal of the History of Biology* 27 (1994): 437–47.

Slater, Leo B. "Max Delbrück: The Physicist and the Phage." Unpublished seminar paper, 1994.

———. "Industry and Academy: The Synthesis of Steroids." *Historical Studies in the Physical and Biological Sciences* 30 (2000): 443–80.

Slotta, K. H., and Heinz Fraenkel-Conrat. "Schlangengifte. [Part 3.] Reinigung und Krystallisation des Klapperschlangen-Giftes." *Berichte der deutschen chemischen Gesellschaft* 71 (1938): 1076–81.

Smadel, Joseph E.; Edward G. Pickels; and Theodore Shedlovsky. "Ultracentrifugation Studies on the Elementary Bodies of Vaccine Virus. Part 2. The

Influence of Sucrose, Glycerol, and Urea Solutions on the Physical Nature of Vaccine Virus." *Journal of Experimental Medicine* 68 (1938): 607–27.

Smadel, Joseph E.; Thomas M. Rivers; and Edward G. Pickels. "Estimation of the Purity of Preparations of Elementary Bodies of Vaccinia." *Journal of Experimental Medicine* 70 (1939): 379–85.

Smith, Crosbie, and M. Norton Wise. *Energy and Empire: A Biographical Study of Lord Kelvin.* Cambridge: Cambridge University Press, 1989.

Smith, Jane S. *Patenting the Sun: Polio and the Salk Vaccine.* New York: William Morrow, 1990.

Sonneborn, Tracy M. "Gene and Cytoplasm. Part 1. The Determination and Inheritance of the Killer Character in Variety 4 of Paramecium Aurealia. Part 2. The Bearing of the Determination and Inheritance of Characters in Paramecium Aurelia on the Problems of Cytoplasmic Inheritance, Pneumococcus Transformations, Mutations, and Development." *Proceedings of the National Academy of Sciences, USA* 29 (1943): 329–38, 338–43.

———. "Beyond the Gene." *American Scientist* 37 (1949): 33–59.

———. "The Cytoplasm in Heredity." *Heredity* 4 (1950): 11–36.

———. "Partner of the Genes." *Scientific American* 183, no. 5 (Nov 1950): 30–39.

———. "Beyond the Gene—Two Years Later." In *Science in Progress,* 7th series, ed. George A. Baitsell. New Haven, CT: Yale University Press, 1951, 167–202.

Sonneborn, Tracy M., and G. H. Beale. "Influence des gènes, des plasmagènes et du milieu dans le déterminisme des caractères antigéniques chez *Paramecium aurelia* (variété 4)." Colloques Internationaux du Centre National de la Recherche Scientifique, 8, *Unités biologiques douées de continuité génétique.* Paris: Publications du Centre National de la Recherche Scientifique, 1949, 25–32.

Spaght, Monroe E. "Basic Research in Industry." *Science* 121 (1955): 784–89.

Spath, Susan. "C. B. van Niels's Conception of the 'Delft School.'" In Bos and Theunissen, *Beijerinck and the Delft School of Microbiology,* 215–20.

Spiegelman, Sol, and Martin D. Kamen. "Genes and Nucleoproteins in the Synthesis of Enzymes." *Science* 104 (1946): 581–84.

Srinivasan, P. R.; Joseph S. Fruton; and John T. Edsall, eds. *The Origins of Modern Biochemistry: A Retrospect on Proteins.* Special issue of *Annals of the New York Academy of Sciences* 325 (1979): 1–373.

Stanley, Wendell M. "The Action of High Frequency Sound Waves on Tobacco Mosaic Virus." *Science* 80 (1934): 339–41.

———. "Chemical Studies on the Virus of Tobacco Mosaic. Part 1. Some Effects of Trypsin." *Phytopathology* 24 (1934): 1055–85.

———. "Chemical Studies on the Virus of Tobacco Mosaic. Part 2. The Proteolytic Action of Pepsin." *Phytopathology* 24 (1934): 1269–89.

———. "Chemical Studies on the Virus of Tobacco Mosaic. Part 3. Rates of Inactivation at Different Hydrogen-Ion Concentrations." *Phytopathology* 25 (1935): 475–92.

———. "Isolation of a Crystalline Protein Possessing the Properties of Tobacco-Mosaic Virus." *Science* 81 (1935): 644–45.

———. "Chemical Studies on the Virus of Tobacco Mosaic. Part 4. Some

Effects of Different Chemical Agents on Infectivity." *Phytopathology* 25 (1935): 899–921.

———. "Chemical Studies on the Virus of Tobacco Mosaic. Part 5. Determination of Optimum Hydrogen-Ion Concentrations for Purification by Precipitation with Lead Acetate." *Phytopathology* 25 (1935): 922–30.

———. "Chemical Studies on the Virus of Tobacco Mosaic. Part 6. The Isolation from Diseased Turkish Tobacco Plants of a Crystalline Protein Possessing the Properties of Tobacco-Mosaic Virus." *Phytopathology* 26 (1936): 305–20.

———. "Chemical Studies on the Virus of Tobacco Mosaic. Part 7. An Improved Method for the Preparation of Crystalline Tobacco Mosaic Virus Protein." *Journal of Biological Chemistry* 115 (1936): 673–78.

———. "Dr. Stanley's Prize Research Described in His Own Words." *Science News Letter* 31 (9 Jan 1937): 20–21.

———. "Chemical Studies on the Virus of Tobacco Mosaic. Part 8. The Isolation of a Crystalline Protein Possessing the Properties of Aucuba Mosaic Virus." *Journal of Biological Chemistry* 117 (1937): 325–40.

———. "Chemical Studies on the Virus of Tobacco Mosaic. Part 9. Correlation of Virus Activity and Protein on Centrifugation of Protein from Solution under Various Conditions." *Journal of Biological Chemistry* 117 (1937): 755–70.

———. "The Reproduction of Virus Proteins." *American Naturalist* 72 (1938): 110–23.

———. "Biochemistry and Biophysics of Viruses." In *Handbuch der Virusforschung,* ed. Robert Doerr and Curt Hallauer. Wien: Julius Springer, 1938, 447–546.

———. "The Isolation and Properties of Tobacco Ring Spot Virus." *Journal of Biological Chemistry* 129 (1939): 405–28.

———. "Recent Advances in the Study of Viruses." In *Science in Progress,* 1st series, ed. George A. Baitsell. New Haven, CT: Yale University Press, 1939, 78–111, 294–301 (chapter 3).

———. "The Architecture of Viruses." *Physiological Reviews* 19 (1939): 524–56.

———. "The Biochemistry of Viruses." *Annual Review of Biochemistry* 9 (1940): 545–70.

———. "Purification of Tomato-Bushy-Stunt Virus by Differential Centrifugation." *Journal of Biological Chemistry* 135 (1940): 437–54.

———. "Some Chemical, Medical, and Philosophical Aspects of Viruses." *Science* 93 (1941): 145–51.

———. "The Concentration and Purification of Tobacco Mosaic Virus by Means of the Sharples Super-Centrifuge." *Journal of the American Chemical Society* 64 (1942): 1804–6.

———. "Chemical Structure and the Mutation of Viruses [from the Messenger Lectures at Cornell by members of the Rockefeller Institute for Medical Research]." In *Virus Diseases.* Ithaca, NY: Cornell University Press, 1943, 35–59.

———. "The Efficiency of Different Sharples Centrifuge Bowls in the Concentration of Tobacco Mosaic and Influenza Viruses." *Journal of Immunology* 53 (1946): 179–89.

————. "Viruses." In *Currents in Biochemical Research,* ed. David E. Green. New York: Interscience, 1946, 13–23.

————. "Virus Composition and Structure—Twenty-five Years Ago and Now." *Federation Proceedings* 15 (1956): 812–18.

————. "The Regulation and Transfer of Biological Information by Viruses." *Proceedings of the Robert A. Welch Foundation Conferences on Chemical Research* 5 (1962): 131–53.

————. "The 'Undiscovered' Discovery." *Archives of Environmental Health* 21 (1970): 256–62.

Stanley, Wendell M., and Thomas F. Anderson. "A Study of Purified Viruses with the Electron Microscope." *Journal of Biological Chemistry* 139 (1941): 325–38.

————. "Electron Micrographs of Protein Molecules." *Journal of Biological Chemistry* 146 (1942): 25–30.

Stanley, Wendell M., and C. Arthur Knight. "The Chemical Composition of Strains of Tobacco Mosaic Virus." *Cold Spring Harbor Symposia on Quantitative Biology* 9 (1941): 255–62.

Stanley, Wendell M., and Max A. Lauffer. "Chemical and Physical Procedures." In *Viral and Rickettsial Infections of Man,* ed. Thomas M. Rivers. Philadelphia: J. B. Lippincott, 1948, 18–66.

Stanley, Wendell M., and Hubert S. Loring. "The Isolation of Crystalline Tobacco Mosaic Virus Protein from Diseased Tomato Plants." *Science* 83 (1936): 85.

Stanley, Wendell M., and Ralph W. G. Wyckoff. "The Isolation of Tobacco Ring Spot and Other Virus Proteins by Ultracentrifugation." *Science* 85 (1937): 181–83.

Star, Susan Leigh, and James R. Griesemer. "Institutional Ecology, 'Translations' and Boundary Objects: Amateurs and Professionals in Berkeley's Museum of Vertebrate Zoology, 1907–1939." *Social Studies of Science* 19 (1989): 387–420.

Starr, Paul. *The Social Transformation of American Medicine.* New York: Basic Books, 1982.

Stent, Gunther S. "That Was the Molecular Biology That Was." *Science* 160 (1968): 345–50.

————. *The Coming of the Golden Age: A View of the End of Progress.* Garden City, NY: Natural History Press for the American Museum of Natural History, 1969.

————. *Molecular Genetics: An Introductory Narrative.* San Francisco, CA: W. H. Freeman, 1971.

————. "Prematurity and Uniqueness in Scientific Discovery." *Scientific American* 227, no. 6 (Dec 1972): 84–93.

Stillwell, Craig R. "Thymectomy as an Experimental System in Immunology." *Journal of the History of Biology* 27 (1994): 379–401.

Stimson, Thomas E., Jr. "Science Probes the Secrets of the Virus." *Popular Mechanics* 106, no. 2 (1956): 92–96.

Strickland, Stephen P. *Politics, Science and Dread Disease: A Short History of United States Medical Research Policy.* Cambridge, MA: Harvard University Press, 1972.

―――. *The Story of the NIH Grants Programs.* Lanham, MD: University Press of America, 1989.

Sullivan, Navin. *Pioneer Germ Fighters.* New York: Scholastic Book Services, 1962.

Summers, William C. "From Culture as Organism to Organism as Cell: Historical Origins of Bacterial Genetics." *Journal of the History of Biology* 24 (1991): 171–90.

―――. "How Bacteriophage Came to Be Used by the Phage Group." *Journal of the History of Biology* 26 (1993): 255–67.

―――. "Concept Migration: The Case of Target Theories in Physics and Biology." Unpublished paper prepared for the History of Science Society Meeting, 26–29 Oct 1995.

―――. *Félix d'Herelle and the Origins of Molecular Biology.* New Haven, CT: Yale University Press, 1999.

Sumner, James B. "The Isolation and Crystallization of the Enzyme Urease." *Journal of Biological Chemistry* 69 (1926): 435–41.

―――. "The Chemical Nature of Enzymes." *Science* 78 (1933): 335.

Svedberg, Theodor. *Die Existenz der Moleküle: Experimentelle Studien.* Leipzig: Akademische Verlagsgesellschaft, 1912.

―――. "Protein Molecules." *Chemical Reviews* 20 (1937): 81–98.

―――. "The Ultra-Centrifuge and the Study of High-Molecular Weight Compounds." *Nature* 139 (1937): 1051–62.

―――. "The Ultracentrifuge and Its Field of Research." *Industrial and Engineering Chemistry, Analytical Edition* 10 (1938): 113–27.

Svedberg, Theodor, and Robin Fåhraeus. "A New Method for the Determination of the Molecular Weight of the Proteins." *Journal of the American Chemical Society* 48 (1926): 430–38.

Svedberg, Theodor, and Kai O. Pedersen. *The Ultracentrifuge.* Oxford: Clarendon Press, 1940.

Svedberg, Theodor, and Herman Rinde. "The Determination of the Distribution of Size of Particles in Disperse Systems." *Journal of the American Chemical Society* 45 (1923): 943–54.

―――. "The Ultra-Centrifuge, a New Instrument for the Determination of Size and Distribution of Size of Particles in Amicroscopic Colloids." *Journal of the American Chemical Society* 46 (1924): 2677–93.

Takahashi, William N., and Mamoru Ishii. "An Abnormal Protein Associated with Tobacco Mosaic Virus Infection." *Nature* 169 (1952): 419–20.

―――. "A Macromolecular Protein Associated with Tobacco Mosaic Virus Infection: Its Isolation and Properties." *American Journal of Botany* 40 (1953): 85–90.

Takahashi, William N., and R. E. Rawlins. "Method for Determining Shape of Colloidal Particles: Application in Study of Tobacco Mosaic Virus." *Proceedings of the Society for Experimental Biology and Medicine* 30 (1932): 155–57.

―――. "Rod-Shaped Particles in Tobacco Mosaic Virus Demonstrated by Stream Double Refraction." *Science* 77 (1933): 26–27.

―――. "Stream Double Refraction Exhibited by Juice from Both Healthy and Mosaic Tobacco Plants." *Science* 77 (1933): 284.

―――. "The Relation of Stream Double Refraction to Tobacco Mosaic Virus." *Science* 81 (1935): 299–300.

Tarbell, D. S.; Ann T. Tarbell; and R. M. Joyce. "The Students of Ira Remsen and Roger Adams." *Isis* 71 (1980): 620–25.

Temin, Howard M., and Harry Rubin, "Characteristics of an Assay for Rous Sarcoma Virus and Rous Sarcoma Cells in Tissue Culture" *Virology* 6 (1958): 669–88.

Thackray, Arnold; Jeffrey L. Sturchio; P. Thomas Carroll; and Robert Bud. *Chemistry in America, 1876–1976: Historical Indicators.* Dordrecht, the Netherlands: D. Reidel, 1985.

Theunissen, Bert. "Martinus Willem Beijerinck and the Beginnings of the Delft Tradition in Microbiology." In Bos and Theunissen, *Beijerinck and the Delft School of Microbiology,* 183–92.

―――. "The Beginnings of the 'Delft Tradition' Revisited: Martinus W. Beijerinck and the Genetics of Microorganisms." *Journal of the History of Biology* 29 (1996): 197–228.

Thieffry, Denis. "Contributions of the 'Rouge-Cloître Group' to the Notion of 'Messenger RNA.' " *History and Philosophy of the Life Sciences* 19 (1997): 89–111.

Thompson, E. P. "The Moral Economy of the English Crowd in the Eighteenth Century." *Past and Present* 50 (1971): 76–136.

Thompson, Emily. "Dead Rooms and Live Wires: Harvard, Hollywood, and the Deconstruction of Architectural Acoustics, 1900–1930." *Isis* 88 (1997): 597–626.

Timoféeff-Ressovsky, N. W.; K. G. Zimmer; and Max Delbrück. "Über die Natur der Genmutation und der Genstruktur." *Nachrichten der Gesellschaft der Wissenschaften in Göttingen,* Fachgr. 6, 1 (1935): 189–245.

Traweek, Sharon. *Beamtimes and Lifetimes: The World of High Energy Physicists.* Cambridge, MA: Harvard University Press, 1988.

Troland, Leonard T. "Biological Enigmas and the Theory of Enzyme Action." *American Naturalist* 51 (1917): 321–50.

Tsugita, A., and Heinz Fraenkel-Conrat. "The Amino Acid Composition and C-Terminal Sequence of a Chemically Evoked Mutant of TMV." *Proceedings of the National Academy of Sciences, USA* 46 (1960): 636–42.

―――. "The Composition of Proteins of Chemically Evoked Mutants of TMV RNA." *Journal of Molecular Biology* 4 (1962): 73–82.

―――. "Contributions from TMV Studies to the Problem of Genetic Information Transfer and Coding." In *Molecular Genetics,* ed. J. Herbert Taylor. New York: Academic Press, 1963, 1: 477–520.

Tsugita, A.; Heinz Fraenkel-Conrat; Marshall W. Nirenberg; and J. Heinrich Matthaei. "Demonstration of the Messenger Role of Viral RNA." *Proceedings of the National Academy of Sciences, USA* 48 (1962): 846–53.

Tsugita, A.; D. T. Gish; J. Young; Heinz Fraenkel-Conrat; C. Arthur Knight; and Wendell M. Stanley. "The Complete Amino Acid Sequence of the Protein of Tobacco Mosaic Virus." *Proceedings of the National Academy of Sciences, USA* 46 (1960): 1463–69.

Turnbull, David, and Terry Stokes, "Manipulable Systems and Laboratory Strategies in a Biomedical Institute." In *Experimental Inquiries: Historical, Philosophical and Social Studies of Experimentation in Science,* ed. H. E. Le Grand. Dordrecht, the Netherlands: Kluwer Academic Publishers, 1990, 167–92.

Turner, Stephen. "Practice in Real Time." *Studies in History and Philosophy of Science* 30A (1999): 149–56.

Turpen, Thomas H. "Tobacco Mosaic Virus and the Virescence of Biotechnology." *Philosophical Transactions of the Royal Society of London* 354B (1999): 665–73.

Twort, Frederick W. "The Ultramicroscopic Viruses." *Journal of State Medicine* 31 (1923): 351–66.

Uchida, Hisao. "Building a Science in Japan: The Formative Decades of Molecular Biology." *Journal of the History of Biology* 26 (1993): 499–518.

Van Helden, Albert, and Thomas L. Hankins, eds. *Instruments. Osiris* (2d ser.) 9 (1994).

van Helvoort, Ton. See Helvoort, Ton van.

Vincent, George E. "Teamplay in Public Health." *American Journal of Public Health* 9 (1919): 14–20.

Vinson, Carl G. "Precipitation of the Virus of Tobacco Mosaic." *Science* 66 (1927): 357–58.

——. "Further Purification of the Virus of Tobacco Mosaic [abstract]." *Phytopathology* 23 (1933): 35.

——. "Purification of the Virus of Tobacco Mosaic." *Phytopathology* 24 (1934): 20.

——. "Possible Chemical Nature of Tobacco Mosaic Virus." *Science* 79 (1934): 548–49.

Vinson, Carl G., and A. W. Petre. "Mosaic Disease of Tobacco." *Botanical Gazette* 87 (1929): 14-38; also published in *Contributions from the Boyce Thompson Institute* 1 (1929): 479–503.

——. "Progress in Freeing the Virus of Mosaic Disease of Tobacco from Accompanying Solids [abstract]." *Phytopathology* 19 (1929): 107–8.

——. "Mosaic Disease of Tobacco. Part 2. Activity of the Virus Precipitated by Lead Acetate." *Contributions from the Boyce Thompson Institute* 3 (1931): 131–45.

Walker, Turnley. *Roosevelt and the Fight Against Polio.* London: Rider, 1954.

Wang, Jessica. "Liberals, the Progressive Left, and the Political Economy of Postwar American Science: The National Science Foundation Debate Revisited." *Historical Studies in the Physical and Biological Sciences* 26 (1995): 139–66.

Waterson, A. P., and Lise Wilkinson. *An Introduction to the History of Virology.* Cambridge: Cambridge University Press, 1978.

Watson, Elizabeth L. *Houses for Science: A Pictorial History of Cold Spring Harbor Laboratory.* Cold Spring Harbor, NY: Cold Spring Harbor Laboratory Press, 1991.

Watson, James D. "The Structure of Tobacco Mosaic Virus. Part 1. X-Ray Evidence of a Helical Arrangement of Sub-Units Around the Longitudinal Axis." *Biochimica et Biophysica Acta* 13 (1954): 10–19.

———. *Molecular Biology of the Gene.* New York: W. A. Benjamin, 1965.

———. *The Double Helix: A Personal Account of the Discovery of the Structure of DNA.* New York: Atheneum Press, 1968.

Watson, James D., and Francis H. C. Crick. "A Structure for Deoxyribose Nucleic Acid." *Nature* 171 (1953): 737–38.

———. "Genetical Implications of the Structure of Deoxyribonucleic Acid." *Nature* 171 (1953): 964–67.

Waugh, John G., and Carl G. Vinson. "Particle Size of the Virus of Tobacco Mosaic in Purified Solutions [abstract]." *Phytopathology* 22 (1931): 29.

Weinberg, Steven. *Dreams of a Final Theory.* New York: Pantheon, 1992.

Weindling, Paul. "Scientific Elites and Laboratory Organisations in Fin de Siècle Paris and Berlin: The Pasteur Institute and Robert Koch's Institute for Infectious Diseases Compared." In *The Laboratory Revolution in Medicine,* ed. Andrew Cunningham and Perry Williams. Cambridge: Cambridge University Press, 1992, 170–88.

Weiner, Jonathon. *Time, Love, Memory: A Great Biologist and His Quest for the Origins of Behavior.* New York: Alfred A. Knopf, 1999.

White, F. H., Jr., and Christian B. Anfinsen. "Some Relationships of Structure to Function in Ribonuclease." *Annals of the New York Academy of Sciences* 81 (1959): 515–23.

White, Ray, and C. Thomas Caskey. "The Human as an Experimental System in Molecular Genetics." *Science* 240 (1988): 1483–88.

Wilkins, M. H. F.; A. R. Stokes; W. E. Seeds; and G. Oster. "Tobacco Mosaic Virus Crystals and Three-Dimensional Microscopic Vision." *Nature* 166 (1950): 127–29.

Wilkinson, Lise. "The Development of the Virus Concept as Reflected in Corpora of Studies on Individual Pathogens. Part 1. Beginnings at the Turn of the Century." *Medical History* 18 (1974): 211–21.

———. "The Development of the Virus Concept as Reflected in Corpora of Studies on Individual Pathogens. Part 3. Lessons of the Plant Viruses—Tobacco Mosaic Virus." *Medical History* 20 (1976): 111–34.

Wilkinson, Lise, and A. P. Waterson. "The Development of the Virus Concept as Reflected in Corpora of Studies on Individual Pathogens. Part 2. The Agent of Fowl Plague—A Model Virus?" *Medical History* 19 (1975): 52–72.

Williams, Greer. *Virus Hunters.* New York: Alfred A. Knopf, 1959.

Williams, Robley C. "The Shapes and Sizes of Purified Viruses as Determined by Electron Microscopy." *Cold Spring Harbor Symposia on Quantitative Biology* 18 (1953): 185–95.

Williams, Robley C.; Robert C. Backus; and Russell L. Steere. "Macromolecular Weights Determined by Direct Particle Counting. Part 2. The Weight of the Tobacco Mosaic Virus Particle." *Journal of the American Chemical Society* 73 (1951): 2062–66.

Williams, Robley C., and Dean Fraser. "Morphology of the Seven T-Bacteriophages." *Journal of Bacteriology* 66 (1953): 458–64.

Williams, Robley C., and Russell L. Steere. "Electron Microscopic Observations on the Unit of Length of the Particles of Tobacco Mosaic Virus." *Journal of the American Chemical Society* 73 (1951): 2057–61.

Williams, Robley C., and Ralph W. G. Wyckoff. "The Thickness of Electron Microscopic Objects." *Journal of Applied Physics* 15 (1944): 712–16.

Wilson, E. B. *The Physical Basis of Life.* New Haven, CT: Yale University Press, 1923.

Winkler, Hans. "Über die Rolle von Kern und Protoplasma bei der Vererbung." *Zeitschrift für induktive Abstammungs- und Vererbungslehre* 33 (1924): 238–53.

Wise, M. Norton. "Mediating Machines." *Science in Context* 2 (1988): 77–113.

Wittgenstein, Ludwig. *Philosophical Investigations.* Trans. G. E. M. Anscombe. Oxford: Basil Blackwell, 1953.

Wittmann, H. G. "Vergleich der Proteine des Normalstammes und einer Nitritmutante des Tabakmosaikvirus." *Zeitschrift für Vererbungslehre* 90 (1959): 463–75.

———. "Comparison of the Tryptic Peptides of Chemically Induced and Spontaneous Mutants of Tobacco Mosaic Virus." *Virology* 12 (1960): 609–12.

Woese, Carl R. "Coding Ratio for the Ribonucleic Acid Viruses." *Nature* 190 (1961): 697–98.

Wolbach, S. B. "The Filterable Viruses: A Summary." *Journal of Medical Research* 27 (1912): 1–25.

Wollman, Elie; F. Holweck; and Salvador Luria. "Effects of Radiation on Bacteriophage C_{16}." *Nature* 145 (1940): 935–36.

Woods, Albert F. "The Destruction of Chlorophyll by Oxidizing Enzymes." *Zentralblatt für Bakteriologie, Parasitenkunde und Infektionskrankheiten*, Abt. 2, 5 (1899): 745–54.

———. "Observations on the Mosaic Disease of Tobacco." *Bulletin of the Bureau of Plant Industry, U.S.D.A.* 18 (1902): 7–24.

Worster, Donald E. *Dust Bowl: The Southern Plains in the 1930s.* Oxford: Oxford University Press, 1979.

———. "Transformations of the Earth: Toward an Agroecological Perspective in History." *Journal of American History* 76 (1990): 1087–1106.

———. "Seeing Beyond Culture." *Journal of American History* 76 (1990): 1142–47.

Wright, Sewell. "The Physiology of the Gene." *Physiological Reviews* 21 (1941): 487–527.

———. "Genes as Physiological Agents: General Considerations." *American Naturalist* 79 (1945): 289–317.

Wright, Susan. *Molecular Politics: Developing American and British Regulatory Policy for Genetic Engineering, 1972–1982.* Chicago: University of Chicago Press, 1994.

Wyckoff, Ralph W. G. "The Ultracentrifugal Purification and Study of Macromolecular Proteins." *Science* 86 (1937): 92–95.

———. "The Ultracentrifugal Study of Virus Proteins." *Proceedings of the American Philosophical Society* 77 (1937): 455–62.

———. "An Ultracentrifugal Analysis of Concentrated Staphylococcus Bacteriophage Preparations." *Journal of General Physiology* 21 (1938): 367–73.

———. "The Ultracentrifugal Study of Macromolecules." *Cold Spring Harbor Symposia on Quantitative Biology* 6 (1938): 361–68.

Wyckoff, Ralph W. G.; J. Biscoe; and Wendell M. Stanley. "An Ultracentrifugal Analysis of the Crystalline Virus Proteins Isolated from Plants Diseased with Different Strains of Tobacco Mosaic Virus." *Journal of Biological Chemistry* 117 (1937): 57–71.

Wyckoff, Ralph W. G., and Robert B. Corey. "The Ultracentrifugal Crystallization of Tobacco Mosaic Virus Protein." *Science* 84 (1936): 513.

Yčas, Martynas. "Correlation of Viral Ribonucleic Acid and Protein Composition." *Nature* 188 (1960): 209–12.

Yoxen, Edward J. "Where Does Schrödinger's *What Is Life?* Belong in the History of Molecular Biology?" *History of Science* 17 (1979): 17–52.

Zaitlin, Milton. "Tobacco Mosaic Virus and Its Contributions to Virology." *American Society for Microbiology News* 65 (1999): 675–80.

Zallen, Doris. "The Rockefeller Foundation and Spectroscopy Research: The Programs at Chicago and Utrecht." *Journal of the History of Biology* 25 (1992): 67–89.

Index

Abir-Am, Pnina, 250
Academy of Sciences (Russia), 24
acid-base effects. *See* pH
acquired immunity, 201–3
ACS. *See* American Cancer Society
Adams, Roger, 49, 123
adrenocorticotropic hormone (ACTH),
    266
agricultural research and TMV, 20–21, 35
air-driven ultracentrifuge: invention of,
    88–92, 89f, 90n24, 91f; use in virus
    studies, 108. *See also* ultracentrifuge
Alexander, Jerome, 35, 73
Allard, H. A., 33, 36
Allen, Ernest, 173
Altenburg, Edgar, 235
American Cancer Society (ACS):
    coordination of research efforts, 178;
    disease-based campaigns' success, 11,
    144, 145; emphasis on funding need,
    175, 176f; establishment of, 143,
    149n39; fund-raising success
    (1944–46), 152. *See also* ASCC
American Chemical Society, 127
American Heart Association, 149n39
American Lung Association. *See* National
    Tuberculosis Association (NTA)
American Phytopathological Society, 38
American Society for the Control of
    Cancer (ASCC), 148, 152
amino acids: carboxypeptidase experiment
    with TMV, 267–69; complete sequence
    in TMV, 304, 305f; RNA composition
    and, 292; size of TMV and, 115; studies
    on virus strain differences and, 217–18,
    219, 220f, 221, 223. *See also* proteins
analytical ultracentrifuge, 81, 83, 85. *See
    also* ultracentrifuge

Anderson, Thomas F.: electron microscope
    work, 120, 122, 205, 206; on shared
    qualities between host and virus, 226
androsterone studies, 76
Anfinsen, Christian, 315
animal viruses: research funding from
    NFIP, 244–45; RNA identification in,
    302, 302n194; Stanley's interest in
    studying, 124, 124n187; ultracentrifuge
    used to isolate, 106, 109n115, 129. *See
    also* influenza virus; papilloma virus;
    poliomyelitis virus
antibiotics, 203
applied research, 172, 172n123, 173. *See
    also* basic research
A-protein, 269
Armstrong, Joanne Karrer, 31, 212
ASCC (American Society for the Control
    of Cancer), 148, 152. *See also*
    American Cancer Society
Astbury, W. T., 110
atomic bomb, 182
Atomic Energy Commission, 174
aucuba mosaic virus, 71, 213, 213n106
autocatalysis: enzyme self-replication and,
    34; Northrop's theory for bacteria,
    198; viruses and, 64

Bacillus coli. See *Escherichia coli*
bacteria: genetic mutation vs. acquired
    immunity, 201–3, 202f, 203n67;
    susceptibility of plants to, 23, 23n22;
    viruses compared to, 31
bacteriophages: biochemistry approach to
    study of, 209–10, 223, 223n137;
    coordination of research efforts, 208;
    electron micrographs, inferences from,
    204–5f, 207–8; evidence of similarities